RANDOM VIBRATION AND SPECTRAL ANALYSIS

SOLID MECHANICS AND ITS APPLICATIONS
Volume 33

Aims and Scope of the Series

The fundamental questions arising in mechanics are: *Why?*, *How?*, and *How much?* The aim of this series is to provide lucid accounts written by authoritative researchers giving vision and insight in answering these questions on the subject of mechanics as it relates to solids.

The scope of the series covers the entire spectrum of solid mechanics. Thus it includes the foundation of mechanics; variational formulations; computational mechanics; statics, kinematics and dynamics of rigid and elastic bodies; vibrations of solids and structures; dynamical systems and chaos; the theories of elasticity, plasticity and viscoelasticity; composite materials; rods, beams, shells and membranes; structural control and stability; soils, rocks and geomechanics; fracture; tribology; experimental mechanics; biomechanics and machine design.

The median level of presentation is the first year graduate student. Some texts are monographs defining the current state of the field; others are accessible to final year undergraduates; but essentially the emphasis is on readability and clarity.

Random Vibration and Spectral Analysis

by

ANDRÉ PREUMONT
Université Libre de Bruxelles, Belgium

KLUWER ACADEMIC PUBLISHERS
DORDRECHT / BOSTON / LONDON

Library of Congress Cataloging-in-Publication Data

Preumont, André.
Random vibration and spectral analysis / by André Preumont.
p. cm. -- (Solid mechanics and its applications)
Includes index.
ISBN 0-7923-3036-6 (alk. paper)
1. Random vibration. 2. Stochastic processes. 3. Spectral theory (Mathematics)--Data processing. I. Title. II. Series.
QA935.P725 1994
531'.32'015192--dc20 94-29946

ISBN 0-7923-3036-6

Published by Kluwer Academic Publishers,
P.O. Box 17, 3300 AA Dordrecht, The Netherlands.

Kluwer Academic Publishers incorporates
the publishing programmes of
D. Reidel, Martinus Nijhoff, Dr W. Junk and MTP Press.

Sold and distributed in the U.S.A. and Canada
by Kluwer Academic Publishers,
101 Philip Drive, Norwell, MA 02061, U.S.A.

In all other countries, sold and distributed
by Kluwer Academic Publishers Group,
P.O. Box 322, 3300 AH Dordrecht, The Netherlands.

Printed on acid-free paper

Translation of the French edition
"Vibrations Aléatoires et Analyse Spectrale"
Première édition
ISBN 2-88074-183-1

Printed in the Netherlands

"...Je n'affirme rien, je me
contente de croire qu'il y a plus
de choses possibles qu'on ne pense."

Voltaire, Micromégas

TABLE OF CONTENTS

Preface

I became interested in Random Vibration during the preparation of my PhD dissertation, which was concerned with the seismic response of nuclear reactor cores. I was initiated into this field through the classical books by Y.K.Lin, S.H.Crandall and a few others. After the completion of my PhD, in 1981, my supervisor M.Geradin encouraged me to prepare a course in Random Vibration for fourth and fifth year students in Aeronautics, at the University of Liege. There was at the time very little material available in French on that subject. A first draft was produced during 1983 and 1984 and revised in 1986. These notes were published by the *Presses Polytechniques et Universitaires Romandes* (Lausanne, Suisse) in 1990.

When Kluwer decided to publish an English translation of the book in 1992, I had to choose between letting Kluwer translate the French text *in-extenso* or doing it myself, which would allow me to carry out a sustantial revision of the book. I took the second option and decided to rewrite or delete some of the original text and include new material, based on my personal experience, or reflecting recent technical advances. Chapter 6, devoted to the response of multi degree of freedom structures, has been completely rewritten, and Chapter 11 on random fatigue is entirely new. The computer programs which have been developed in parallel with these chapters have been incorporated in the general purpose finite element software SAMCEF, developed at the University of Liege. All the chapters have been supplemented with a set of problems.

I am deeply endebted to Prof. G.M.L.Gladwell from the University of Waterloo, who read the manuscript and corrected many mistakes and misuses of the English language. His comments have been invaluable in improving the text.

I take this opportunity to thank Prof. Michel Geradin from the University of Liege for his advice, encouragement, and long friendship.

I dedicate this book to Prof. André Jaumotte, Honorary Rector of the University of Brussels. His enthusiastic response to new ideas, and tireless action to promote research, his supportive friendship and his humanism have been a constant stimulus and example.

André Preumont
Bruxelles, November 1993.

Chapter 1

Introduction

1.1 Overview

Structural dynamics aims at predicting the response of structures in a given loading environment. It has enjoyed a tremendous development since the late 60's, mainly because of the availability of high-speed computers and general purpose finite-element programs. Numerical (finite-element) and experimental modal analysis have become standard engineering tools.

Random vibration has emerged from the need to analyse and assess the reliability of structures operating in a *random environment*. Examples are numerous: wind on tall buildings and bridges; sea waves on off-shore structures and ships; earthquakes on civil engineering structures; atmospheric turbulence on aircrafts; launch acceleration on satellites; acoustic fatigue of aircraft panels, etc...

One of the standard assumptions in random vibration is that the structure is *known and deterministic*, that is, not subject to random variations of its properties.

The theory of stationary random processes has been well established for several decades; for *linear* systems, the input-output relationship can be expressed in a very simple manner in the frequency domain (Fig.1.1). Most of the difficulties in predicting the structural response statistics are related to defining the physical excitation acting on the structure in a way which is compatible with the structural modelling, on the one hand, and handling the large number of degrees of freedom of finite-element models, on the other. A drastic reduction of the dimension of the system can often be achieved by using normal modes, but the computational effort needed to determine the modal excitations for complicated loadings with frequency-dependent spatial coherence remains high.

In frequency domain analysis, it is assumed implicitly that the excitation process is *Gaussian*. This is in general justified by the *central limit theorem*

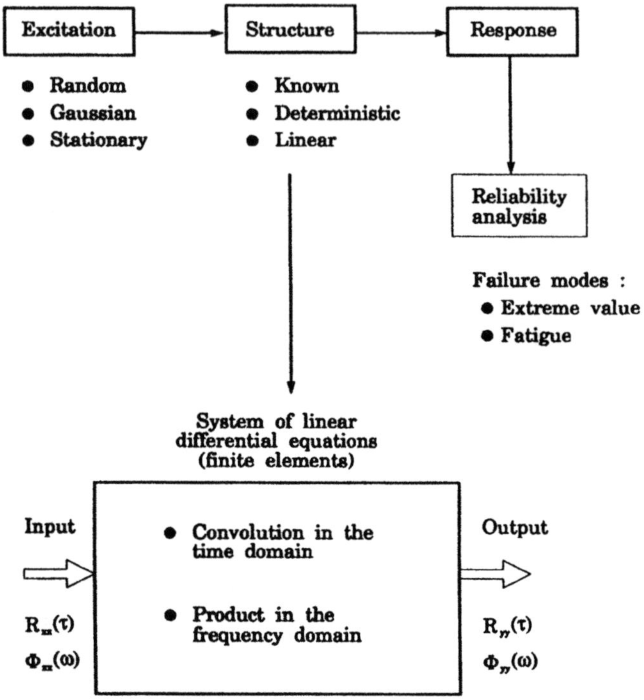

Figure 1.1: Stationary random vibration analysis of a linear structure.

which states that a phenomenon resulting from the superposition of a large number of statistically independent contributions is Gaussian, independently of the distribution of the elementary contributions. The Gaussian character is preserved by linear transformation.

Once the statistics of the random response have been calculated, the reliability of the system can be assessed. The reliability can be expressed in several forms, depending on what failure mode we are concerned with. If response amplitudes exceeding some threshold can jeopardize the normal operation of the system (e.g. vibration amplitude of a rotor exceeding the gap with the casing), one is interested in the probabilty distribution of the ***extreme value*** of the response over the lifetime of the system. Alternatively, one may wish to avoid ***fatigue*** failure of structural components. Random fatigue for uniaxial stress has been known for some time, based on the linear damage theory (***Palmgren-Miner*** criterion); an extension to multidimensional stress states is proposed in this text, based on the quadratic ***von Mises*** criterion.

There may be other harmful effects of vibrations, as for example the discomfort associated with the swaying of tall buildings. Research into human response to acceleration at building frequencies of 0.1 to 0.5 *Hz* indicates that the threshold lies in the range 1-10 *mg*. The designer will often be more concerned with limiting the maximum acceleration than reducing the fatigue damage induced by the wind loading. Noise generation can also be a major nuisance associated with vibrations.

The spectral analysis of non-stationary but slowly varying random processes is a natural extension of that of stationary processes. However, a major difficulty is that there does not exist a *local* decomposition of the energy in the process in the time-frequency plane comparable to the power spectral density function for a stationary random process. Only local averages can be defined and, according to the *uncertainty principle*, a good resolution in one domain (e.g. frequency) can only be achieved at the expense of a poor resolution in the dual domain (time). *Priestley's evolutionary spectrum* has emerged as the most effective analytical spectral representation of non-stationary random processes. The input-output relationship for time-invariant multi degree of freedom systems excited by evolutionary processes can be written simply in state variable form.

Many physical random processes which have a correlation time smaller than the time constant of the system to which they are applied can be approximated by *purely random* processes, also called *white noises* (the values at different times of a purely random process are statistically independent). When a white noise process is applied at the input of a system governed by a differential equation of order n, linear or not, the state vector (of dimension n) constitutes a vector *Markov* process, whose transitional probability is governed by the *Fokker-Planck* equation. Unlike the frequency domain analysis, the Fokker-Planck approach applies to *nonlinear* systems as well. However, it rapidly becomes cumbersome as the order of the system increases.

Many aspects of the spectral decomposition of stationary random processes are essential in the *estimation* of power spectral density functions from experimental data. This is used extensively for frequency response determination in modal testing. The modern digital spectral analysers make extensive use of the *Fast Fourier Transform*, which constitutes a digital approximation to the continuous Fourier transform. The quality of the approximation, however, is strongly dependent of the choice of some critical parameters such as the sampling frequency, the record length, and the type of *window* used to reduce the *leakage* associated with the truncation of the record. A sound understanding of these basic concepts is necessary to ensure a correct use of the spectral analyser. This is the justification for the final chapter of this book.

This textbook is adressed to mechanical engineers already trained in linear structural dynamics; only a very limited background is assumed in stochastic processes. This, more or less, reflects the realities of the engineering education in Belgium and, likely, in other parts of Europe and North America.

1.1.1 Organization

The book is organized as follows:

Chapters 2 to 4 cover the basic theory of random processes. These chapters can be skipped by the reader already trained in random processes.

Chapters 5 and 6 cover the input-output relationship for single and multi degree of freedom (discrete and continuous) linear structures subjected to a stationary random excitation. Both external forces and support excitation are considered.

Chapter 7 is devoted to the effect of noise on the input-output relationship.

Chapter 8 covers the spectral description and the input-output relationship for non-stationary processes. This chapter may be skipped during a first reading.

Chapter 9 studies the Markov processes; it may be skipped during a first reading.

Chapters 10 and 11 are devoted to the failure modes for Gaussian processes. Chapter 10 covers the probability distribution of the threshold crossings, the maxima, the envelope and the extreme value (peak factor); while chapter 11 deals with the prediction of fatigue life for uniaxial and multiaxial stress fields.

Chapter 12 is devoted to the Fast Fourier Transform. This chapter can be read independently of the other chapters.

Each chapter is followed by problems and a limited set of references suggested as additional reading; a more comprehensive biblography is given at the end of the book.

1.1.2 Notations

Most of the notations used in this text follow the customary usage: Capital letters are used to designate random quantities but no special care is taken to distinguish scalars from vectors and matrices; this will generally be clear from the context. For example,

$$m\ddot{x} + c\dot{x} + kx = f$$

is the equation of motion of a single degree of freedom oscillator subjected to the deterministic load f;

$$m\ddot{X} + c\dot{X} + kX = F$$

is the equation governing the random response of the same oscillator excited by the random force $F(t)$;

$$M\ddot{X} + C\dot{X} + KX = F$$

is the equation governing the random response of a multi degree of freedom system subjected to the random vector force $F(t)$.

1.2 The Fourier transform

The Fourier transform will be used extensively throughout the text. In this section, we recall some of its main properties. More will be said in chapter 12, when we study the Discrete Fourier transform.

The Fourier transform of $h(t)$ is defined by

$$H(\omega) = \int_{-\infty}^{\infty} h(t)e^{-j\omega t}dt \tag{1.1}$$

if this integral exists for all real ω. $H(\omega)$ is a complex valued function of the real parameter ω. A sufficient condition for $H(\omega)$ to exist is that $h(t)$ be absolutely integrable:

$$\int_{-\infty}^{\infty} |h(t)|\, dt < \infty \tag{1.2}$$

The inverse transform is given by

$$\frac{1}{2\pi}\int_{-\infty}^{\infty} H(\omega)e^{j\omega t}d\omega = h(t) \tag{1.3}$$

wherever $h(t)$ is continuous. At a point of discontinuity, the integral tends to the average value

$$\frac{h(t^-) + h(t^+)}{2}$$

Condition (1.2), however, is too restrictive for many functions of practical interest such as the periodic functions. If one considers the above integral in the sense of principal value, the existence can be extended to functions of the form

$$\frac{\sin at}{at}$$

and, using the theory of distribution, to periodic functions, constant, and Dirac impulse functions. A short table of Fourier transform pairs which are particularly important for our discussion is provided at the beginning of chapter 12; more elaborate and illustated tables are available in (Bracewell, 1978; Brigham, 1974).

The following theorems are used extensively throughout the text:

1.2.1 Differentiation theorem

If $h(t)$ and $H(\omega)$ constitute a Fourier transform pair, then

$$\frac{\mathrm{d}^n h(t)}{\mathrm{d}t^n} \Longleftrightarrow (j\omega)^n H(\omega) \tag{1.4}$$

also constitute a Fourier transform pair.

1.2.2 Translation theorem

If $h(t)$ and $H(\omega)$ constitute a Fourier transform pair, then

$$h(t-t_0) \Longleftrightarrow H(\omega)e^{-j\omega t_0} \tag{1.5}$$

$$h(t)e^{j\omega_0 t} \Longleftrightarrow H(\omega-\omega_0) \tag{1.6}$$

also constitute Fourier transform pairs.

1.2.3 Parseval's theorem

Parseval's theorem consists of the following identity between the energy distribution in the time and frequency domains

$$E = \int_{-\infty}^{\infty} h^2(t)\,dt = \frac{1}{2\pi}\int_{-\infty}^{\infty} |H(\omega)|^2 d\omega \tag{1.7}$$

$|H(\omega)|^2/2\pi$ is called the ***energy spectrum*** of the signal; it describes the density of energy in the vicinity of the frequency ω.

1.2.4 Symmetry, change of scale, duality

If $h(t)$ and $H(\omega)$ constitute a Fourier transform pair, then

$$H(t) \Longleftrightarrow 2\pi h(-\omega) \tag{1.8}$$

$$h(at) \Longleftrightarrow \frac{1}{|a|}H(\frac{\omega}{a}) \tag{1.9}$$

also constitute Fourier transform pairs.

A consequence of the foregoing theorems is illustrated by the following example: Consider the rectangular function (Fig.1.2):

$$\Pi_T(t) = \left\{ \begin{array}{ll} 1 & |t| \le T \\ 0 & |t| > T \end{array} \right\} \Longleftrightarrow 2\frac{\sin \omega T}{\omega} \tag{1.10}$$

One observes that $H(\omega)$ is an oscillating function of the frequency. The width of the central lobe is equal to $2\pi/T$; the amplitude of the side lobes attenuate as $1/(\omega T)$ and their width is π/T. Thus, as the duration T of the rectangle increases, the amplitude of $H(\omega)$ increases at the origin and the width of the lobes decreases. At the limit, as $T \to \infty$, $H(\omega)$ tends towards the Dirac function:

$$1 \Longleftrightarrow 2\pi\,\delta(\omega) \tag{1.11}$$

This is one of the manifestation of the ***duality*** between the time and frequency domains: ***The shorter a signal is in one domain, the longer it is in the dual domain.***

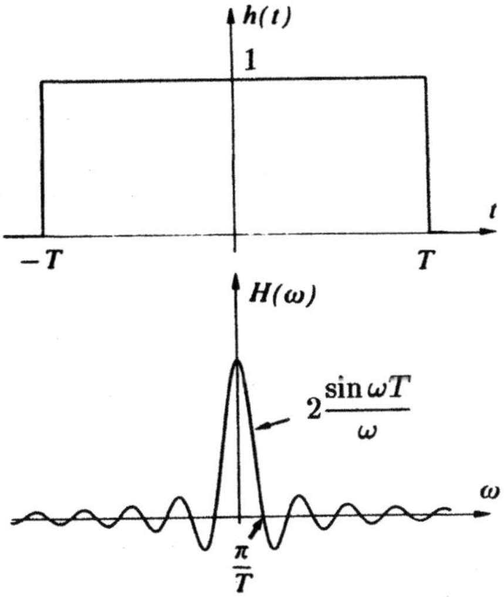

Figure 1.2: Rectangular function $\Pi_T(t)$ and its Fourier transform.

The following Fourier transform pair can be derived from Equ.(1.10) by application of theorem (1.8).

$$\frac{\sin at}{\pi t} \Longleftrightarrow \Pi_a(\omega) \tag{1.12}$$

This relationship is dual of (1.10). The reader is strongly encouraged to examine pictorial tables, where many other illustrations of the duality can be found.

1.2.5 Harmonic functions

From Equ.(1.11) and the frequency translation theorem, the following Fourier transform pairs are readily established

$$e^{j\omega_0 t} \Longleftrightarrow 2\pi\,\delta(\omega - \omega_0) \tag{1.13}$$

$$\cos\omega_0 t \Longleftrightarrow \pi[\delta(\omega+\omega_0) + \delta(\omega-\omega_0)] \tag{1.14}$$

$$\sin\omega_0 t \Longleftrightarrow j\pi[\delta(\omega+\omega_0) - \delta(\omega-\omega_0)] \tag{1.15}$$

1.3 Convolution, correlation

1.3.1 Convolution integral

If $x(t)$ is applied to the input of a linear single input single output system of impulse response $h(t)$, the output $y(t)$ is given by the ***convolution integral***:

$$y(t) = x(t) \star h(t) = \int_{-\infty}^{\infty} x(\tau)h(t-\tau)\,d\tau = \int_{-\infty}^{\infty} h(\tau)x(t-\tau)\,d\tau \tag{1.16}$$

This result can be readily established by replacing the actual input by a set of elementary impulses of intensity $x(\tau)\,d\tau$ and using the principle of superposition (Problem P.1.6). The convolution integral can be visualized as illustrated in Fig.1.3: The value $y(t)$ is obtained after the following sequence of operations:
(1) *Folding* $h(\tau)$ gives $h(-\tau)$.
(2) *Translating* $h(-\tau)$ by t produces $h(t-\tau)$.
(3) Next, $x(\tau)$ and $h(t-\tau)$ must be multiplied and the convolution integral is equal to the area under the curve $x(\tau)h(t-\tau)$ (shaded in Fig.1.3).

The convolution theorem states that if $X(\omega)$ and $H(\omega)$ are the Fourier transforms of $x(t)$ and $h(t)$, then the Fourier transform of the convolution $y(t)$ is

$$Y(\omega) = X(\omega)H(\omega) \tag{1.17}$$

Thus, *a convolution in the time domain corresponds to a product in the frequency domain.* Conversely, it can be shown that *a product in the time domain corresponds to a convolution in the frequency domain*

$$x(t)y(t) \Longleftrightarrow \frac{1}{2\pi}\int_{-\infty}^{\infty} X(\nu)Y(\omega-\nu)\,d\nu \tag{1.18}$$

This is another manifestation of the duality. The convolution integral is commutative, associative and distributive over addition (Problem P.1.7).

1.3.2 Correlation integral

In a similar manner, the *correlation integral* is defined by

$$z(t) = \int_{-\infty}^{\infty} x(\tau)h(t+\tau)\,d\tau \tag{1.19}$$

Its Fourier transform is related to those of the contributing functions by

$$Z(\omega) = X^{*}(\omega)H(\omega) \tag{1.20}$$

1.3.3 Example: The leakage

Consider the finite duration cosine wave

$$f(t) = \cos\omega_0 t.\Pi_T(t) \tag{1.21}$$

where $\Pi_T(t)$ is the rectangular function of duration $2T$. The Fourier transforms of the contributing functions are given by Equ.(1.14) and (1.10), respectively. According to the frequency convolution theorem (1.18),

$$F(\omega) = \frac{1}{2\pi}\{2\frac{\sin\omega T}{\omega}\} \star \{\pi[\delta(\omega-\omega_0)+\delta(\omega+\omega_0)]\}$$

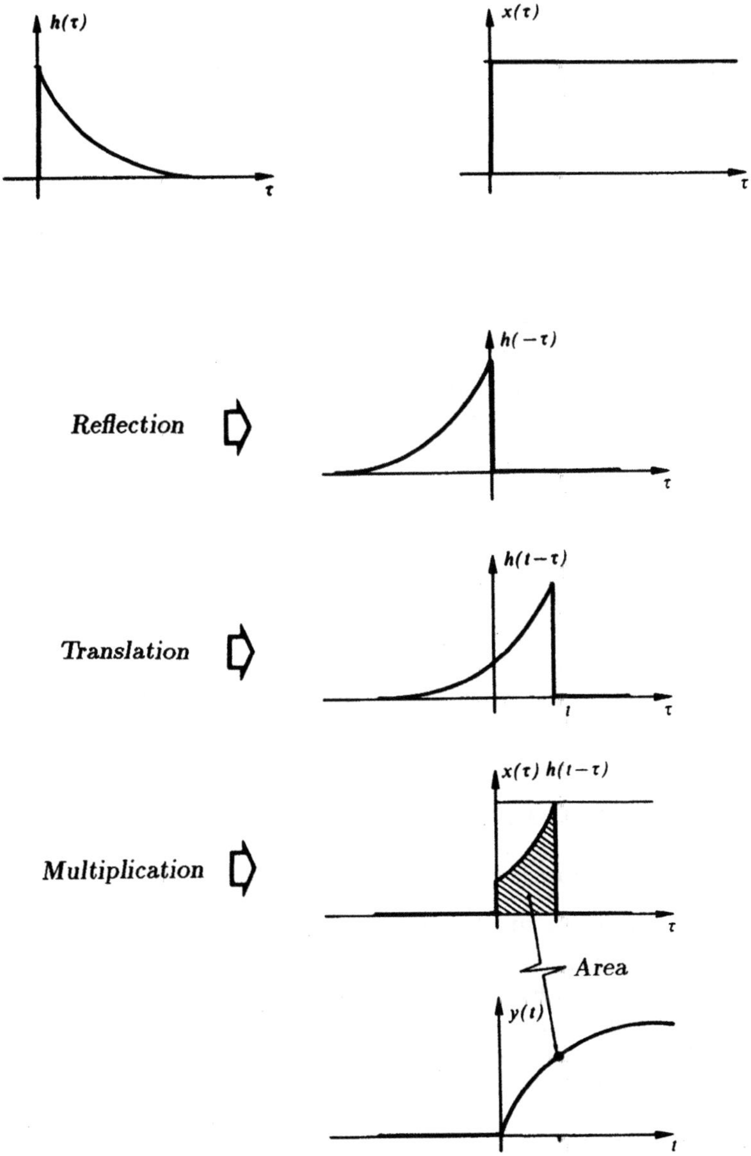

Figure 1.3: Graphical illustration of the convolution.

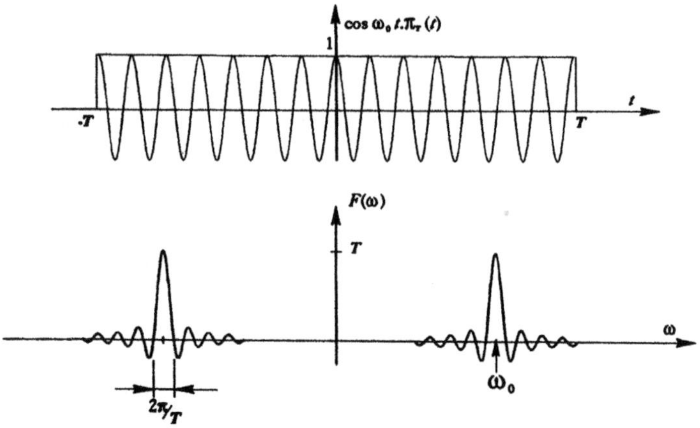

Figure 1.4: Fourier transform of a cosine wave of finite duration.

or

$$F(\omega) = \frac{\sin(\omega - \omega_0)T}{\omega - \omega_0} + \frac{\sin(\omega + \omega_0)T}{\omega + \omega_0} \tag{1.22}$$

$F(\omega)$ is represented in Fig.1.4. One notices that, because of the truncation of the signal, its original energy content spreads at frequencies different from ω_0. The spreading is larger if T is shorter. This phenomenon is known as *leakage*; it constitutes a major difficulty in spectral analysis. It can be alleviated by using an appropriate window of observation with smaller side lobes than the rectangular function. Such windows will be discussed in chapter 12.

1.4 References

Structural Dynamics

R.W.CLOUGH & J.PENZIEN, *Dynamics of Structures*, McGraw-Hill, 1975.
R.R.CRAIG, Jr. *Structural Dynamics*, Wiley, 1981.
M.GERADIN & D.RIXEN, *Mechanical Vibrations, Theory and Application to Structural Dynamics*, Wiley, 1993.
L.MEIROVITCH, *Computational Methods in Structural Dynamics*, Sijthoff & Noordhoff, 1980.

Modal Testing

D.J.EWINS, *Modal Testing: Theory and Practice*, Wiley, 1984.

Fourier Integral

R.N.BRACEWELL, *The Fourier Transform and its Applications*, McGraw-Hill, 1978.

E.O.BRIGHAM, *The Fast Fourier Transform*, Prentice Hall, 1974.
A.PAPOULIS, *The Fourier Integral and its Applications*, McGraw-Hill, 1962.

Other books on Random Vibration

S.H.CRANDALL & W.D.MARK, *Random Vibration in Mechanical Systems*, Academic Press, 1963.
I.ELISHAKOFF, *Probabilistic Methods in the Theory of Structures*, Wiley, 1982.
Y.K.LIN, *Probabilistic Theory of Structural Dynamics*, McGraw-Hill, 1967.
D.E.NEWLAND, *Random Vibrations and Spectral Analysis*, Longmans, 1975.
N.C.NIGAM, *Introduction to Random Vibration*, MIT Press, 1983.

1.5 Problems

P.1.1 *Using the graphical interpretation of the convolution, show that a triangular function*

$$q_T(t) = \begin{cases} 1 - \frac{|t|}{T} & |t| \leq T \\ 0 & |t| > T \end{cases}$$

can be constructed by convolving two identical rectangular functions.
P.1.2 *Based on the result of the previous problem, show that*

$$q_T(t) \Longleftrightarrow \frac{4\sin^2(\omega T/2)}{T\omega^2}$$

Sketch the result and compare it to that of a rectangular function.
P.1.3 *Compute the Fourier transform of*

$$\cos \omega_0 t . q_T(t)$$

where $q_T(t)$ is the triangular function defined at problem P.1.1. With respect to the leakage, compare the situation with that of the rectangular window.
P.1.4 *Show that the following functions constitute Fourier transform pairs*

$$e^{-\alpha|t|} \Longleftrightarrow \frac{2\alpha}{\alpha^2 + \omega^2}$$

$$e^{-\alpha t^2} \Longleftrightarrow \sqrt{\frac{\pi}{\alpha}} e^{-\omega^2/4\alpha}$$

P.1.5 *Consider the Fourier transform pair*

$$h(t) = -\frac{1}{\pi t} \Longleftrightarrow j\,\mathrm{sign}(\omega)$$

Show that

$$\sin \omega_0 t \star h(t) = \cos \omega_0 t$$

$h(t)$ defines the impulse response of a system which produces a phase shift of $\pi/2$. This is called the Hilbert transform (chapter 10).

P.1.6 *Show that the input and the output of a single input single output, linear, time invariant system are related by the convolution integral (1.16).*

P.1.7 *Show that the convolution integral is commutative, associative, and distributive over addition:*

$$f \star g = g \star f$$

$$f \star (g \star h) = (f \star g) \star h$$

$$f \star (g + h) = f \star g + f \star h$$

P.1.8 *Show that the impulse response of an ideal narrow-band filter of bandwidth $\Delta\omega$ centered on ω is*

$$h(\tau) = \frac{2}{\pi\tau} \sin\frac{\Delta\omega\,\tau}{2} \cos\omega\tau$$

Sketch the impulse response for two values of the bandwidth $\Delta\omega$. Is this filter physically realizable?

P.1.9 *Show that the Fourier transform satisfies the following properties:*

$$h(t) \in \{real \cap even\} \Longleftrightarrow H(\omega) \in \{real \cap even\}$$

$$h(t) \in \{real \cap odd\} \Longleftrightarrow H(\omega) \in \{imaginary \cap odd\}$$

$$h(t) \in \{real\} \Longleftrightarrow \mathrm{Re}[H(\omega)] \in \{even\} \;\cap\; \mathrm{Im}[H(\omega)] \in \{odd\}$$

Chapter 2

Random Variables

2.1 Axioms of probability theory

2.1.1 Bernoulli's law of large numbers

Consider the random experiment consisting of tossing a coin. The only possible outcomes of the experiment are ***head (h)*** and ***tail (t)***. Repeating the experiment $n = 50$ times, assume we obtained the following sequence of results:

$$h\ h\ t\ t\ h\ t\ t\ t\ h\ h\ t\ t\ t\ t\ t\ h\ t\ h\ h\ t\ t\ h\ t\ h\ t$$

$$h\ h\ h\ h\ t\ t\ t\ t\ h\ t\ h\ t\ t\ t\ h\ h\ h\ t\ t\ t\ t\ h\ h\ t\ t$$

The outcome ***head*** occured $n_h = 21$ times while ***tail*** occured $n_t = 29$ times. The relative frequencies are respectively

$$r_h = \frac{n_h}{n} = \frac{21}{50} = 0.42 \qquad r_t = \frac{n_t}{n} = \frac{29}{50} = 0.58$$

In fact, we know that if the number n of repetitions of the experiment increases indefinitely, the relative frequencies tend towards fixed limits which, if the coin is fair, are both equal to 0.5. This is known as ***statistical regularity***. The ***probability*** of an event E can be defined as the limit of the relative frequency $r_E = n_E/n$ when the number of repetitions of the random experiment goes to infinity:

$$P(E) = \lim_{n\to\infty} r_E = \frac{n_E}{n} \tag{2.1}$$

This is called ***Bernoulli's law of large numbers.***

2.1.2 Alternative interpretation

The above experiment was not necessary to establish the probabilty of ***head*** and ***tail***, because they are known ***a priori*** from the symmetry of the coin. The same thing applies to the toss of a dice: The probabilities associated with the 6 faces of the dice are all the same, if the dice is perfectly symmetrical. Being the only possible outcomes and being mutually exclusive, their sum must be equal to 1; the elementary probability is therefore 1/6.

The foregoing argument sounds quite attractive, particularly in view of the slow convergence of the law of large numbers. However, except for very simple situations like those described above, things become rapidly more difficult, as illustrated by the following example:

Consider the random experiment consisting of tossing two dices with the faces numbered from 1 to 6; we attempt to evaluate the probability that the sum of the results be 7. This sum can be achieved with the 3 following pairs:

$$(3,4) \quad (5,2) \quad (6,1)$$

If one does not distinguish the result of the first toss from that of the second one, one has to consider a total of 21 possibilities:

$$(1,1) \quad (1,2) \quad (1,3) \quad \ldots \quad (6,6)$$

One would be tempted to conclude that the probabilty that the sum be 7 is 3/21. This is not true, because the 21 elementary results which are considered above are not equally probable: The pair (1,1) can only be achieved in one way while the pair (1,2) can also be achieved as (2,1).

The argument requires that the elementary results be ***equally probable***. This requires that the outcomes of the two dices be considered in their order, keeping (1,2) distinct from (2,1). If we do that, there are 6 favourable results out of a total of 36 with equal probability; the probability is therefore $p = 6/36$.

In accordance with the foregoing discussion, the probability of the event E is defined as

$$P(E) = \frac{N_E}{N} \tag{2.2}$$

where N is the total number of possible outcome with ***equal probability*** and N_E is the number of these outcomes which are favourable to the event. Note that this definition is very different from the previous one, in that the probability is obtained ***a priori***, without experimentation. However, as we illustrated on an example, it must be used with care, because it is not always easy to decide what are equally probable outcomes.

2.1.3 Axioms

During a random experiment (trial) one observes the result R_i. The event E is said to have occured if R_i belongs to E. For example if the toss of a dice produced a 6, one can say that the event "6" has occured, but also the event "***even***", or the event "***larger than*** 3", etc.

Let Ω be a sample space representing the totality of the possible outcomes of the experiment. The event E is a subset of Ω to which a probability $P(E)$ is assigned. The event Ω is certain.

Probability theory is entirely based on the three following axioms:

- *Axiom 1:*

$$0 \leq P(E) \leq 1$$

- *Axiom 2:*

$$P(\Omega) = 1$$

- *Axiom 3:* If E_i are mutually exclusive (disjoint) events, that is

$$E_i \cap E_k = 0, \quad i \neq k$$

in number finite or infinite, then,

$$P[\bigcup_{i=1}^{n} E_i] = \sum_{i=1}^{n} P(E_i) \tag{2.3}$$

The notion of probability resulting from the foregoing axioms essentially agrees with the definitions given in the previous sections. One can therefore expect that the theory which is derived from them provides a satisfactory representation of the physical phenomena involved. The events being defined as sets, they are ruled by the set theory (Boolian algebra). To visualize the results, it is often useful to use Venn diagrams which, without being a formal proof, provide a physical insight which is more useful to engineers.

In the remainder of this chapter, we shall recall the main theorems of probability theory for future use in the subsequent chapters. Most of them are already familiar to the reader and we shall not attempt to demonstrate them. The demonstrations can be found in the references given at the end of the chapter. We emphasize that all the theorems follow from the axioms given above.

2.2 Theorems and definitions

- *Complement:*

The set of sample points of the sample space Ω which are not in the event E is called the *complement* of E; it is denoted by $\bar{E}$.

$$P(\bar{E}) = 1 - P(E) \tag{2.4}$$

- *Theorem of the Total Event:*

If E_1 and E_2 are two arbitrary events of a sample space,

$$P(E_1 \cup E_2) = P(E_1) + P(E_2) - P(E_1 \cap E_2) \tag{2.5}$$

This is known as the *Theorem of the Total Event*. It can be visualized with a Venn diagram. If the events are disjoint, that is, mutually exclusive, then $P(E_1 \cap E_2) = 0$ and one recovers the third axiom. For three events, by successive applications of Equ.(2.5), one easily gets

$$P(E_1 \cup E_2 \cup E_3) = P(E_1) + P(E_2) + P(E_3) - P(E_1 \cap E_2)$$
$$-P(E_2 \cap E_3) - P(E_3 \cap E_1) + P(E_1 \cap E_2 \cap E_3)$$

The generalization to more than two events reads

$$P(E_1 \cup E_2 \cup ... \cup E_n) = \sum P(E_i) - \sum P(E_i \cap E_j)$$

$$+\sum P(E_i \cap E_j \cap E_k) - ... + (-1)^{n-1} P(E_1 \cap E_2 \cap ... \cap E_n) \tag{2.6}$$

where each summation includes all distinct combinations of distinct events.

- *Conditional probability:*

The probability of the event E_2 *conditional to the occurence of event* E_1 is defined as

$$P(E_2|E_1) = \frac{P(E_1 \cap E_2)}{P(E_1)} \qquad [\text{if } P(E_1) > 0] \tag{2.7}$$

This equation can be rewritten

$$P(E_1 \cap E_2) = P(E_1)P(E_2|E_1) = P(E_2)P(E_1|E_2)$$

Two events E_1 and E_2 are *independent* if the occurence of one of them does not change the probability of occurence of the other:

$$P(E_2|E_1) = P(E_2) \qquad P(E_1|E_2) = P(E_1)$$

The probability of the joint occurence of the n events $E_1, ..., E_n$ reads

$$P(E_1 \cap E_2 \cap ... \cap E_n) = P(E_1)P(E_2|E_1)P(E_3|E_1 \cap E_2)$$
$$...P(E_n|E_1 \cap E_2 \cap ... \cap E_{n-1}) \tag{2.8}$$

If the events are independent,

$$P(E_1 \cap E_2 \cap ... \cap E_n) = P(E_1)P(E_2)...P(E_n)$$

Note that this condition is necessary, but not sufficient, for the events E_i to be independent. The events E_i are independent if the following relationships hold for all $i, j, k, ...$

$$P(E_i \cap E_j) = P(E_i)P(E_j)$$

$$P(E_i \cap E_j \cap E_k) = P(E_i)P(E_j)P(E_k)$$

$$...$$

$$P(E_1 \cap E_2 \cap ... \cap E_n) = P(E_1)P(E_2)...P(E_n)$$

Let $E_1, ..., E_n$ be a partition of the sample space Ω, that is

$$E_i \cap E_j = 0, \quad i \neq j$$

$$\Omega = E_1 \cup E_2 \cup ... \cup E_n$$

with $P(E_i) > 0$. If A is an arbitrary event of Ω,

$$P(A) = \sum_i P(A|E_i)P(E_i) \tag{2.9}$$

- *Bayes' theorem:*

If $E_1, ..., E_n$ partition the sample space Ω with $P(E_i) > 0$ for all i, and if A is an event in Ω, then

$$P(E_i|A) = \frac{P(A|E_i)P(E_i)}{\sum_{k=1}^{n} P(A|E_k)P(E_k)} \tag{2.10}$$

The unconditional probabilities $P(E_i)$ are called *a priori*, while the conditional probabilities $P(E_i|A)$ are called *a posteriori.*

As an example of the use of Bayes' theorem, suppose that the conditional probabilities $P(A|E_i)$ that a given event A occurs, given that certain causes E_i occur, is known, and that the *a priori* probabilities $P(E_i)$ are also known. If the event A is then observed to occur, Equ.(2.10) may be used to calculate the *a posteriori* probabilities $P(E_i|A)$ and thus update the probabilities of the causes from the observation of the event A.

2.3 Random variable

For most of the physical phenomena of interest to engineers, the results of random experiments consist of numerical values, such as the daily ambiant temperature, the atmospheric pressure, or the mean wind velocity. For random experiments which do not possess that property (e.g. the toss of a coin), one can always associate a numerical value to every outcome of the experiment (e.g. the value 1 is associated with the outcome *head* and 0 is associated with the outcome *tail*). It is therefore quite general to say that one can represent a random phenomenon by a random number or *random variable*, X, whose value depends on the outcome ω of the random experiment. $X(\omega)$, $\omega \in \Omega$ is a function defined on the sample space Ω.

$X(\omega)$ is a mapping of the sample space Ω on the real axis.

A random variable can be *discrete* or *continuous*. An example of the former is provided by the toss of a coin where only two values (0 and 1) are allowed, while numerous examples of the latter are met in engineering applications.

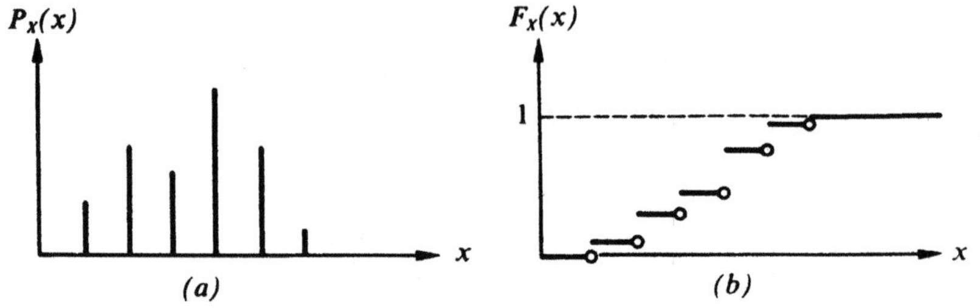

Figure 2.1: Description of a discrete random variable. (a) Probability function. (b) Probability distribution function.

2.3.1 Discrete random variable

The simplest way to describe a discrete random variable X is to supply the probability associated with each discrete value x_i: $P_X(x_i)$. This is called the ***probability function*** (Fig.2.1). For reasons which will become clearer in the next section, it is often more convenient to use an alternative representation called the ***probability distribution function***, $F_X(x)$, defined as the probability that the random variable be smaller or equal to x.

$$F_X(x) = P(X \leq x) = \sum_{x_i \leq x} P_X(x_i) \tag{2.11}$$

This function has the appearance of a staircase, with steps of amplitude $P_X(x_i)$ located at the discrete values x_i (Fig.2.1.b). Note that, because of the $\leq$ sign, $F_X(x)$ is continuous from the right at a point of discontinuity (the circled points are excluded in Fig.2.1.b). Some comments on the notations used in Equ.(2.11) are appropriate: The subscript refers to the random variable under consideration. Although, according to our conventions, capital letters are used to represent random variables, there is no ambiguity in using lower case letters instead of capital ones in subscripts. This will be frequently done in later chapters for aesthetic reasons. As a result, $F_x(x)$ is completely equivalent to $F_X(x)$.

By definition, a probability distribution function is a monotonically non decreasing function of its argument satisfying

$$F_X(-\infty) = 0 \qquad F_X(+\infty) = 1 \tag{2.12}$$

2.3.2 Continuous random variable

The probability associated with any specific value of a continuous random variable is zero. The probability function is therefore inappropriate to represent

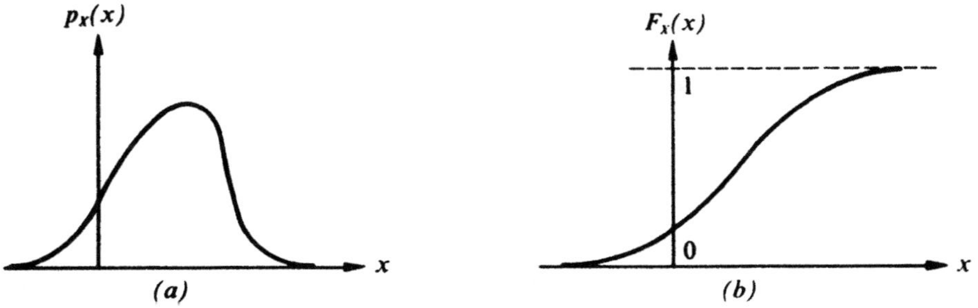

Figure 2.2: Description of a continuous random variable. (a) Probability density function. (b) Probability distribution function.

continuous random variables. The probability distribution function, $F_X(x)$, represents the probability that X be smaller or equal to x. It is therefore appropriate for the representation of continuous random variables as well as discrete ones. The most frequent characterization of continuous random variables, however, is the ***probability density function*** (One often uses the acronym *PDF*) (Fig.2.2):

$$p_X(x) = \frac{\mathrm{d}F_X(x)}{\mathrm{d}x} \tag{2.13}$$

Clearly,

$$p_X(x)\mathrm{d}x = F_X(x + \mathrm{d}x) - F_X(x) = P(x < X \leq x + \mathrm{d}x) \tag{2.14}$$

represents the probability that the random variable X takes a value between x and $x + \mathrm{d}x$.The inverse relation is

$$F_X(x) = \int_{-\infty}^{x} p_X(y)\mathrm{d}y \tag{2.15}$$

with the initial condition $F_X(-\infty) = 0$. It is a continuous, non decreasing function of x. From the second condition (2.12), one gets the normalization condition

$$\int_{-\infty}^{+\infty} p_X(x)\mathrm{d}x = 1 \tag{2.16}$$

If one uses Dirac delta functions, the probability density function can be extended to continuous random variables with a countable number of discrete values x_i with nonzero probability:

$$p_X(x) = p_X^c(x) + \sum_i P_X(x_i)\delta(x - x_i) \tag{2.17}$$

where $p_X^c(x)$ refers to the continuous part of the random variable and $P_X(x_i)$ are the probabilities associated with the discrete values x_i. Integrating this equation, one gets

$$F_X(x) = \int_{-\infty}^{x} p_X^c(y)\mathrm{d}y + \sum_{x_i \le x} P_X(x_i) \tag{2.18}$$

$F_X(x)$ is continuous non decreasing between the discrete values x_i, with steps of amplitude $P_X(x_i)$ at x_i. The normalization condition reads

$$\int_{-\infty}^{+\infty} p_X^c(x)\mathrm{d}x + \sum_i P_X(x_i) = 1 \tag{2.19}$$

2.4 Jointly distributed random variables

Let us first consider two random variables, X_1 and X_2. Their individual behaviour can be characterized as in the foregoing section. This decription, however, does not provide any information on their interrelationship. That additional information is available from the *joint* distribution of X_1 and X_2. The joint probability distribution function is defined as

$$F_{X_1X_2}(x_1, x_2) = P[\{X_1 \le x_1\} \cap \{X_2 \le x_2\}] \tag{2.20}$$

Just as in the case of a single random variable, it is a non decreasing function of both of its arguments and, because of the $\le$ signs, it is continuous from the right. The joint probability distribution function satisty the following conditions which are straightforward deductions from the definition (2.20):

$$F_{X_1X_2}(-\infty, x_2) = F_{X_1X_2}(x_1, -\infty) = 0$$

$$F_{X_1X_2}(\infty, \infty) = 1 \tag{2.21}$$

$$F_{X_1X_2}(x_1, \infty) = F_{X_1}(x_1)$$

$$F_{X_1X_2}(\infty, x_2) = F_{X_2}(x_2)$$

Thus, one sees that the probability distribution function of the individual random variables can be recovered from the joint probability distribution function. Just as for a single random variable, the *joint* probability density function is defined as

$$p_{X_1X_2}(x_1, x_2) = \frac{\partial^2}{\partial x_1 \partial x_2} F_{X_1X_2}(x_1, x_2) \tag{2.22}$$

It is a non negative function of its arguments. $p_{X_1X_2}(x_1, x_2)\mathrm{d}x_1\mathrm{d}x_2$ represents the probability that $[\{x_1 < X_1 \le x_1 + \mathrm{d}x_1\} \cap \{x_2 < X_2 \le x_2 + \mathrm{d}x_2\}]$. The inverse relationship is

$$F_{X_1X_2}(x_1, x_2) = \int_{-\infty}^{x_1} \int_{-\infty}^{x_2} p_{X_1X_2}(y_1, y_2)\mathrm{d}y_1\mathrm{d}y_2 \tag{2.23}$$

From the third equation (2.21),

$$F_{X_1}(x_1) = P(X_1 \le x_1) = \int_{-\infty}^{x_1} \mathrm{d}y_1 \int_{-\infty}^{\infty} p_{X_1X_2}(y_1, y_2)\,\mathrm{d}y_2 \tag{2.24}$$

Comparing this equation to (2.15), one sees that the probability density function of an individual random variable can be obtained by partial integration of the joint probability density function over the whole range of values of the other random variable:

$$p_{X_1}(x_1) = \int_{-\infty}^{\infty} p_{X_1X_2}(x_1, x_2)\mathrm{d}x_2 \tag{2.25}$$

A direct consequence of (2.16) is that the normalization condition for the joint probability density function is

$$\int_{-\infty}^{\infty}\int_{-\infty}^{\infty} p_{X_1X_2}(x_1, x_2)\mathrm{d}x_1\mathrm{d}x_2 = 1 \tag{2.26}$$

The concept of joint probability distribution function and joint probability density function can be extended to more than two random variables without difficulty. For n random variables $X_1, \ldots, X_n$, they are defined as

$$F_{X_1\ldots X_n}(x_1, \ldots, x_n) = P[\{X_1 \le x_1\} \cap \ldots \cap \{X_n \le x_n\}] \tag{2.27}$$

$$p_{X_1\ldots X_n}(x_1, \ldots, x_n) = \frac{\partial^n}{\partial x_1 \ldots \partial x_n} F_{X_1\ldots X_n}(x_1, \ldots, x_n) \tag{2.28}$$

The joint distribution of order n contains all the information contained in all the distributions of lower orders; the latter can always be recovered by partial integration as in (2.26).

2.5 Conditional distribution

Let X and Y be two discrete random variables with joint probability function $P_{XY}(x, y)$. The probability that $X = x$, under the condition that $Y = y$, is given by

$$P_{X|Y}(x|y) = \frac{P_{XY}(x, y)}{P_Y(y)} \tag{2.29}$$

provided $P_Y(y) > 0$. This equation is a direct consequence of (2.7). X and Y are independent random variables if $P_{X|Y}(x|y) = P_X(x)$.

For continuous random variables, the *conditional* probability density function of X is defined as

$$p_{X|Y}(x|y) = \frac{p_{XY}(x, y)}{p_Y(y)} = \frac{p_{XY}(x, y)}{\int_{-\infty}^{\infty} p_{XY}(x, y)\mathrm{d}x} \tag{2.30}$$

$p_{X|Y}(x|y)\mathrm{d}x$ represents the probability that $\{x < X \leq x + \mathrm{d}x\}$, under the condition $\{y < Y \leq y + \mathrm{d}y\}$. From this equation, one sees that the conditional probability density is completely defined by the joint probability density. Two continuous random variables x and y are independent if the condition does not affect the probability density:

$$p_{X|Y}(x|y) = p_X(x) \tag{2.31}$$

Equation (2.30) shows that this is equivalent to

$$p_{XY}(x,y) = p_X(x)\,p_Y(y) \tag{2.32}$$

2.6 Functions of Random variables

A random variable has been defined as a mapping of the sample space on the real axis. The choice of that mapping may not be unique. For example, consider a particle with a random velocity; the magnitude of the velocity, V, can be taken as the random variable. However, depending on the problem, it may be just as relevant to use the kinetic energy, K, as the random variable. These two random variables constitute two different mappings of the same random experiment. Of course, they are related by the equation $K = \frac{1}{2}mV^2$. K can be considered as a function of V or vice versa. How can we determine the probability distribution of K from that of V?

2.6.1 Function of one random variable

Let $Y = f(X)$ be a function of the random variable X. Y is also a random variable. Its probability distribution function is defined as the mass of probability attached to the value $Y \leq y$. It is equal to the mass of probability associated to the values of X belonging to the domain D_x such that $f(X) \leq y$:

$$F_Y(y) = P[Y \leq y] = P(X \in D_x) \tag{2.33}$$

where

$$D_x \equiv \{x : f(x) \leq y\}$$

Let us illustrate this with an example. Consider the function $Y = 1/X^2$: Clearly,

$$D_x \equiv \{(-\infty, -y^{-1/2}] \cup [y^{-1/2}, \infty)\}$$

From (2.33),

$$\begin{aligned} F_Y(y) &= P(X \leq -y^{-1/2}) + P(X \geq y^{-1/2}) \\ &= F_X(-y^{-1/2}) + 1 - F_X(y^{-1/2}) \end{aligned}$$

where F_X is the probability distribution function of X. Differentiating this equation would give the probabilty density function.

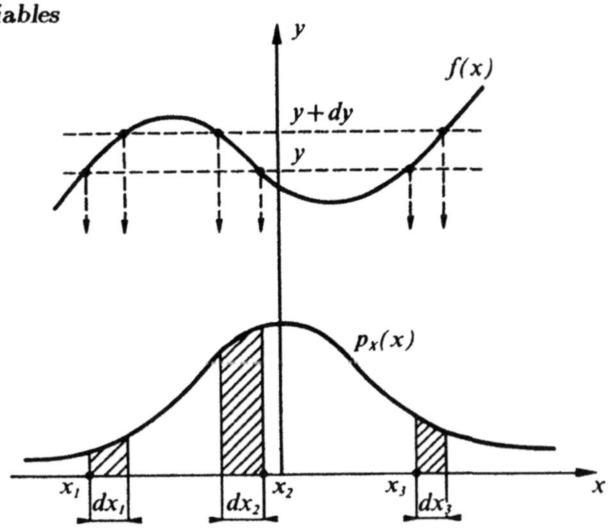

Figure 2.3: Function of one random variable of probability density $p_X(x)$.

As an alternative to the foregoing procedure, assume that $y = f(x)$ is continuous and has, for each value of y, a countable number of real roots $x_1, x_2, \ldots$ ($y = f(x_1) = f(x_2) = \ldots$). From Fig.2.3, one notices that Y is within $(y, y + \mathrm{d}y]$ if X is within either of the intervals $(x_i, x_i + \mathrm{d}x_i]$. The corresponding probability equality is

$$\begin{aligned} p_Y(y)\mathrm{d}y &= P[\{y < Y \le y + \mathrm{d}y\}] \\ &= P[\{\{x_1 < X \le x_1 + \mathrm{d}x_1\} \cup \{x_2 < X \le x_2 + \mathrm{d}x_2\} \cup \ldots] \\ &= P[\{x_1 < X \le x_1 + \mathrm{d}x_1\}] + P[\{x_2 < X \le x_2 + \mathrm{d}x_2\}] + \ldots \end{aligned}$$

or

$$p_Y(y)\mathrm{d}y = p_X(x_1)\mathrm{d}x_1 + p_X(x_2)\mathrm{d}x_2 + \ldots \tag{2.34}$$

where we have used the fact that the intervals are disjoint (mutually exclusive events). It remains to relate the increments $\mathrm{d}x_i$ to $\mathrm{d}y$. To do that, we use the functional relationship between x and y. Figure 2.3 shows that

$$\mathrm{d}x_i = \frac{\mathrm{d}y}{|f'(x_i)|} = \frac{\mathrm{d}y}{|(\frac{\mathrm{d}y}{\mathrm{d}x})_{x=x_i}|} \tag{2.35}$$

where the absolute sign takes into account the fact that the mass of probability relative to each interval must be taken with a positive sign. Upon substituting into Equ.(2.34), one gets

$$p_Y(y) = \sum_i \frac{p_X(x_i)}{|f'(x_i)|} + \ldots \tag{2.36}$$

where the sum extends to all the roots of $y = f(x_i)$. A couple of examples will illustrate the procedure.

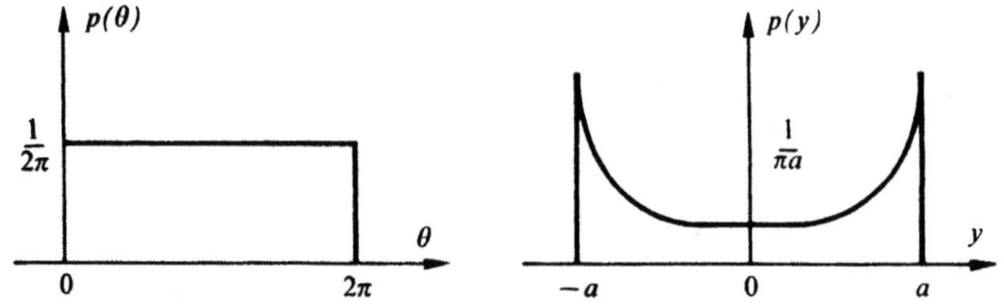

Figure 2.4: (a) Uniform distribution $p_\theta(\theta)$. (b) $p_Y(y)$ for the transformation $Y = a\sin\theta$.

First, consider the linear transformation $Y = aX + b$. For any value of y, there is a single solution $x = (y-b)/a$. Since $f'(x) = a$, one gets from (2.36)

$$p_Y(y) = \frac{1}{|a|}p_X(\frac{y-b}{a}) \tag{2.37}$$

Next, consider the transformation

$$Y = a\sin\theta \tag{2.38}$$

where a is a positive constant and θ is a random variable with uniform distribution between 0 and 2π (Fig.2.4). Any value of y between $-a$ and $+a$ corresponds to two possible values of θ. For each of them, the slope satisfies

$$|\frac{dy}{d\theta}| = |a\cos\theta| = a\sqrt{1-\sin^2\theta} = \sqrt{a^2-y^2}$$

(the absolute value of the slope is the same for the two roots). From Equ.(2.36), one gets that, within $[-a, a]$,

$$p_Y(y) = \frac{1}{\pi}\frac{1}{\sqrt{a^2-y^2}} \qquad (-a \le y \le a) \tag{2.39}$$

$p_Y(y) = 0$ outside. The distribution is represented in Fig.2.4.

2.6.2 Function of two random variables

Consider the random variable $Z = g(X,Y)$ function of the two random variables X and Y with joint probability density function $p_{XY}(x,y)$. The probability distribution function of Z reads

$$F_Z(z) = P(Z \le z) = P[\{(X,Y) \in D_z\}] = \int\int_{D_z} p_{XY}(x,y)\,dx dy \tag{2.40}$$

where D_z represents the region in the plane (x,y) corresponding to $g(x,y) \le z$. The probability density function is obtained by differentiation with respect to z.

2.6.3 The sum of two independent random variables

Consider $Z = X + Y$. The domain D_z in Equ.(2.40) is defined by $x + y \leq z$. The integral can therefore be written as

$$F_Z(z) = \int_{-\infty}^{\infty} dy \int_{-\infty}^{z-y} p_{XY}(x,y)\, dx$$

Upon differentiating with respect to z, one gets

$$p_Z(z) = \int_{-\infty}^{\infty} p_{XY}(z-y,y)\, dy \tag{2.41}$$

So far, we have not made any assumption about the joint probability density of X and Y. If they are independent, $p_{XY}(x,y) = p_X(x)\, p_Y(y)$. Introducing this into the above equation, one gets the interesting result that the probabilty density function of z is the ***convolution*** of the probability densities of the random variables contributing to the sum:

$$p_Z(z) = \int_{-\infty}^{\infty} p_X(z-y) p_Y(y)\, dy = \int_{-\infty}^{\infty} p_X(x) p_Y(z-x)\, dx = p_X(z) \star p_Y(z) \tag{2.42}$$

This result can be extended to the sum of an arbitrary number of independent random variables. If $Z = \sum X_i$,

$$p_Z(z) = p_{X_1}(z) \star p_{X_2}(z) \star \ldots p_{X_n}(z) \tag{2.43}$$

If the function has the form $Z = aX + bY$, it is easy to combine the above result with that of Equ.(2.37). As a first step, the random variables $X_1 = aX$ and $Y_1 = bY$ are introduced; their probability density functions are derived from Equ.(2.37). The foregoing result is then used with $Z = X_1 + Y_1$.

2.6.4 Rayleigh distribution

Consider the random variable $Z = \sqrt{X^2 + Y^2}$. The domain D_z is defined as the region of the plane (x, y) such that $x^2 + y^2 \leq z^2$. It consists of a circle of radius z. Equation (2.40) reads

$$F_Z(z) = \int\int_{x^2+y^2 \leq z^2} p_{XY}(x,y)\, dx dy \qquad (z \geq 0) \tag{2.44}$$

Now, anticipating chapter 4, let us assume that the random variables X and Y are Gaussian, independent, with zero mean and the same standard deviation σ. As we shall see, this implies that their joint probability density function is

$$p_{XY}(x,y) = \frac{1}{2\pi\sigma^2} \exp\left(-\frac{x^2+y^2}{2\sigma^2}\right) \tag{2.45}$$

This form must be introduced in Equ.(2.44). Since the domain of integration is a circle, it is convenient to change into polar coordinates:

$$x = r\cos\theta$$

$$y = r\sin\theta$$

The determinant of the Jacobian is r. With this new set of coordinates, Equ.(2.44) becomes nicely decoupled:

$$F_Z(z) = \frac{1}{2\pi\sigma^2}\int_0^{2\pi} d\theta \int_0^z r\exp(-\frac{r^2}{2\sigma^2})\,dr \qquad (z \geq 0)$$

$$= 1 - \exp(-\frac{z^2}{2\sigma^2})$$

Upon differentiating with respect to z, one gets

$$p_Z(z) = \frac{z}{\sigma^2}\exp(-\frac{z^2}{2\sigma^2}) \qquad (z \geq 0) \tag{2.46}$$

This distribution is known as the ***Rayleigh distribution***; it will be of great interest later in this book, because it describes the distribution of the envelope of a narrow band process.

2.6.5 n functions of n random variables

Let $Y_1, \dots, Y_n$ be a set of n functions of the n random variables $X_1, \dots, X_n$ with joint probability density function $p_{X_1 \dots X_n}(x_1, \dots, x_n)$. One wishes to find the joint probability density function $p_{Y_1 \dots Y_n}(y_1, \dots, y_n)$. We assume that the mapping between X_i and Y_k is one-to-one, so that the transformation

$$Y_k = f_k(X_1, \dots, X_n) \tag{2.47}$$

can be inverted

$$X_i = g_i(Y_1, \dots, Y_n) \tag{2.48}$$

If $\mathcal{D}$ represents an arbitrary domain in the space X_i and $\mathcal{D}'$ is the corresponding domain in the space Y_i, the conservation of the mass of probability implies

$$\int_{\mathcal{D}} p_{X_1 \dots X_n}(x_1, \dots, x_n)dx_1 \dots dx_n = \int_{\mathcal{D}'} p_{Y_1 \dots Y_n}(y_1, \dots, y_n)dy_1 \dots dy_n \tag{2.49}$$

Taking into account the theorem of change of variables in integrals, one gets

$$p_{Y_1 \dots Y_n}(y_1, \dots, y_n) = p_{X_1 \dots X_n}(x_1, \dots, x_n)|\det(\frac{\partial x_i}{\partial y_j})| \tag{2.50}$$

where $J = \frac{\partial x_i}{\partial y_j}$ is the Jacobian of the transformation.

When the number of random variables $Y_1, \ldots, Y_m$ is lower than that of $X_1, \ldots, X_n$, the above procedure can still be applied by first introducing the dummy variables $Y_{m+1} = X_{m+1}, \ldots, Y_n = X_n$. Once $p_{Y_1 \ldots Y_n}(y_1, \ldots, y_n)$ has been obtained, $p_{Y_1 \ldots Y_m}(y_1, \ldots, y_m)$ can be recovered by partial integration on $y_{m+1}, \ldots, y_n$.

2.7 Moments

2.7.1 Expected value

The ***expected value*** or ***expectation*** or ***mean*** of a random variable X is defined as

$$\mu_X = \mathrm{E}[X] = \int_{-\infty}^{\infty} x\, p_X(x)\, \mathrm{d}x \tag{2.51}$$

If $Y = f(X)$ is a function of the random variable X, its expected value can be calculated ***without prior determination of its probability density function***:

$$\mathrm{E}[Y] = \int_{-\infty}^{\infty} y\, p_Y(y)\, \mathrm{d}y = \int_{-\infty}^{\infty} f(x)\, p_X(x)\, \mathrm{d}x \tag{2.52}$$

This result can be generalized to functions of several random variables

$$\mathrm{E}[f(X_1, \ldots, X_n)] = \int_{-\infty}^{\infty} \cdots \int f(x_1, \ldots, x_n)\, p_{X_1 \ldots X_n}(x_1, \ldots, x_n)\, \mathrm{d}x_1 \ldots \mathrm{d}x_n \tag{2.53}$$

If X and Y are jointly distributed random variables, the ***conditional expectation*** of X, subject to the condition that $Y = y$, is defined as in (2.51), except that the conditional *PDF* is used instead of the *PDF*:

$$\mathrm{E}[X|Y = y] = \int_{-\infty}^{\infty} x\, p_{X|Y}(x|y)\, \mathrm{d}x \tag{2.54}$$

2.7.2 Moments

The expected value $\mathrm{E}[X]$ is the ***first order moment*** of the random variable X. The ***moment of order*** n of X and the ***joint moment of order*** $n + m$ of X and Y are defined respectively as

$$\mathrm{E}[X^n] \qquad \mathrm{E}[X^n Y^m] \tag{2.55}$$

A moment of arbitrary order can be calculated from the probability density function. Conversely, except for special situations like for the Gaussian distribution, the moments of all orders are necessary to characterize completely the probability distribution.

The *central moment* of order n is defined as the moment of order n of the deviation of X with respect to its mean value μ_X. Similarly, the *joint central moment* of order $n+m$ of X and Y is defined as the moment of order $n+m$ of the deviations with respect to the means

$$E[(X-\mu_X)^n] \qquad E[(X-\mu_X)^n(Y-\mu_Y)^m] \tag{2.56}$$

The central moments of order 2 are especially important

$$\sigma_X^2 = E[(X-\mu_X)^2] = E[X^2] - \mu_X^2 \tag{2.57}$$

$$\kappa_{XY} = E[(X-\mu_X)(Y-\mu_Y)] = E[XY] - \mu_X\mu_Y \tag{2.58}$$

σ_X^2 is called the *variance*, σ_X is the *standard deviation*, and κ_{XY} is called the *covariance* of X and Y.

2.7.3 Schwarz inequality

Consider the quadratic form

$$E[(aX-Y)^2] = a^2E[X^2] - 2aE[XY] + E[Y^2] \geq 0$$

where a is an arbitrary real number. This second order polynomial in a, being non-negative, must have a non-positive discriminant. This implies

$$E[XY]^2 \leq E[X^2]E[Y^2] \tag{2.59}$$

This important relation is known as the *Schwarz inequality*. A direct consequence is that the *coefficient of correlation* ϱ_{XY} satisfies the following inequality:

$$-1 \leq \varrho_{XY} = \frac{\kappa_{XY}}{\sigma_X\sigma_Y} \leq 1 \tag{2.60}$$

If $\varrho_{XY} = 0$, the random variables are *uncorrelated* or *linearly independent*. Note that this is not in general sufficient for being independent in the sense of Equ.(2.32). As we shall see later, being uncorrelated is a necessary and sufficient condition for *Gaussian* random variables to be independent.

If $Y = X_1 + \ldots + X_n$ is the sum of uncorrelated random variables, it is easily established that the variance of Y is the sum of the variances of the contributing random variables (problem P.2.7):

$$\sigma_Y^2 = \sum_{i=1}^{n} \sigma_{X_i}^2 \tag{2.61}$$

2.7.4 Chebyshev's inequality

The standard deviation σ_X measures the ***dispersion*** of the random variable X about its mean μ_X. A beautiful interpretation is provided by ***Chebyshev's inequality***, which states an upper bound on the probability that X be at some distance from its mean. More precisely, the probability that the deviation with respect to the mean, $|X - \mu_X|$, be larger than h times σ_X, has the following upper bound

$$P[|X - \mu_X| \geq h\sigma_X] \leq \frac{1}{h^2} \qquad (h > 0) \tag{2.62}$$

The reader already familiar with the Gaussian distribution will observe that this inequality is fairly conservative. Indeed, for $h = 3$, it provides a probability of exceedance of 1/9, while it is well known that, for the Gaussian distribution, this probability is as low as 0.003. Note, however, that Chebyshev's inequality applies to ***any*** probability distribution.

2.8 Characterstic function, Cumulants

2.8.1 Single random variable

The ***characteristic function*** of the random variable X is defined as

$$M_X(\theta) = \mathrm{E}[e^{j\theta X}] = \int_{-\infty}^{\infty} e^{j\theta x} p_X(x)\, \mathrm{d}x \tag{2.63}$$

It is the Fourier transform of the probability density function and, as a result, contains the same information. It is a complex function of the real parameter θ; it always exists, because the *PDF* is absolutely integrable. The inverse relation reads

$$p_X(x) = \frac{1}{2\pi} \int_{-\infty}^{\infty} e^{-j\theta x} M_X(\theta)\, \mathrm{d}\theta \tag{2.64}$$

From Equ.(2.63), it is readily verified that the moments are related to the characteristic function by

$$\mathrm{E}[X] = \frac{1}{j}(\frac{\mathrm{d}M_X}{\mathrm{d}\theta})_{\theta=0}$$

$$\mathrm{E}[X^n] = \frac{1}{j^n}(\frac{\mathrm{d}^n M_X}{\mathrm{d}\theta^n})_{\theta=0} \tag{2.65}$$

The Maclaurin series expansion of the characteristic function reads

$$M_X(\theta) = 1 + \sum_{n=1}^{\infty} \frac{(j\theta)^n}{n!} \mathrm{E}[X^n] \tag{2.66}$$

This shows that, in general, the moments of all orders are necessary to specify the probability distribution of a random variable. As we shall see, the first two moments are enough to specify a ***Gaussian*** random variable, because the higher order moments can be expressed in terms of the first two.

The following alternative series expansion is often preferable to (2.66):

$$M_X(\theta) = \exp\{\sum_{n=1}^{\infty} \frac{(j\theta)^n}{n!} \kappa_n[X]\} \tag{2.67}$$

where $\kappa_n[X]$ is defined as the ***cumulant of order n***. Upon expanding the exponential and identifying the power of θ of increasing orders, one gets the relationship between the cumulants and the moments:

$$\kappa_1 = \mathrm{E}[X] = \mu_X$$

$$\kappa_2 = \mathrm{E}[X^2] - \mu_X^2 = \sigma_X^2$$

$$\kappa_3 = \mathrm{E}[X^3] - 3\mu_X \mathrm{E}[X^2] + 2\mu_X^3$$

$$\kappa_4 = \mathrm{E}[X^4] - 3\mathrm{E}[X^2]^2 - 4\mu_X \mathrm{E}[X^3] + 12\mathrm{E}[X^2]\mu_X^2 - 6\mu_X^4$$

$$\cdots$$

Observe that κ_1 is the mean and κ_2 is the variance. It may be checked that κ_3 is equal to the third central moment too, but this is no longer true for higher orders. From the above equations, one sees that the cumulant of order n can be expressed in terms of the moments up to order n and vice versa; this means that the cumulants of order up to n contain the same information as the moments up to the same order:

$$\kappa_1, \ldots \kappa_n \leftrightarrow \mathrm{E}[X], \ldots \mathrm{E}[X^n]$$

However, to understand the difference, consider the second cumulant $\kappa_2 = \sigma_X^2$ and the second moment $\mathrm{E}[X^2]$. The former describes the magnitude of the deviation with respect to the mean and does not contain any information on the magnitude of the random variable as the mean square value does. The information on the magnitude is supplied by the cumulant of the first order, κ_1. Thus, the cumulants present the information in such a way that their importance decreases as their order increases. Unlike Equ.(2.66), Equ.(2.67) can be truncated after a limited number of terms without losing much signficance. We shall see later that having the cumulants of orders higher than two equal to zero is a necessary and sufficient condition for the random variable to be *Gaussian*. From Equ.(2.67), one sees that

$$\kappa_n[X] = \frac{1}{j^n}\{\frac{\mathrm{d}^n}{\mathrm{d}\theta^n} \ln M_X(\theta)\}_{\theta=0} \tag{2.68}$$

Since the characteristic function and the probability density function are a Fourier transform pair, they constitute completely equivalent characterizations of the random variable X. Whether we use one rather than the other will depend on circumstances. One situation where the characteristic function is more appealing than the density function is for the sum of independent random variables. In fact, Equ.(2.43) tells us that the probabilty density function of the sum of independent random variables is equal to the ***convolution*** of the probability density functions of the contributing randon variables. From the properties of the Fourier transform, we know that the corresponding characteristic function is equal to the ***product*** of the characteristic functions of the contributing random varibles. Equation (2.43) is therefore equivalent to

$$M_Z(\theta) = M_{X_1}(\theta)\ldots M_{X_n}(\theta) \tag{2.69}$$

It is also often simpler to calculate the moments from Equ.(2.65) than from the integrals involving the probability density function.

2.8.2 Jointly distributed random variables

The characteristic function of n jointly ditributed random variables is defined as

$$M_{X_1\ldots X_n}(\theta_1,\ldots,\theta_n) = \mathrm{E}[e^{j(\theta_1 X_1+\ldots+\theta_n X_n)}] \tag{2.70}$$

$$= \int_{-\infty}^{\infty}\ldots\int p_{X_1\ldots X_n}(x_1,\ldots,x_n)e^{j(\theta_1 x_1+\ldots+\theta_n x_n)}\mathrm{d}x_1\ldots\mathrm{d}x_n$$

It is the n–fold Fourier transform of the joint probability density function. As in the discussion in the previous section, one may readily verify that the joint moments are related to the partial derivatives by

$$\mathrm{E}[X_1\ldots X_n] = \frac{1}{j^n}\{\frac{\partial^n}{\partial\theta_1\ldots\partial\theta_n}M_{X_1\ldots X_2}\}_{\theta_1=\ldots=\theta_n=0} \tag{2.71}$$

The Maclaurin series expansion reads

$$M_{X_1\ldots X_n}(\theta_1,\ldots,\theta_n) = 1 + j\theta_k\mathrm{E}[X_k] + \frac{1}{2!}(j\theta_k)(j\theta_l)\mathrm{E}[X_k X_l] + \ldots \tag{2.72}$$

where the repeated indices $k, l, \ldots$ indicate summations from 1 to n. All this is a direct extension of the previous section. Similarly, the ***joint cumulants*** are defined by the following expansion

$$M_{X_1\ldots X_n}(\theta_1,\ldots,\theta_n) = \exp\{j\theta_k\kappa_1[X_k] + \frac{1}{2!}(j\theta_k)(j\theta_l)\kappa_2[X_k, X_l] + \ldots\} \tag{2.73}$$

$\kappa_m[X_1,\ldots,X_m]$ is the joint cumulant of order m; it measures the multiple correlation between the random variables $X_1,\ldots,X_m$; it vanishes if at least one of the random variables is linearly independent of all the others; it is related to the characteristic function by

$$\kappa_m[X_1,\ldots,X_m] = \frac{1}{j^n}\{\frac{\partial^m}{\partial\theta_1\ldots\partial\theta_m}\ln M_{X_1\ldots X_2}\}_{\theta_1=\ldots=\theta_n=0} \tag{2.74}$$

Just as for the case of a single random variable, the first three joint cumulants are identical to the joint central moments

$$\kappa_2[X_1,X_2] = \mathrm{E}[(X_1-\mu_{X_1})(X_2-\mu_{X_2})] \tag{2.75}$$

$$\kappa_3[X_1,X_2,X_3] = \mathrm{E}[(X_1-\mu_{X_1})(X_2-\mu_{X_2})(X_3-\mu_{X_3})]$$

If the random variables $X_1,\ldots,X_n$ are independent, their joint probability density can be partitioned as the product of the first order densities:

$$p_{X_1\ldots X_n}(x_1,\ldots,x_n) = \prod_{i=1}^{n} p_{X_i}(x_i) \tag{2.76}$$

If one substitutes this in the definition (2.70), one gets

$$M_{X_1\ldots X_n}(\theta_1,\ldots,\theta_n) = \prod_{i=1}^{n} M_{X_i}(\theta_i) \tag{2.77}$$

Either of these equations is a necessary and sufficient condition for the random variables $X_1,\ldots,X_n$ to be independent.

2.9 References

W.B.DAVENPORT, *Probability and Random Processes*, McGraw Hill, 1970.
Y.K.LIN, *Probabilistic Theory of Structural Dynamics*, McGraw Hill, 1967.
A.PAPOULIS, *Probability, Random Variables and Stochastic Processes*, McGraw Hill, 1965.
E.PARZEN, *Stochastic Processes*, Holden Day, 1962.
R.L.STRATONOVICH, *Topics in the Theory of Random Noise*, 1, Gordon & Breach, N-Y, 1963.

2.10 Problems

P.2.1 *If $B \subset A$ (B implies A), show that*

$$P[B|A] \geq P[B]$$

P.2.2 *Consider n independent tosses of a fair coin [equal probability of occurence of head (H) and tail (T)]. Show that the probability that k heads occur is*

$$P[kH] = \frac{1}{2^n}\frac{n!}{k!(n-k)!}$$

Plot this distribution for $n = 4$ and $n = 8$. This result is known as the binomial distribution; *it will be used in chapter 4.*

P.2.3 *Consider the* exponential *probability density function*

$$p_X(x) = ae^{-ax} \qquad (x \geq 0)$$

where a is a positive number. Compute the mean μ_X and standard deviation σ_X. Compute the characteristic function and check that that the previous results can be recovered simply by application of formula (2.65).

P.2.4 *Consider the* unity Gaussian distribution

$$p_X(x) = \frac{1}{\sqrt{2\pi}}\exp[-\frac{x^2}{2}]$$

Show that the characteristic function is

$$M_X(\theta) = \exp[-\frac{\theta^2}{2}]$$

P.2.5 *Consider the function $f(X_1,\dots,X_n) = \prod f_i(X_i)$ of the independent random variables X_i. Show that*

$$\mathrm{E}[f(X_1,\dots,X_n)] = \prod \mathrm{E}[f_i(X_i)]$$

P.2.6 *The conditional expectation $\mathrm{E}[X|Y = y]$ can be considered as a function of the random variable Y. Show that*

$$\mathrm{E}[\mathrm{E}[X|Y = y]] = \mathrm{E}[X]$$

P.2.7 *If X_i are uncorrelated random variables, show that the variance of $Y = X_1 + \dots + X_n$ is*

$$\sigma_Y^2 = \sum_{i=1}^{n} \sigma_{X_i}^2$$

P.2.8 *Let X be a random variable uniformly distributed over $[-\pi,\pi]$*

$$p_X(x) = \frac{1}{2\pi} \qquad -\pi \leq x \leq \pi$$

and Y a random variable with a Rayleigh distribution

$$p_Y(y) = y\exp[-\frac{y^2}{2}]$$

Show that, if X and Y are independent, the random variable $Z = Y \sin X$ has a Gaussian distribution

$$p_Z(z) = \frac{1}{\sqrt{2\pi}} \exp[-\frac{z^2}{2}]$$

This interesting fact is closely related to the definition of the envelope of a narrow band random process *(chapters 5 and 10).*

P.2.9 *Show that the mean and the mean square value of the Rayleigh distribution (2.46) are respectively*

$$\mu_X = \sqrt{\frac{\pi}{2}}\sigma \qquad E[X^2] = 2\sigma^2$$

P.2.10 *Find the probability density function $p_Y(y)$ of $Y = X^2$, if X is uniformly distributed in $[-1, 1]$.*

P.2.11 *Show that the characteristic function of $Y = aX + b$ is*

$$M_Y(\theta) = e^{j\theta b} M_X(a\theta)$$

Chapter 3

Random Processes

3.1 Introduction

Consider a random experiment where the outcomes are no longer numbers (as in the case of a random random variable), but functions of time, $X(\omega, t)$, or other parameters. *A random process is a parametrized family of random variables.* When there are several parameters, as for example the space coordinates, it is called a *random field*. A sea wave is an example of a random field: at a fixed point, the sea level is a random process with the time as parameter; at a fixed time, the sea surface is a random field with the space coordinates as parameters. The random process $X(\omega, t)$ can represent four different things:

- A family of functions of time (t and ω variables);
- A function of time (t variable, ω fixed);
- A random variable (t fixed, ω variable);
- A number (t fixed, ω fixed).

If the parameter is discontinuous, a random process is called a *random sequence*. A random process is called *discrete* if it can take only discrete values. The terminology is therefore as follows

- *Continuous process* if t and X are continuous;
- *Discrete process* if t is continuous and X is discrete;
- *Continuous sequence* if X is continuous and t is discrete;
- *Discrete sequence* if t and X are discrete.

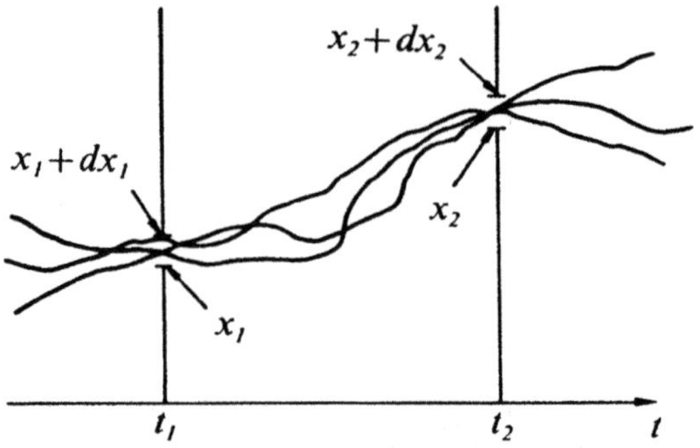

Figure 3.1: Interpretation of the second order probability density function.

3.2 Specification of a random process

There are many different ways of specifying a random process; they are reviewed below.

3.2.1 Probability density functions

Just as for random variables, the most natural way to specify a random process is by its probability density functions of increasing orders: The first density

$$p_X(x,t)$$

supplies the probability structure of the random variables $X(t)$ for every fixed value of the parameter t. It does not reflect the interdependency between the values of the random function at different times. To specify this, one needs the joint probability densities of higher orders

$$p_X(x_1,t_1;x_2,t_2)$$

$$\ldots$$

$$p_X(x_1,t_1;\ldots;x_n,t_n)$$

They are all non-negative functions, symmetric with respect to their arguments. They satisfy the normalization condition

$$\int_{-\infty}^{\infty}\ldots\int_{-\infty}^{\infty} p_X(x_1,t_1;\ldots;x_n,t_n)dx_1\ldots dx_n = 1 \tag{3.1}$$

The meaning of the second order probability density function is illustrated in Fig.3.1:

$$p_X(x_1,t_1;x_2,t_2)dx_1dx_2$$

represents the probability that the value of the process belongs to $(x_1, x_1+dx_1]$ at t_1 and to $(x_2, x_2+dx_2]$ at t_2. Although it contains more information than the density of the first order, the second order density is not in general sufficient to characterize completely a random process; this requires the probability densities of all orders ($n = 1, 2, 3, \ldots$). Note that, as we have seen in chapter 2, the probability densities of lower orders can always be recovered from higher order ones by partial integration

$$p_X(x_1, t_1; \ldots; x_n, t_n) = \int_{-\infty}^{\infty} \cdots \int_{-\infty}^{\infty} p_X(x_1, t_1; \ldots; x_{n+k}, t_{n+k}) dx_{n+1} \ldots dx_{n+k} \tag{3.2}$$

When there are two random processes, $X(t)$ and $Y(t)$, their interdependency must be specified in addition to their individual behaviour. This can be done by the joint densities of increasing orders:

$$p_{XY}(x, t; y, s)\, dx\, dy$$

represents the probability that $X(t)$ belongs to $(x, x + dx]$ at time t and $Y(t)$ belongs to $(y, y + dy]$ at time s. Higher order joint density functions are defined in a similar manner.

In general, the probability densities of all orders are necessary to specify a random process completely. Two special cases are very important:

A ***purely random process*** is such that the values of the process for different times are ***statistically independent***. It is entirely specified by its first order density. The higher densities can be factorized into that of the first order according to

$$p_X(x_1, t_1; x_2, t_2) = p_X(x_1, t_1) p_X(x_2, t_2) \tag{3.3}$$

etc.

A ***Markov process*** is completely specified by its second order probabilty density function. It is also called ***process with one-step memory***. Markov processes possess very nice properties and are extremely useful in practice; they are treated in detail in chapter 9.

3.2.2 Characteristic function

We have seen in the previous chapter that the characteristic function is the Fourier transform of the probability density function. As a result, it contains the same information and constitutes an alternative specification of a random process. It is sometimes easier to manipulate. The sequence of characteristic functions is

$$M_X(\theta_1, t_1) = \mathrm{E}[e^{j\theta_1 X(t_1)}]$$

$$\ldots$$

$$M_X(\theta_1, t_1; \ldots; \theta_n, t_n) = \mathrm{E}[e^{j\theta_1 X(t_1) + \ldots + j\theta_n X(t_n)}]$$

Just as for the probability density function, the characteristic function of order n repeats all the information contained in the characteristic functions of lower orders. Indeed, it follows from the definition that

$$M_X(\theta_1, t_1; \dots; \theta_n, t_n) = M_X(\theta_1, t_1; \dots; \theta_n, t_n; 0, t_{n+1}; \dots; 0, t_{n+k}) \tag{3.4}$$

3.2.3 Moment functions

We have seen that the characteristic functions can be expanded in terms of the joint moments [Equ.(2.72)]. The moments can therefore be used as an alternative specification of a random process:

$$\mathrm{E}[X(t)] = \int x\, p_X(x,t)\, dx$$

$$\mathrm{E}[X(t_1)X(t_2)] = \int\int x_1 x_2\, p_X(x_1, t_1; x_2, t_2)\, dx_1\, dx_2$$

$$\dots$$

The moment functions of the first and second order are especially important; they are called the *mean* and the *autocorrelation function*; they have received the special notations

$$\mu_x(t) = \mathrm{E}[X(t)]$$

$$\phi_{xx}(t_1, t_2) = \mathrm{E}[X(t_1)X(t_2)] \tag{3.5}$$

$\phi_{xx}(t,t) = \mathrm{E}[X^2(t)]$ is the *mean-square value* at t. Although the moments of all orders are necessary to specify a random process in the general case, the mean and the autocorrelation functions contain the most important information about the process; they describe a Gaussian process *completely*, as we shall see in the next chapter.

For two random processes, we define the *cross-correlation* as

$$\phi_{xy}(t_1, t_2) = \mathrm{E}[X(t_1)Y(t_2)]$$

3.2.4 Cumulant functions

Since the series expansion of the logarithm of the characteristic function involves the cumulants [see Equ.(2.73)], they constitute an alternative specification of a random process. As we already discussed in chapter 2, the cumulant of order n can be expressed in terms of the moments of orders up to n and vice versa. However, unlike the moment of order n, the cumulant of order n does not reproduce the information already contained in the cumulants of lower orders. The second order cumulant is called the *autocovariance function*

$$\kappa_{xx}(t_1, t_2) = \kappa_2[X(t_1), X(t_2)] = \phi_{xx}(t_1, t_2) - \mu_x(t_1)\mu_x(t_2) \tag{3.6}$$

For $t_1 = t_2 = t$, one gets the *variance* $\kappa_{xx}(t,t) = \sigma_x^2(t)$.

For two different processes, we define the *cross-covariance function* as

$$\kappa_{xy}(t_1,t_2) = \kappa_2[X(t_1), Y(t_2)] = \mathrm{E}\{[X(t_1) - \mu_x(t_1)][Y(t_2) - \mu_y(t_2)]\}$$
$$= \phi_{xy}(t_1,t_2) - \mu_x(t_1)\mu_y(t_2) \tag{3.7}$$

Just as for random variables, the *correlation coefficient function* is the normalized covariance

$$\varrho_{xx}(t_1,t_2) = \frac{\kappa_{xx}(t_1,t_2)}{\sigma_x(t_1)\sigma_x(t_2)} \qquad \varrho_{xy}(t_1,t_2) = \frac{\kappa_{xy}(t_1,t_2)}{\sigma_x(t_1)\sigma_y(t_2)} \tag{3.8}$$

The definition implies that $\varrho_{xx}(t,t) = 1$, and it follows from the Schwarz inequality that $-1 \le \varrho_{xy}(t_1,t_2) \le 1$.

3.2.5 Characteristic functional

For a continuous process, it is useful to define the *characteristic functional*

$$M_X[\theta(t)] = \mathrm{E}\{\exp[j\int_T \theta(t)X(t)dt]\} \tag{3.9}$$

This functional can be seen as the limit of the characteristic function as the times $t_1, \dots, t_n$ become infinitely close to each other. The fact that the characteristic functional completely specifies the process is assessed by the fact that the characteristic function of an arbitrary order n can be recovered from it by choosing

$$\theta(t) = \sum_{i=1}^{n} \theta_i \delta(t - t_i)$$

The characteristic functional can be expanded as (Stratonovich,1963)

$$M_X[\theta(t)] = 1 + \sum_{n=1}^{\infty} \frac{j^n}{n!} \int .. \int \mathrm{E}[X(t_1)\dots X(t_n)]\theta(t_1)\dots\theta(t_n)\, dt_1 \dots dt_n \tag{3.10}$$

or

$$M_X[\theta(t)] = \exp\{1 + \sum_{n=1}^{\infty} \frac{j^n}{n!} \int .. \int \kappa_n[X(t_1), \dots, X(t_n)]\theta(t_1)\dots\theta(t_n)\, dt_1 \dots dt_n\} \tag{3.11}$$

and it is readily checked that the expansions (2.72) and (2.73) are recovered by the above special choice of $\theta(t)$. Equation (3.11) can be regarded as giving the definitions of the cumulants. Because the cumulants of order larger than 2 vanish for Gaussian processes, this expansion of the characteristic functional is certainly the simplest way to define a Gaussian process.

3.3 Stationary random process

A random process is ***strongly stationary*** or ***stationary in the strict sense*** if its probability structure is independent of a shift of the time origin. This implies

$$p_X(x_1,t_1;x_2,t_2) = p_X(x_1,t_1+a;x_2,t_2+a)$$

$$\dots \tag{3.12}$$

$$p_X(x_1,t_1;\dots;x_n,t_n) = p_X(x_1,t_1+a;\dots;x_n,t_n+a)$$

The first order probability density is independent of the time, and the higher order densities depend only on the difference between the time arguments. Substituting into Equ.(3.5) and (3.6), one finds that the process has a constant mean and its autocorrelation function depends only on the time difference:

$$\mu_x = \int_{-\infty}^{\infty} x\, p_X(x,t)\, dx = Constant$$

$$\phi_{xx}(t_1,t_2) = \int_{-\infty}^{\infty}\int x_1 x_2\, p_X(x_1;x_2,t_2-t_1)\, dx_1 dx_2 \tag{3.13}$$

$$= R_{xx}(t_1-t_2) = R_{xx}(\tau)$$

$$\kappa_{xx}(t_1,t_2) = \phi_{xx}(t_1,t_2) - \mu_x(t_1)\mu_x(t_2)$$

$$= R_{xx}(\tau) - \mu_x^2 = \Gamma_{xx}(t_1-t_2) = \Gamma(\tau)$$

The notations $R_{xx}(\tau)$ and $\Gamma_{xx}(\tau)$ are the most frequently used for stationary processes.

Thus, we have established that the strong stationarity implies that the mean is constant and that the autocorrelation function depends only on the difference of its arguments. The converse is not true in general, except for Gaussian processes, because they are completely characterized by their mean and autocovariance functions. As a result, the above conditions make their entire probability structure independent of the time origin and imply that they are strongly stationary.

Because of the practical importance of the Gaussian process and also because, even for non-Gaussian processes, their analysis is often limited to the moments up to the second order, we define a ***weakly stationary process*** or a ***process stationary in the wide sense*** as one for which the conditions (3.13) are satisfied. A weakly stationary Gaussian process is also strongly stationary.

3.4 Properties of the correlation functions

So far, we have only considered real random processes. It is sometimes convenient to consider complex random functions. In that case, the correlation functions are defined as

$$\phi_{xx}(t_1,t_2) = \mathrm{E}[X(t_1)X^\star(t_2)] \tag{3.14}$$

$$\phi_{xy}(t_1,t_2) = \mathrm{E}[X(t_1)Y^\star(t_2)] \tag{3.15}$$

where $X^\star$ is the complex conjugate of X. These definitions are identical to the previous ones if $X(t)$ is real. The correlation functions enjoy the following properties

- **Symmetry**: It immediately follows from the definition that

$$\phi_{xx}(t_1,t_2) = \phi^\star_{xx}(t_2,t_1)$$

$$\phi_{xy}(t_1,t_2) = \phi^\star_{yx}(t_2,t_1) \tag{3.16}$$

For a real valued stationary process,

$$R_{xx}(\tau) = R_{xx}(-\tau)$$

$$R_{xy}(\tau) = R_{yx}(-\tau) \tag{3.17}$$

The autocorrelation function of a real valued weakly stationary random process is an even function of the time delay.

- **Inequalities**: Consider the quadratic form

$$|X(t_1) \pm Y(t_2)|^2 = [X(t_1) \pm Y(t_2)][X(t_1) \pm Y(t_2)]^\star$$
$$= X(t_1)X^\star(t_1) + Y(t_2)Y^\star(t_2) \pm 2\mathrm{Re}[X(t_1)Y^\star(t_2)] \geq 0$$

Upon taking the expectation, one gets

$$\phi_{xx}(t_1,t_1) + \phi_{yy}(t_2,t_2) \geq 2|\mathrm{Re}\,\phi_{xy}(t_1,t_2)| \tag{3.18}$$

For real valued stationary processes,

$$R_{xx}(0) + R_{yy}(0) \geq 2|R_{xy}(\tau)| \tag{3.19}$$

A direct consequence of the Schwarz inequality is that

$$\sqrt{R_{xx}(0)R_{yy}(0)} \geq |R_{xy}(\tau)| \quad \text{and} \quad R_{xx}(0) \geq |R_{xx}(\tau)| \tag{3.20}$$

Note that, since the geometric mean of two numbers cannot exceed their arithmetic mean, Equ.(3.20) implies (3.19). It states that *the autocorrelation function of a stationary process has its maximum at* $\tau = 0$. The two

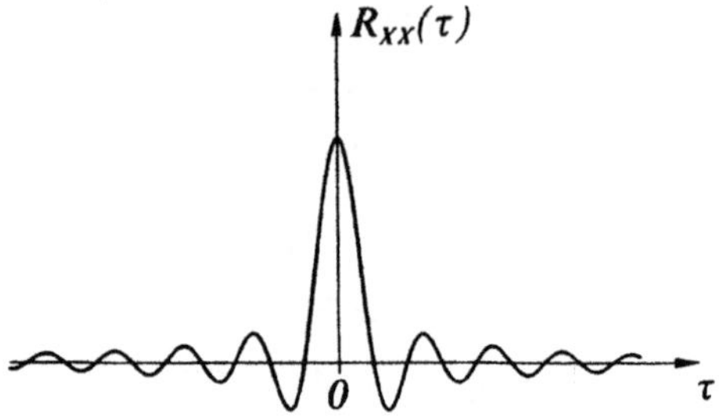

Figure 3.2: Autocorrelation function of a weakly stationary random process. $R_{xx}(\tau)$ is symmetric with its maximum at $\tau = 0$.

foregoing properties imply that the autocorrelation function of a weakly stationary process looks like that represented in Fig.3.2. However, these properties are by no means sufficient for a given function to qualify for being an autocorrelation function. In addition, the following property must be satisfied.

- **Positive Fourier transform**: $\phi_{xx}(t_1, t_2)$ is such that, for an arbitrary function $h(t)$ defined on the domain $[a, b]$,

$$\int_a^b \int_a^b \phi_{xx}(t_1, t_2) h(t_1) h^\star(t_2)\, dt_1 dt_2 \geq 0 \tag{3.21}$$

Such a function is called *positive definite* (strictly speaking *non-negative* definite). In the weakly stationary case, this equation becomes

$$\int_a^b \int_a^b R_{xx}(t_1 - t_2) h(t_1) h^\star(t_2)\, dt_1 dt_2 \geq 0 \tag{3.22}$$

It can be shown that this implies that $R_{xx}(\tau)$ has a non-negative Fourier transform (Bochner's theorem):

$$\Phi_{xx}(\omega) = \frac{1}{2\pi} \int_{-\infty}^{\infty} R_{xx}(\tau) e^{-j\omega\tau} d\tau \geq 0 \tag{3.23}$$

$\Phi_{xx}(\omega)$ is called the *power spectral density (PSD)*. It provides a frequency decomposition of the power in the process. It will be discussed extensively later in the text.

Since the covariance functions are special cases of correlation functions, they enjoy the same properties. Besides, if the process does not contain any periodic component, the autocovariance function goes to zero as the time delay increases:

$$\lim_{|t_1 - t_2| \to \infty} \kappa_{xx}(t_1, t_2) = 0$$

3.5 Differentiation

3.5.1 Convergence

Before discussing the derivative of a random process, it is necessary to consider the concept of convergence of a random sequence. We know that, in the deterministic case, the sequence of numbers x_n converges towards the limit x if, for any positive ε, one can find n_0 such that

$$|x_n - x| < \varepsilon, \qquad n > n_0$$

Now, consider a random sequence. Every random experiment $\omega \in \Omega$ supplies a sequence of results $X_n(\omega)$. If each of them satisfies the above relationship, the random sequence is said to converge ***everywhere***. The limit X is in general a random variable. Less restrictive modes of convergence can be defined, accepting a limited set of experiments invalidating the above relation. Here, we shall restrict ourselves to the convergence ***in the mean-square sense***: The random sequence X_n tends to the limit X in the mean-square sense if

$$\mathrm{E}[|X_n - X|^2] \rightarrow 0 \qquad \text{as } n \rightarrow \infty \tag{3.24}$$

According to Chebyshev's inequality, Equ.(2.62), the probability that the difference $|X_n - X|$ exceeds ε can be made arbitrarily small when n increases.

3.5.2 Continuity

According to the foregoing section, the process $X(t)$ is continuous in the mean-square sense at t if

$$\mathrm{E}\{[X(t+\varepsilon) - X(t)]^2\} \rightarrow 0 \qquad (\varepsilon \rightarrow 0) \tag{3.25}$$

Since

$$\mathrm{E}\{[X(t+\varepsilon) - X(t)]^2\} =$$
$$\phi_{xx}(t+\varepsilon, t+\varepsilon) - \phi_{xx}(t+\varepsilon, t) - \phi_{xx}(t, t+\varepsilon) + \phi_{xx}(t,t)$$

$X(t)$ is continuous in the mean-square sense if $\phi_{xx}(t_1, t_2)$ is continuous with respect to both of its arguments at $t_1 = t_2 = t$. If the process is stationary,

$$\mathrm{E}\{[X(t+\varepsilon) - X(t)]^2\} = 2[R_{xx}(0) - R_{xx}(\varepsilon)]$$

$X(t)$ is continuous in the mean-square sense if the autocorrelation function is continuous at the origin.

Since, for any random variable Z,

$$\mathrm{E}[Z^2] = \sigma_z^2 + \mathrm{E}^2[Z] \geq \mathrm{E}^2[Z]$$

we have

$$\mathrm{E}\{[X(t+\varepsilon)-X(t)]^2\} \geq \mathrm{E}^2[X(t+\varepsilon)-X(t)]$$

As a result, the mean-square continuity implies that

$$\mathrm{E}[X(t+\varepsilon)-X(t)] \to 0 \qquad (\varepsilon \to 0)$$

or

$$\lim_{\varepsilon\to 0} \mathrm{E}[X(t+\varepsilon)] = \mathrm{E}[X(t)] \tag{3.26}$$

Accordingly, *one can interchange the order of limit and expected value* if the process is continuous in the mean-square sense.

3.5.3 Stochastic differentiation

The derivative of the random process $X(t)$ is defined by

$$\dot{X}(t) = \lim_{\varepsilon\to 0} \frac{X(t+\varepsilon)-X(t)}{\varepsilon}$$

If this limit exists for every single sample of the process, it has the usual meaning of a derivative. If it exists in the mean-square sense, one says that the process $X(t)$ has a derivative in this sense. A random process has a derivative in the mean-square sense if one can find another process $\dot{X}(t)$ such that

$$\lim_{\varepsilon\to 0} \mathrm{E}\{[\frac{X(t+\varepsilon)-X(t)}{\varepsilon} - \dot{X}(t)]^2\} = 0 \tag{3.27}$$

One can show that the existence of this process is guaranteed if

$$\frac{\partial^2}{\partial t_1 \partial t_2} \phi_{xx}(t_1, t_2)$$

exists at $t_1 = t_2$. *A stationary process is differentiable in the mean-square sense if its autocorrelation function $R_{xx}(\tau)$ has derivatives of order up to 2 at $\tau = 0$.* Under that condition,

$$\mathrm{E}[\dot{X}(t)] = \mathrm{E}[\lim_{\varepsilon\to 0} \frac{X(t+\varepsilon)-X(t)}{\varepsilon}] = \lim_{\varepsilon\to 0} \frac{\mathrm{E}[X(t+\varepsilon)]-\mathrm{E}[X(t)]}{\varepsilon}$$

or

$$\mathrm{E}[\dot{X}(t)] = \frac{\mathrm{d}}{\mathrm{d}t} \mathrm{E}[X(t)] \tag{3.28}$$

One can therefore *interchange the order of derivative and expected value* if the process is differentiable in the mean-square sense. As a result, the following relations are easily derived

$$\frac{\partial}{\partial t} \phi_{xx}(t,s) = \frac{\partial}{\partial t} \mathrm{E}[X(t)X(s)] = \mathrm{E}[\dot{X}(t)X(s)] = \phi_{\dot{x}x}(t,s) \tag{3.29}$$

$$\frac{\partial^2}{\partial t \partial s}\phi_{xx}(t,s) = \frac{\partial}{\partial s}\mathrm{E}[\dot{X}(t)X(s)] = \mathrm{E}[\dot{X}(t)\dot{X}(s)] = \phi_{\dot{x}\dot{x}}(t,s) \tag{3.30}$$

For a stationary random process, these relations become

$$R_{\dot{x}x}(\tau) = R'_{xx}(\tau) \tag{3.31}$$

$$R_{\dot{x}\dot{x}}(\tau) = -R''_{xx}(\tau) \tag{3.32}$$

Since $R_{xx}(\tau)$ is an even function of τ, one must have, if the process is differentiable,

$$R_{\dot{x}x}(0) = 0 \tag{3.33}$$

A weakly stationary process is orthogonal to its first derivative (evaluated at the same time). If the process is not differentiable, $R'_{xx}(\tau)$ may be discontinuous at $\tau = 0$ where $R''_{xx}(\tau)$ does not exist.

3.6 Stochastic integrals, Ergodicity

3.6.1 Integration

Consider the stochastic integral

$$Y = \int_a^b X(t)\,dt \tag{3.34}$$

If this integral exists in the Riemann sense (limit sum) for every sample $X(\omega,t)$, it defines a random variable which represents the random area defined by the curve $X(\omega,t)$ over the interval $[a,b]$. Just as in the previous section, the integral may not exist for *every* sample, but in a weaker sense. The integral exists in the mean-square sense if

$$\lim_{n\to\infty} \mathrm{E}\{[Y - \sum_{i=1}^{n} X(t_i)\Delta t_i]^2\} = 0 \tag{3.35}$$

As before, this allows a set with zero probability of non integrable functions. It can be shown that a necessary and sufficient condition for $X(t)$ to be integrable in the mean-square sense is that its autocorrelation function $\phi_{xx}(t_1,t_2)$ be twice integrable over the domain $[a,b]$. Under that condition, one can interchange the integral and the expected value. The mean-square value of the integral reads

$$\mathrm{E}[Y^2] = \int_a^b \int_a^b \mathrm{E}[X(t_1)X(t_2)]\,dt_1 dt_2$$

or

$$\mathrm{E}[Y^2] = \int_a^b \int_a^b \phi_{xx}(t_1,t_2)\,dt_1 dt_2 \tag{3.36}$$

An interesting generalization is

$$Y(\nu) = \int_a^b X(t)h(t,\nu)\,dt \tag{3.37}$$

where $h(t,\nu)$ is a complex function of two arguments t and ν. This integral includes the Fourier transform and the convolution integral as particular cases. In the former case, $h(t,\nu) = e^{-j\nu t}$ while in the latter, $h(\nu - t)$ is the impulse response of the system. The integral $Y(\nu)$ is now a random process of parameter ν. The mean and autocorrelation functions of $Y(\nu)$ are :

$$\mu_y(\nu) = \int_a^b \mu_x(t)h(t,\nu)\,dt$$

$$\phi_{yy}(\nu_1,\nu_2) = \mathrm{E}[Y(\nu_1)Y^\star(\nu_2)] = \int_a^b \int_a^b \phi_{xx}(t_1,t_2)h(t_1,\nu_1)h^\star(t_2,\nu_2)\,dt_1 dt_2 \tag{3.38}$$

The integral (3.37) exists in the mean-square sense if and only if the foregoing integral is bounded for all ν_1 and ν_2.

3.6.2 Temporal mean

Consider the real valued stationary process $X(t)$. The temporal mean is defined by the integral

$$S = \frac{1}{2T}\int_{-T}^{T} X(t)\,dt \tag{3.39}$$

S is a random variable with mean and variance

$$\mathrm{E}[S] = \mu_x$$

$$\sigma_s^2 = \frac{1}{4T^2}\int_{-T}^{T}\int_{-T}^{T} \Gamma_{xx}(t_1 - t_2)\,dt_1 dt_2 \tag{3.40}$$

After a few manipulations, this latter expression can be transformed into [e.g. see (Papoulis, 1965), p.325]

$$\sigma_s^2 = \frac{1}{T}\int_0^{2T} (1 - \frac{\tau}{2T})[R_{xx}(\tau) - \mu_x^2]\,d\tau \tag{3.41}$$

Note that, because of the T at the denominator, $\sigma_s \to 0$ as $T \to \infty$, if the integral is bounded. This is essential in the discussion on ergodicity.

3.6.3 Ergodicity theorem

Consider a real valued stationary random process $X(t)$. The ergodicity theorem deals with the issue of determining the statistics of $X(t)$ ***from a single sample*** of the process. The ergodicity property allows us to replace ***ensemble averages*** by ***time averages on a single sample.*** The most general form of ergodicity is concerned with all the statistics of the process; We shall restrict ourselves to the mean and autocorrelation function.

Let $\hat{x}(t)$ be a sample of the stationary process $X(t)$ [$\hat{x}(t)$ is a simple function of time]. Consider the limits

$$\hat{\mu} = \lim_{T\to\infty} \frac{1}{2T} \int_{-T}^{T} \hat{x}(t)\, dt$$

$$\hat{R}(\tau) = \lim_{T\to\infty} \frac{1}{2T} \int_{-T}^{T} \hat{x}(t+\tau)\hat{x}(t)\, dt \tag{3.42}$$

The above integral is closely related to the *correlation integral* (1.19).

The property of *ergodicity* implies that

$$\hat{\mu} = \mu_X = \mathrm{E}[X(t)]$$

and

$$\hat{R}(\tau) = R_{xx}(\tau) = \mathrm{E}[X(t+\tau)X(t)] \tag{3.43}$$

The mean and the autocorrelation functions which are ensemble averages are replaced by time averages on a single sample of the process. Obviously $\hat{\mu}$ and $\hat{R}(\tau)$ are samples of random variables $\bar{\mu}$ and $\bar{R}(\tau)$. According to Chebyshev's inequality, if a random variable has a zero variance, it is equal to its mean with probability 1. Therefore, the foregoing relationships will be true if

$$\mathrm{E}[\bar{\mu}] = \mu_x \qquad \text{and} \qquad \sigma_{\bar{\mu}}^2 = 0$$

and

$$\mathrm{E}[\bar{R}(\tau)] = R_{xx}(\tau) \qquad \text{and} \qquad \sigma_{\bar{R}}^2 = 0$$

Thus, a stationary random process is ergodic with respect to the mean if it is such that the variance of the temporal mean vanishes at the limit, when $T \to \infty$. Referring to Equ.(3.41), one sees that the ergodicity of the mean is guaranteed if the integral is bounded.

Similarly, it is clear that

$$\mathrm{E}[\bar{R}(\tau)] = \frac{1}{2T} \int_{-T}^{T} \mathrm{E}[X(t+\tau)X(t)]\, dt = R_{xx}(\tau)$$

The condition under which $\sigma_{\bar{R}}^2 = 0$ follows a development parallel to that for the mean; it involves higher order moments of $X(t)$.

The assumption of ergodicity is always implicit in the experimental estimation of power spectral density functions.

3.7 Spectral decomposition

The frequency distribution of the energy in a nonstationary random process will be studied in detail in chapter 8. In this section, we discuss the existence of the Fourier transform and introduce the power spectral density of a weakly stationary random process.

3.7.1 Fourier transform

Consider the Fourier transform stochastic integral

$$\mathcal{X}(\omega) = \int_{-\infty}^{\infty} X(t)\, e^{-j\omega t} dt \tag{3.44}$$

It has the general form (3.37) and defines another random process of parameter ω. According to section 3.6, this integral exists in the mean-square sense if and only if

$$\mathrm{E}[\mathcal{X}(\omega_1)\mathcal{X}^\star(\omega_2)] = \int_{-\infty}^{\infty}\int_{-\infty}^{\infty} \phi_{xx}(t_1,t_2)\, e^{-j(\omega_1 t_1 - \omega_2 t_2)} dt_1 dt_2 \tag{3.45}$$

exists for all ω_1 and ω_2. Since this is the 2-fold Fourier transform of the autocorrelation function, a sufficient condition is that $\phi_{xx}(t_1,t_2)$ be absolutely integrable over the complete domain. Under that condition, $X(t)$ and $\mathcal{X}(\omega)$ can be regarded as a Fourier transform pair.

Condition (3.45) is not satisfied by a weakly stationary random process, because the autocorrelation function $\phi_{xx}(t_1,t_2) = R_{xx}(t_1 - t_2)$ does not even vanish at infinity along straight lines corresponding to constant values of $t_1 - t_2$. Thus, just as ordinary functions which do not vanish at infinity, a stationary random process does not have a Fourier transform.

3.7.2 Power spectral density

Now, consider a *stationary* random process. The integral of finite duration

$$\mathcal{X}(\omega,T) = \int_{-T/2}^{T/2} X(t)\, e^{-j\omega t} dt \tag{3.46}$$

is the truncated Fourier transform of the process. Its existence condition in the mean-square sense is easily derived from (3.38). It can be shown [e.g. (Papoulis, 1965), p.343] that, if

$$\int_{-\infty}^{\infty} |\tau R_{xx}(\tau)|\, d\tau < \infty \tag{3.47}$$

then,

$$\lim_{T\to\infty} \frac{1}{2\pi T}\, \mathrm{E}[|\mathcal{X}(\omega,T)|^2] = \Phi_{xx}(\omega) \tag{3.48}$$

where the ***power spectral density (PSD)*** function is defined as the Fourier transform of the autocorrelation function (to a constant factor):

$$\Phi_{xx}(\omega) = \frac{1}{2\pi}\int_{-\infty}^{\infty} R_{xx}(\tau)\, e^{-j\omega\tau}\, d\tau \tag{3.49}$$

$$R_{xx}(\tau) = \int_{-\infty}^{\infty} \Phi_{xx}(\omega)\, e^{j\omega\tau}\, d\omega \tag{3.50}$$

Condition (3.47) is met by most processes of practical interest and Equ.(3.48) is the starting point for the estimation of the *PSD* from records of finite duration. Being the limit of a positive quantity, it also proves that $\Phi_{xx}(\omega) \geq 0$, as already stated in section 3.4.

Equations (3.49) and (3.50) are known as the ***Wiener-Khintchine theorem.*** The reason why the *PSD* is often used in practice is related to the fact that the input-output relationship for linear systems is a convolution in the time domain, while it becomes a product in the frequency domain. Thus, there is no coupling between the components of the response relative to distinct frequencies. At $\tau = 0$, Equ.(3.50) reads

$$R_{xx}(0) = \mathrm{E}[X^2(t)] = \int_{-\infty}^{\infty} \Phi_{xx}(\omega)\, d\omega \tag{3.51}$$

This equation shows that $\Phi_{xx}(\omega)$ is a ***frequency decomposition of the mean-square of the process.***

Since the autocorrelation is an even function of τ, the power spectral density is an even function of ω (Problem P.1.9) and Equ.(3.49) and (3.50) can be written

$$\Phi_{xx}(\omega) = \frac{1}{\pi}\int_{0}^{\infty} R_{xx}(\tau)\, \cos\omega\tau\, d\tau$$

$$R_{xx}(\tau) = 2\int_{0}^{\infty} \Phi_{xx}(\omega)\, \cos\omega\tau\, d\omega$$

The power spectral density is defined for positive as well as for negative circular frequencies ω. According to Equ.(3.51), $\Phi_{xx}(\omega)$ is expressed in

$$\frac{(\text{unit of } X)^2 \times \text{sec}}{\text{rad}}$$

In the literature, one often meets a ***one-sided*** power spectral density, $G_x(f)$, defined only for positive frequencies in *Hertz* ($f = \omega/2\pi$). We must understand it in the sense

$$R_{xx}(0) = \mathrm{E}[X^2] = \int_{0}^{\infty} G_x(f)\, df \tag{3.52}$$

Comparing this to Equ.(3.51), one gets

$$G_x(f) = 4\pi\, \Phi_{xx}(2\pi f) \tag{3.53}$$

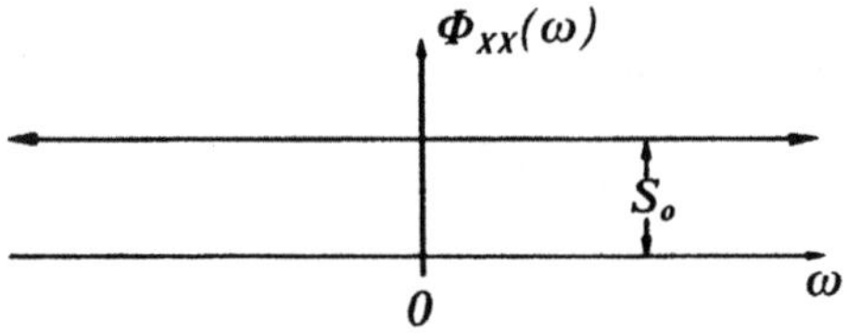

Figure 3.3: White noise. The power spectral density is uniform.

$G_x(f)$ is expressed in

$$\frac{(\text{unit of } X)^2}{\text{Hertz}}$$

3.8 Examples

3.8.1 White noise

A *white noise* is a stationary random process with a uniform power spectral density (Fig.3.3):

$$\Phi_{xx}(\omega) = S_0 \qquad -\infty < \omega < \infty \tag{3.54}$$

Its correlation function is

$$R_{xx}(\tau) = 2\pi S_0\, \delta(\tau) \tag{3.55}$$

where $\delta(\tau)$ is the Dirac function. It is readily observed that this process is not physically realizable, because there is an infinite area under the spectrum, which implies that the mean-square value of such a process would be infinite. Although not physically acceptable, the white noise process is often a convenient approximation in system analysis, whenever the correlation time of the excitation is small with respect to the time constant of the system. In particular, the statistics of the response of a lightly damped oscillator subjected to a wide-band excitation can be evaluated quite accurately using the white noise approximation (chapter 5).

3.8.2 Ideal low-pass process

A white noise can be approximated by an *ideal low-pass process*, also called *band-limited white noise*, defined as:

$$\Phi_{xx}(\omega) = S_0 \qquad (|\omega| \le \omega_c) \tag{3.56}$$

The corresponding autocorrelation function is

$$R_{xx}(\tau) = 2S_0 \frac{\sin \omega_c \tau}{\tau} \tag{3.57}$$

The area under the spectrum is now finite (Fig.3.4.a) and this process becomes a white noise at the limit, when $\omega_c \rightarrow \infty$. Comparing Equ.(3.55) and (3.57), one gets the following limiting form of the Dirac function:

$$\delta(\tau) = \lim_{\omega_c \rightarrow \infty} \frac{\sin \omega_c \tau}{\pi \tau} \tag{3.58}$$

3.8.3 Process with exponential correlation

A *process with exponential correlation* (Fig.3.4.b) is defined by

$$\Phi_{xx}(\omega) = S_0 \frac{\beta^2}{\beta^2 + \omega^2} \tag{3.59}$$

$$R_{xx}(\tau) = \pi S_0 \, \beta \, e^{-\beta|\tau|} \tag{3.60}$$

This process also becomes a white noise at the limit, when $\beta \rightarrow \infty$. Comparing Equ.(3.60) and (3.55), one gets another limiting form of the Dirac function:

$$\delta(\tau) = \lim_{\beta \rightarrow \infty} \frac{\beta}{2} e^{-\beta|\tau|} \tag{3.61}$$

Although the two processes of Fig.3.4 constitute approximations of a white noise, one observes a major difference in the behaviour of their autocorrelation functions in the vicinity of the origin. $R''_{xx}(0)$ does not exist for the process with exponential correlation. According to section 3.5, this means that this process is not differentiable in the mean-square sense. Indeed, we shall see that it can be seen as the response to a white noise of a system governed by a first order differential equation.

3.8.4 Construction of a random process with specified power spectral density

Consider the random process

$$X(t) = a \, e^{j\Omega t} \tag{3.62}$$

where a is a constant and Ω is a random variable of probability density function $p(\omega)$. The autocorrelation function reads

$$R_{xx}(\tau) = \mathrm{E}[X(t+\tau)X^*(t)] = |a|^2 \int_{-\infty}^{\infty} e^{j\omega\tau} \, p(\omega) \, d\omega \tag{3.63}$$

Comparing this with Equ.(3.50), one gets the power spectral density

$$\Phi_{xx}(\omega) = |a|^2 p(\omega) \tag{3.64}$$

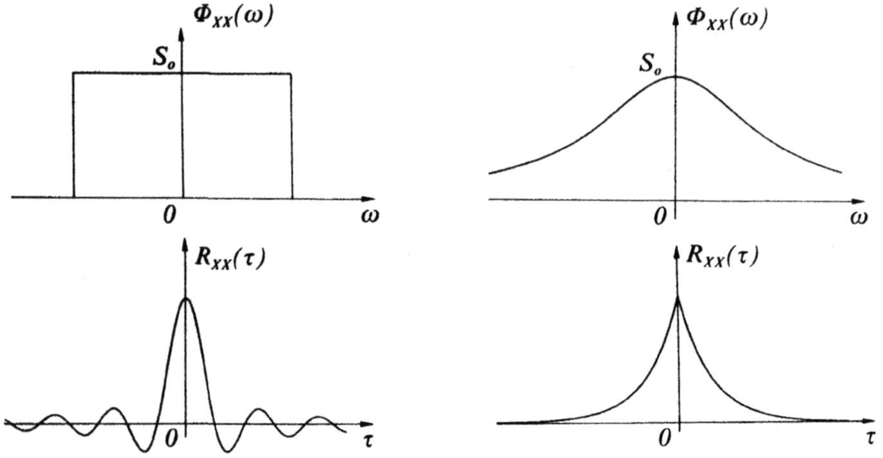

Figure 3.4: Approximations of a white noise. (a) Ideal low-pass process. (b) Process with exponential correlation.

Thus, the shape of the power spectral density duplicates that of the probability density function. The reader will observe that each sample of the process consists of an exponential function with a fixed frequency. Obviously, the time averages on a single sample cannot be equivalent to ensemble averages, in this case, and the process is not ergodic. In chapter 12, we shall see how samples of an ergodic process of arbitrary *PSD* can be generated.

3.9 Cross power spectral density

The *cross power spectral density* and the cross-correlation functions of two random processes $X(t)$ and $Y(t)$ constitute a Fourier transform pair:

$$\Phi_{xy}(\omega) = \frac{1}{2\pi} \int_{-\infty}^{\infty} R_{xy}(\tau)\, e^{-j\omega\tau} d\tau \tag{3.65}$$

$$R_{xy}(\tau) = \int_{-\infty}^{\infty} \Phi_{xy}(\omega)\, e^{j\omega\tau} d\omega \tag{3.66}$$

$\Phi_{xy}(\omega)$ exists if both $\Phi_{xx}(\omega)$ and $\Phi_{yy}(\omega)$ exist. A consequence of Equ.(3.17) is that

$$\Phi_{xy}(\omega) = \Phi^{\star}_{yx}(\omega) \tag{3.67}$$

In a similar manner to Equ.(3.48), it can be demonstrated that the cross-power spectral density is related to the truncated Fourier transform by

$$\lim_{T\to\infty} \frac{1}{2\pi T} \mathrm{E}[\mathcal{X}(\omega,T)\mathcal{Y}(\omega,T)^{\star}] = \Phi_{xy}(\omega) \tag{3.68}$$

Since

$$|\mathrm{E}[\mathcal{X}(\omega,T)\mathcal{Y}(\omega,T)^\star]|^2 \leq \mathrm{E}[|\mathcal{X}(\omega,T)\mathcal{Y}(\omega,T)^\star|^2]$$
$$\leq \mathrm{E}[|\mathcal{X}(\omega,T)|^2]\,\mathrm{E}[|\mathcal{Y}(\omega,T)|^2]$$

taking the limit for $T \to \infty$, one gets the inequality:

$$|\Phi_{xy}(\omega)|^2 \leq \Phi_{xx}(\omega)\Phi_{yy}(\omega) \tag{3.69}$$

3.10 Periodic process

A stationary random process $X(t)$ is periodic in the mean-square sense if there exists a period T such that

$$\mathrm{E}\{[X(t) - X(t+T)]^2\} = 0 \tag{3.70}$$

Developing this expression, one gets

$$\mathrm{E}\{[X(t) - X(t+T)]^2\} = \mathrm{E}[X(t)^2] - 2\mathrm{E}[X(t)X(t+T)] + \mathrm{E}[X(t+T)^2]$$
$$= 2[R_{xx}(0) - R_{xx}(T)]$$

It follows that a necessary and sufficient condition for a stationary random process to be periodic in the mean-square sense is that

$$R_{xx}(0) = R_{xx}(\tau) \tag{3.71}$$

Moreover, since

$$\mathrm{E}^2\{X(t).[X(t+\tau)-X(t+T+\tau)]\} \leq \mathrm{E}[X^2(t)].\mathrm{E}\{[X(t+\tau)-X(t+T+\tau)]^2\} = 0$$

one concludes that

$$\mathrm{E}\{X(t).[X(t+\tau) - X(t+T+\tau)]\} = 0$$

or

$$R_{xx}(\tau) = R_{xx}(\tau + T) \tag{3.72}$$

The autocorrelation function of a mean-square periodic random process is also periodic, with the same period T.

Since a periodic function is not absolutely integrable, its Fourier transform does not exist and the power spectral density does not exist in the sense of Equ.(3.49). However, $R_{xx}(\tau)$ can be expanded in a Fourier series

$$R_{xx}(\tau) = \sum_{n=-\infty}^{\infty} \alpha_n e^{jn\omega_0\tau} \tag{3.73}$$

where $\omega_0 = 2\pi/T$ and

$$\alpha_n = \frac{1}{T}\int_0^T R_{xx}(\tau)e^{-jn\omega_0\tau}\,d\tau \tag{3.74}$$

Introducing formally Equ.(3.73) into (3.49) and interchanging the order of integration and summation, one gets

$$\Phi_{xx}(\omega) = \frac{1}{2\pi}\sum_{n=-\infty}^{\infty}\alpha_n\int_{-\infty}^{\infty}e^{j(n\omega_0-\omega)\tau}\,d\tau = \sum_{n=-\infty}^{\infty}\alpha_n\,\delta(\omega - n\omega_0) \tag{3.75}$$

because of the orthogonality of the exponential functions. The mean-square value reads

$$R_{xx}(0) = \int_{-\infty}^{\infty}\Phi_{xx}(\omega)\,d\omega = \sum_{n=-\infty}^{\infty}\alpha_n = \alpha_0 + 2\sum_{0}^{\infty}\alpha_n \tag{3.76}$$

where we have used the fact that $\Phi_{xx}(\omega)$ is even for a real valued process. Thus, α_n describe the power distribution at the various harmonics of the periodic process.

A mean-square periodic random process can be expanded in a Fourier series:

$$X(t) = \sum_{n=-\infty}^{\infty}A_n e^{jn\omega_0\tau} \tag{3.77}$$

where A_n are *orthogonal* random variables such that

$$E[A_0] = E[X(t)], \qquad E[A_n] = 0 \quad (n \neq 0)$$

$$E[A_k A_n^*] = \alpha_n\,\delta(k-n) \tag{3.78}$$

The proof of this statement is based on the orthogonality of the exponential functions [see e.g. (Papoulis, 1965), p.368].

3.11 References

A.BLANC-LAPIERRE & R.FORTET, *Théorie des Fonctions Aléatoires*, Masson, Paris, 1953.

W.B.DAVENPORT, *Probability and Random Processes*, McGraw-Hill, 1970.

Y.K.LIN, *Probabilistic Theory of Structural Dynamics*, McGraw-Hill, 1967.

J.H.LANING & R.H.BATTIN, *Random Processes in Automatic Control*, McGraw-Hill, 1956.

A.PAPOULIS, *Probability, Random Variables and Stochastic Processes*, McGraw-Hill, 1965.

E.PARZEN, *Stochastic Processes*, Holden Day, 1962.

R.L.STRATONOVICH, *Topics in the Theory of Random Noise*, 1, Gordon & Breach, N-Y, 1963.
A.A.SVESHNIKOV, *Applied Methods of the Theory of Random Functions*, Pergamon Press, 1966.

3.12 Problems

P.3.1 *Consider a stationary random process with a triangular autocorrelation function:*

$$R_{xx}(\tau) = \frac{2\pi S_0}{T}(1 - \frac{|\tau|}{T}) \qquad |\tau| \le T$$

($R_{xx}(\tau) = 0$ if $|\tau| > T$). Show that its PSD is

$$\Phi_{xx}(\omega) = 4S_0 \frac{\sin^2(\omega T/2)}{\omega^2 T^2}$$

Check that this process becomes a white noise at the limit $T \to 0$. Is this process differentiable in the mean-square sense?
P.3.2 *Show that the following pair of functions satisfy the Wiener-Khintchine theorem:*

$$R_{xx}(\tau) = \frac{\sqrt{2\pi}}{\varepsilon} S_0 \, e^{-\tau^2/2\varepsilon^2}$$

$$\Phi_{xx}(\omega) = S_0 \, e^{-\varepsilon^2\omega^2/2}$$

Show that this process tends towards a white noise as $\varepsilon \to 0$. Is it differentiable in the mean-square sense?
P.3.3 *Sketch the PSD corresponding to the following autocorrelation function:*

$$R_{xx}(\tau) = e^{-\beta|\tau|} \cos \omega_0 \tau$$

[Hint: Start from the process with exponential correlation and use the translation theorem of the Fourier transform.]
P.3.4 *Consider the random process*

$$X(t) = A \cos \omega t + B \sin \omega t$$

where A and B are independent random variables of zero mean and variance σ^2, and ω is a constant. Show that $X(t)$ is stationary with zero mean and the autocorrelation function

$$R_{xx}(\tau) = \sigma^2 \cos \omega \tau$$

P.3.5 *If the $2n$ random variables A_i and B_i are uncorrelated with zero mean and the same standard deviation σ_i, show that the process*

$$X(t) = \sum_{i=1}^{n} A_i \cos \omega_i t + B_i \sin \omega_i t$$

is wide sense stationary with zero mean and autocorrelation function

$$R_{xx}(\tau) = \sum_{i=1}^{n} \sigma_i^2 \cos \omega_i \tau$$

P.3.6 *Show that a band-limited white noise*

$$\Phi_0(\omega) = S_0 \qquad (\omega_1 \leq |\omega| \leq \omega_2)$$

has the following autocorrelation function

$$R_{xx}(\tau) = 2S_0 \frac{\sin \omega_2 \tau - \sin \omega_1 \tau}{\tau}$$

[Hint: Start from the ideal low-pass process and use the convolution theorem.]
P.3.7 *Consider the random process $X(t)$ constructed by sampling independent random variables at regular times:*

$$X(t) = Y_k \qquad t \in (k\Delta t, (k+1)\Delta t]$$

where Y_k are independent random variables with the same probability distribution and mean square value $\mathrm{E}[Y^2] = \sigma^2$. *Show that the PSD of this process is*

$$\Phi_{xx}(\omega) = \frac{\sigma^2 \Delta t}{2\pi} \cdot \frac{\sin^2(\omega \Delta t/2)}{(\omega \Delta t/2)^2}$$

Under what conditions does $X(t)$ tend to a white noise? [Hint: Start from Equ(3.48).] Note that this result does not depend on the probability distribution of Y_k.

Chapter 4

Gaussian Process, Poisson Process

4.1 Gaussian random variable

A random variable X is *Gaussian* or *normal* if its probability density function can be written as

$$p_X(x) = \frac{1}{\sqrt{2\pi}\sigma} \exp[-\frac{(x-\mu)^2}{2\sigma^2}] \qquad -\infty < x < \infty \tag{4.1}$$

μ is the mean and σ is the standard deviation of the distribution:

$$E[X] = \int_{-\infty}^{\infty} x\, p_X(x) dx = \mu$$

$$E[(X-\mu)^2] = \int_{-\infty}^{\infty} (x-\mu)^2 p_X(x) dx = \sigma^2$$

Since μ and σ are the only parameters in the specification of the distribution, the two moments $E[X]$ and $E[(X-\mu)^2]$ characterize it completely. The *unity* Gaussian distribution is that corresponding to $\sigma = 1$ and $\mu = 0$:

$$p_X(x) = \frac{1}{\sqrt{2\pi}} \exp[-\frac{x^2}{2}] \tag{4.2}$$

It is represented in Fig.4.1. The mass of probability between -1 and +1 is 0.683, that between -2 and +2 is 0.954 and that between -3 and +3 is 0.997. The characteristic function corresponding to (4.1) reads

$$M_X(\theta) = \exp(j\mu\theta - \frac{\sigma^2\theta^2}{2}) \tag{4.3}$$

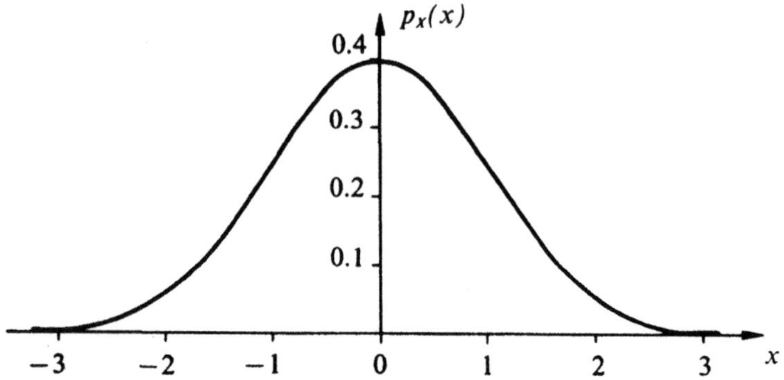

Figure 4.1: Unity Gaussian distribution.

Being the Fourier transform of the probability density function, it conveys the same information. Comparing Equ.(4.3) with the exponential form (2.67), one observes that ***the cumulants of order larger than 2 vanish.*** The moments of order larger than 2 do not vanish, but the following recursive formula can be established from Equ.(2.65):

$$E[X^n] = \mu E[X^{n-1}] + (n-1)\sigma^2 E[X^{n-2}] \qquad (n > 2) \tag{4.4}$$

This implies that the central moments of odd orders vanish. Any linear function of Gaussian random variables is also Gaussian.

4.2 The central limit theorem

This theorem establishes that, under mild conditions, the distribution of the sum of ***independent*** random variables tends to be Gaussian when the number of contributions goes to infinity, irrespective to the distribution of the individual contributions. There are many forms of this theorem; the simplest one is that where the contributing random variables have identical distributions. In that case, it reads:

The probability distribution function of the normalized sum

$$Z_n = \frac{Y_n - E[Y_n]}{\sigma(Y_n)} \quad \text{where} \quad Y_n = \sum_{k=1}^{n} X_k \tag{4.5}$$

of mutually independent random variables X_k with the same, but arbitrary, probability distribution function, tends to that of a unity Gaussian random variable as $n \to \infty$.

In this form, the theorem can be established easily, using the characteristic function [e.g. see (W.B.Davenport, 1970), p.440]. Note that the distribution of

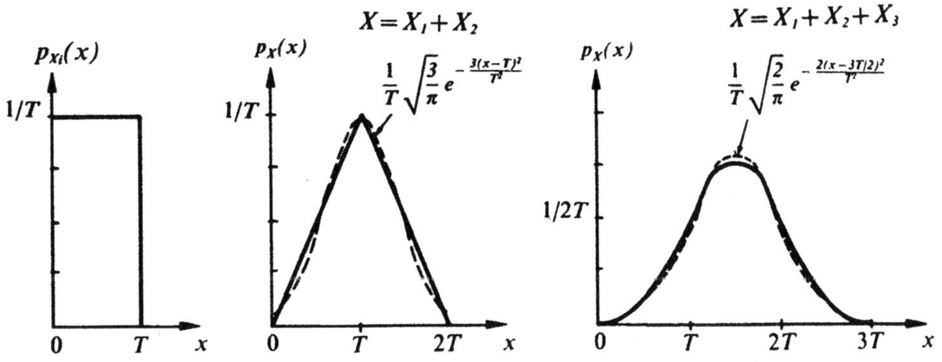

Figure 4.2: Distribution of the sum of independent random variables of rectangular distribution.

the contributing random variables X_k can be arbitrary, but they must be independent. The theorem can be extended to the case where the independent random variables have different probability distributions, providing none of them dominates the others in its contribution to the sum. The following examples illustrate the convergence of the process.

4.2.1 Example 1

Consider the contributing independent random variables X_i uniformly distributed in $[0, T]$ (Fig.4.2.a). Their mean and variance are

$$\mathrm{E}[X_i] = \frac{T}{2} \qquad \sigma_i^2 = \frac{T^2}{12}$$

Consider the sum of two random variables, $X = X_1 + X_2$. Because X_1 and X_2 are independent, the probability density function of X is given by the convolution of the probability density functions of X_1 and X_2:

$$p_X(x) = p_{X_1}(x) \star p_{X_2}(x) = \int_{-\infty}^{\infty} p_{X_1}(y) p_{X_2}(x-y) dy$$

We know from the graphical interpretation of the convolution that convolving a rectangle with itself provides a triangle with a double width (Fig.4.2.b). The mean and variance are

$$\mathrm{E}[X] = T \qquad \sigma_X^2 = \frac{T^2}{6}$$

The Gaussian distribution corresponding to these parameters is represented in dotted line in Fig.4.2.b. Proceeding to the sum of three independent random variables, $X = X_1 + X_2 + X_3$, the probability density function is obtained by convolving the triangular distribution (that of $X_1 + X_2$) with the rectangular one (of X_3). One gets the distribution of Fig.4.2.c. The mean and variance are

$$\mathrm{E}[X] = \frac{3T}{2} \qquad \sigma_X^2 = \frac{T^2}{4}$$

The corresponding Gaussian distribution is represented in dotted line in Fig.4.2.c. One observes that the curves are very close, which indicates that the convergence is extremely fast in the central part of the distribution. This is not the case in the tails of the distribution, because the convolution of n rectangular functions will always be zero outside the interval $[0, nT]$.

4.2.2 Example 2: Binomial distribution

Let X_i be independent random variables with discrete values 0 and 1. Their probability function is

$$P_{X_i}(0) = 1 - p \qquad P_{X_i}(1) = p$$

The mean and standard deviation are respectively

$$\mu_i = \mathrm{E}[X_i] = \sum x_i P_{X_i}(x_i) = p$$

$$\sigma_i^2 = \mathrm{E}[(X_i - \mu_i)^2] = \mathrm{E}[X_i^2] - \mu_i^2 = p(1-p)$$

Now consider the sum

$$X = \sum_{i=1}^{N} X_i$$

Its probabilty distribution is known as the *binomial distribution*:

$$P[X = k] = \binom{N}{k} p^k (1-p)^{N-k} = \frac{N!}{k!(N-k)!} p^k (1-p)^{N-k} \tag{4.6}$$

In fact, $p^k(1-p)^{N-k}$ is the probability associated with each sequence of N samples x_i containing k times $x_i = 1$ and $N - k$ times $x_i = 0$. This probability is multiplied by the binomial coefficient which represents the number of different ways one can construct such a sequence. X being the sum of *independent* random variables of mean μ_i and standard deviation σ_i, its mean and standard deviation are N times that of the contributing random variables:

$$\mu_x = N\mu_i, \qquad \sigma_x^2 = N\sigma_i^2$$

Since the mean and the variance of X increase linearly with N, consider the reduced variable

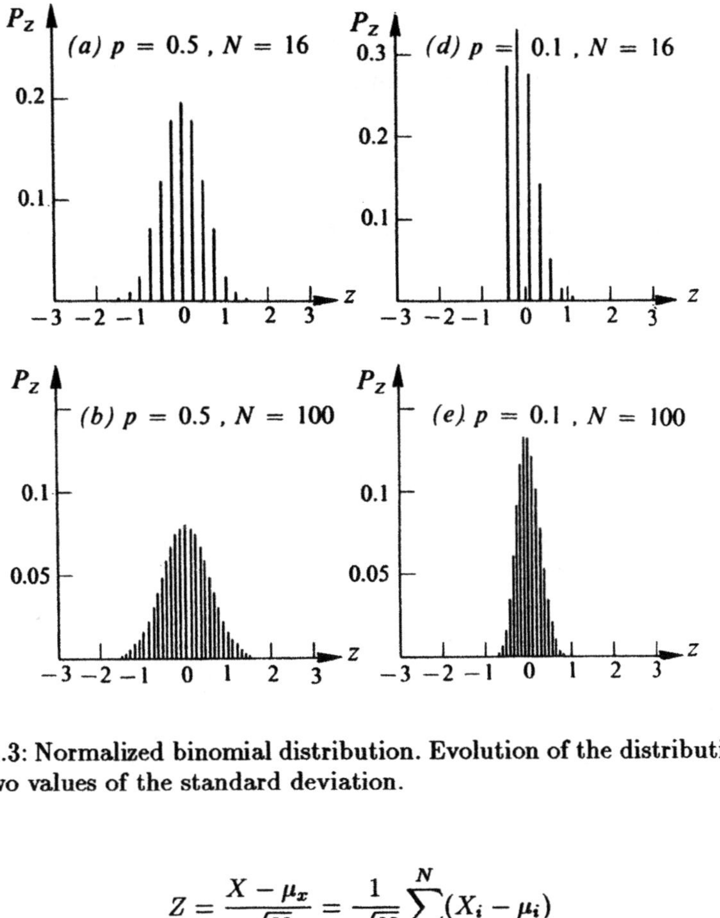

Figure 4.3: Normalized binomial distribution. Evolution of the distribution with N for two values of the standard deviation.

$$Z = \frac{X - \mu_x}{\sqrt{N}} = \frac{1}{\sqrt{N}} \sum_{i=1}^{N} (X_i - \mu_i) \tag{4.7}$$

It has a zero mean ($\mu_z = 0$), and its standard deviation is that of X_i ($\sigma_z = \sigma_i$). Apart from the scaling, the distribution of Z is similar to that of X; it is illustrated in Fig.4.3 for various values of N and p $[\sigma_z^2 = p(1-p)]$. One sees that the envelope of the distribution tends to the normal distribution. Obviously, the contributing random variables X_i being discrete, so is Z. The probability density function $p_Z(z)$ consists of a set of Dirac delta functions of intensity given by (4.6) at the discrete values of z; it does not become identical to (4.1) as N goes to infinity. However, the probability distribution function does converge towards that of a Gaussian random variable as N increases:

$$\lim_{N\to\infty} F_Z(z) = \lim_{N\to\infty} \int_{-\infty}^{z} p_Z(x)dx = \int_{-\infty}^{z} \frac{1}{\sqrt{2\pi}\sigma_z} \exp[-\frac{x^2}{2\sigma_z^2}]dx \tag{4.8}$$

The central limit theorem is the justification of the frequent assumption that the physical phenomena arising from the superposition of a large number of independent random contributions are Gaussian. An earthquake, for example, results from a large number of elementary fractures occuring along the fault line at random times; each of them generates seismic waves which propagate in the various layers of the soil, undergoing multiple refractions and reflections. The ground motion at a given point which results from the superposition of all these independent contributions can be considered as Gaussian. This has been confirmed by statistical analysis of actual earthquake records. The same considerations apply to other physical phenomena like atmospheric turbulence, etc...

Two nice features of the Gaussian random processes are that *(i)* they are completely characterized by their first and second order moments and *(ii)* the Gaussian character is preserved by *linear* transformations.

4.3 Jointly Gaussian random variables

Two random variables X and Y are *jointly Gaussian* if their joint probability density function reads

$$p_{XY}(x,y) = \frac{1}{2\pi\sigma_x\sigma_y\sqrt{1-\varrho_{xy}^2}}$$

$$.\exp\{-\frac{1}{2(1-\varrho_{xy}^2)}[\frac{(x-\mu_x)^2}{\sigma_x^2} - \frac{2\varrho_{xy}(x-\mu_x)(y-\mu_y)}{\sigma_x\sigma_y} + \frac{(y-\mu_y)^2}{\sigma_y^2}]\} \tag{4.9}$$

It can be checked that

$$E[X] = \mu_x, \qquad E[Y] = \mu_y$$

$$E[(X-\mu_x)^2] = \sigma_x^2, \qquad E[(Y-\mu_y)^2] = \sigma_y^2$$

$$E[(X-\mu_x)(Y-\mu_y)] = \sigma_x\sigma_y\varrho_{xy}$$

These five quantities define completely the joint probability density function. ϱ_{xy} is the *correlation coefficient* ($|\varrho| \leq 1$). If $\varrho_{xy} = 0$, X and Y are uncorrelated; it is readily seen from Equ.(4.9) that, in that case,

$$p_{XY}(x,y) = p_X(x).p_Y(y)$$

This equation shows that *uncorrelated Gaussian random variables are independent.*

In order to illustrate the meaning of the correlation coefficient, consider the case where X and Y are both distributed according to (4.2). Since $\mu_x = \mu_y = 0$

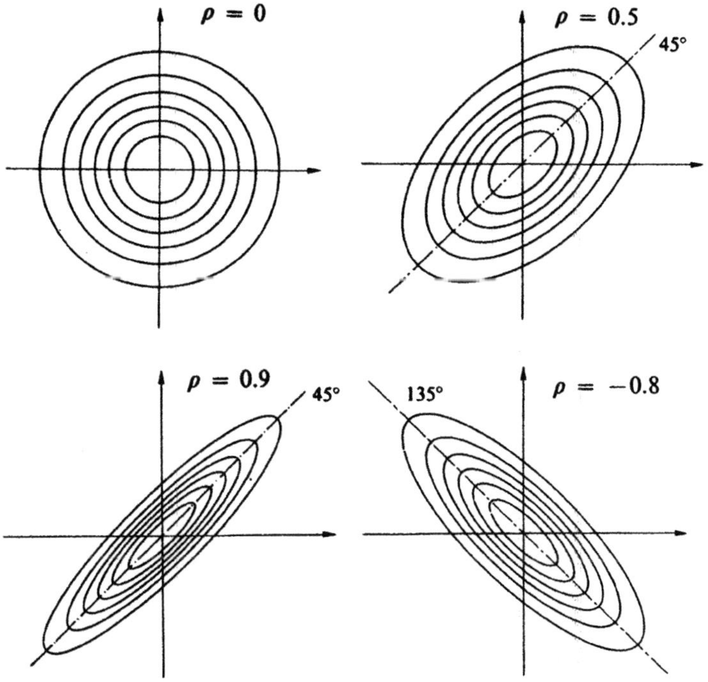

Figure 4.4: Influence of the correlation coefficient ϱ on the joint distribution of unity Gaussian random variables.

and $\sigma_x = \sigma_y = 1$, ϱ_{xy} is the only remaining parameter in the joint distribution which reads

$$p_{XY}(x,y) = \frac{1}{2\pi\sqrt{1-\varrho^2}} \exp[-\frac{x^2 - 2\varrho xy + y^2}{2(1-\varrho^2)}] \tag{4.10}$$

This distribution is represented for various values of ϱ in Fig.4.4. The contours of $p_{XY}(x,y)$ consist of ellipses having their major principal axis at 45^0 of the x axis if $\varrho > 0$ or at 135^0 if $\varrho < 0$. The ellipses become more elongated along their major axis as $|\varrho|$ increases; they become circles when the variables are uncorrelated. The volume under the distribution is 1 and the amplitude at the origin is $(2\pi\sqrt{1-\varrho^2})^{-1}$.

The conditional probability density function of X under the condition $Y = y$ is readily obtained from Equ.(4.10):

$$p_{X/Y}(x/y) = \frac{p_{XY}(x,y)}{p_Y(y)} = \frac{1}{\sqrt{2\pi(1-\varrho^2)}} \exp[-\frac{(x-\varrho y)^2}{2(1-\varrho^2)}] \tag{4.11}$$

It is a Gaussian distribution of [conditional] mean ϱy and variance $1-\varrho^2$; it is illustrated in Fig.4.5. Graphically, the conditional density can be visualized

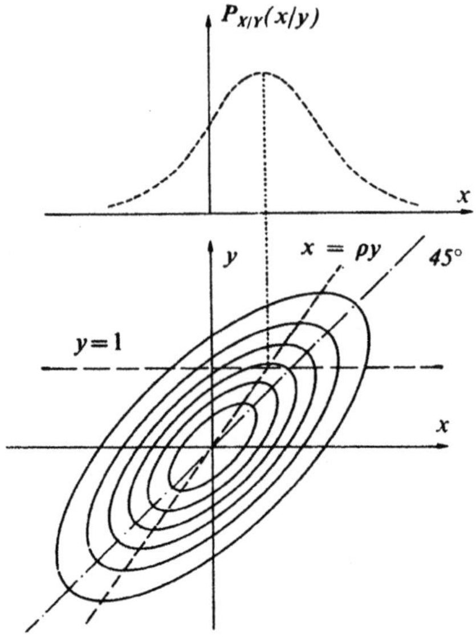

Figure 4.5: Conditional probability density function $p_{X/Y}(x/y)$ ($\varrho = 0.7$ and $y = 1$).

as the intersection between the joint probability density function and a vertical plane parallel to the x axis at the prescribed value of y. As one would expect, the scatter of the distribution from the conditional mean decreases when the correlation coefficient increases.

4.3.1 Remark

Two random variables are said to be

- *Orthogonal* if $\quad E[XY] = 0.$
- *Uncorrelated* or *linearly independent* if $\quad E[(X - \mu_x)(Y - \mu_y)] = 0$ or $\varrho_{xy} = 0.$
- *Independent* if $\quad p_{XY}(x, y) = p_X(x).p_Y(y)$

Orthogonal random variables of zero mean are uncorrelated. Independent random variables are uncorrelated, but being uncorrelated does not imply independence. If they are Gaussian, uncorrelated random variables are also independent.

4.4 Gaussian random vector

When there are more than two random variables, it is convenient to use vector notations. The components of the vector $X = (X_1, ..., X_n)^T$ are jointly Gaussian if their joint probability density function reads

$$p_X(x) = \frac{1}{(2\pi)^{n/2}|S|^{1/2}} \exp[-\frac{1}{2}(x-\mu_x)^T S^{-1}(x-\mu_x)] \tag{4.12}$$

where S stands for the ***covariance matrix***

$$S = \mathrm{E}[(X-\mu_x)(X-\mu_x)^T] \tag{4.13}$$

S is symmetric non-negative definite ($S \geq 0$).

If one performs a linear transformation $Y = AX$, where A is non singular, then $\mu_y = A\mu_x$ and the Jacobian of the transformation reads $(\partial x_i/\partial y_k) = A^{-1}$. According to Equ.(2.50), the probability density function of the new random vector is

$$p_Y(y) = \frac{1}{(2\pi)^{n/2}|S|^{1/2}} \cdot \exp[-\frac{1}{2}(y-\mu_y)^T A^{-T} S^{-1} A^{-1}(y-\mu_y)].|A^{-1}| \tag{4.14}$$

One observes that the distribution of Y is also Gaussian, with a covariance matrix ASA^T. Because S is symmetric non-negative definite, it is always possible to find an orthogonal transformation ($A^{-1} = A^T$) such that the covariance matrix is diagonalized:

$$D = ASA^T = \mathrm{diag}(\lambda_1^2, ..., \lambda_n^2) \tag{4.15}$$

From Equ.(4.14), it is readily established that the components of the transformed vector have the following joint probability density function

$$p_Y(y) = \prod_{i=1}^{n} \frac{1}{\sqrt{2\pi}\lambda_i} \exp[-\frac{(y_i-\mu_{y_i})^2}{2\lambda_i^2}] = \prod_{i=1}^{n} p_{Y_i}(y_i) \tag{4.16}$$

The new random variables Y_i are therefore ***independent***. This result is not surprising, since a diagonal covariance matrix means that the various components of the transformed vector Y are uncorrelated, and we established in the previous section that uncorrelated Gaussian random variables are independent.

From Equ.(2.77) and (4.3), the characteristic function of Y reads

$$\begin{aligned} M_Y(\theta) &= M_{Y_1 \dots Y_n}(\theta_1, ..., \theta_n) = \prod_{i=1}^{n} M_{Y_i}(\theta_i) \\ &= \exp[\sum_{i=1}^{n} (j\mu_{y_i}\theta_i - \frac{\lambda_i^2\theta_i^2}{2})] = \exp(j\mu_y^T\theta - \frac{1}{2}\theta^T D\theta) \end{aligned} \tag{4.17}$$

Returning to the original random vector $X = A^T Y$,

$$\begin{aligned} M_X(\theta) &= \mathrm{E}[e^{j\theta^T X}] = \mathrm{E}[e^{j\theta^T A^T Y}] = M_Y(A\theta) \\ &= \exp(j\mu_y^T A\theta - \frac{1}{2}\theta^T A^T D A\theta) \end{aligned}$$

After substituting the mean $\mu_x = A^T\mu_y$ and the covariance matrix $S = A^T DA$, one gets

$$M_X(\theta) = \exp(j\mu_x^T\theta - \frac{1}{2}\theta^T S\theta) \tag{4.18}$$

Thus, with matrix notation, the joint probability density function and the characteristic function have the same form, whether the random variables are correlated or not. The correlation is reflected only in the fact that the covariance matrix has off-diagonal components. Comparing (4.12) to (4.16) or (4.17) to (4.18), one sees that ***being uncorrelated is a necessary and sufficient condition for Gaussian random variables to be independent.*** Besides, comparing Equ.(4.18) with the expansion (2.73), one notices that ***the joint cumulants of order greater than 2 vanish.*** This property is a necessary and sufficient condition for a random vector to be Gaussian and it can be used as an alternative definition.

4.5 Gaussian random process

A random process $X(t)$ is Gaussian if the random vector constituted by the values of the process at arbitrary times $t_1, ..., t_n$ is Gaussian, for any value of n. Using Equ.(3.9), a Gaussian process can be defined as a random process with the following ***characteristic functional***:

$$M_X[\theta(t)] = \mathrm{E}\{\exp[j\int_T \theta(t)X(t)dt]\}$$

$$= \exp[j\int_T \mu_x(t)\theta(t)dt - \frac{1}{2}\int\int_T \kappa_{xx}(t_1,t_2)\theta_1(t_1)\theta_2(t_2)dt_1dt_2] \tag{4.19}$$

where $\mu_x(t)$ is the mean and $\kappa_{xx}(t_1,t_2)$ the autocovariance function of $X(t)$. Indeed, using

$$\theta(t) = \sum_{i=1}^{n} \theta_i\delta(t - t_i)$$

as we did in section 3.2, we find that this characteristic functional supplies the following characteristic function for the random vector $X = (X(t_1), ..., X(t_n))^T$:

$$M_X(\theta_1, ..., \theta_n) = \exp[j\sum_{i=1}^{n}\theta_i\mu_x(t_i) - \frac{1}{2}\sum_{i,k=1}^{n}\kappa_{xx}(t_i,t_k)\theta_i\theta_k] \tag{4.20}$$

Being identical to (4.18), this equation shows that the random variables $X(t_i)$ are jointly Gaussian. Comparing Equ.(4.19) to the general form (3.11) of the characteristic functional, one notices that

- The cumulants $\kappa_n[X(t_1), ..., X(t_n)]$ of order n larger than 2 vanish for a Gaussian process. Just as for the random vector, this can be used as an alternative definition of a Gaussian process.

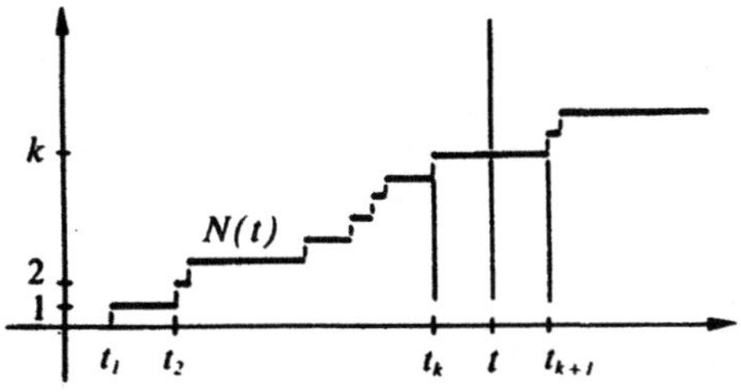

Figure 4.6: Sample of a counting process $N(t)$.

- ***A Gaussian random process is completely defined by its mean $\mu_x(t)$ and its autocovariance function $\kappa_{xx}(t_1,t_2)$.***
- A direct consequence of the foregoing property is that *a weakly* stationary Gaussian process is also *strongly* stationary. Indeed, if the mean is independent of t and if the autocorrelation function depends only on the difference of its arguments, $\kappa_{xx}(t_1 - t_2)$, the characteristic functional is invariant with respect to a change of the time origin.

We have seen that any linear transformation of a Gaussian random vector is also Gaussian. Similarly, ***any linear operator transforms a Gaussian process into another Gaussian process.*** Thus, the Gaussian character is preserved by the operations of differentiation, integration and filtering; the response of a linear system to a Gaussian excitation is also Gaussian.

4.6 Poisson process

4.6.1 Counting process

Consider an event occuring repeatedly at random times (e.g. telephone calls through a switchboard, customers at a gas station, users of a computer system,...). The number of times $N(t)$ the event occurs in the observation period $[0,t]$ constitutes a ***counting process.*** A typical sample is represented in Fig.4.6 ($N(0) = 0$). These processes have been studied extensively by Stratonovich (1963). In the general case, a counting process may be characterized by its probability functions of increasing order:

$$P_N(n,t) = P[N(t) = n]$$

$$P_N(n_1,t_1;n_2,t_2) = P[N(t_1) = n_1 \cap N(t_2) = n_2] \tag{4.21}$$

etc...

The first one expresses the probability of n occurrences in $[0,t]$, while the second expresses the probabilty of n_1 occurrences in $[0,t_1]$ ***and*** n_2 occurrences in

$[0, t_2]$, etc... In general, all these probability functions are necessary to define the counting process. If the probability of occurrence relative to disjoint time intervals are independent, the counting process has ***independent increments***; the Poisson process belongs to this class. If, in addition to that, the probabilty distribution is invariant with respect to a change of the time origin, the process has ***stationary increments.*** In that case, the probability functions can be factorized in terms of that of the first order only:

$$P_N(n_1, t_1; n_2, t_2) = P_N(n_1, t_1).P_N(n_2 - n_1, t_2 - t_1) \qquad (t_1 < t_2) \qquad (4.22)$$

4.6.2 Uniform Poisson process

A ***uniform Poisson process*** is a counting process with stationary increments which satisfies the following conditions:

- The events occur ***independently***;
- the ***arrival rate*** λ is uniform:

$$P_N(1, dt) = \lambda.dt \qquad (\lambda > 0)$$

- the probability of simultaneous arrival is negligible:

$$P_N(n, dt) = 0 \qquad (n > 1, \; dt \to 0)$$

Under these assumptions, the probabilty function of $N(t)$ is

$$P_N(n, t) = e^{-\lambda t} \frac{(\lambda t)^n}{n!} \qquad (4.23)$$

To prove that, consider the event $N(t + dt) = n$; it may result of either of the following mutually exclusive events:

$$\{N(t) = n\} \quad \cap \quad \{\text{no arrival in } (t, t + dt]\}$$

or

$$\{N(t) = n - 1\} \quad \cap \quad \{1 \text{ arrival in } (t, t + dt]\}$$

More than one arrival in $(t, t+dt]$ is impossible because of the third assumption. Since the arrivals are independent, the corresponding probability relationship is

$$P_N(n, t + dt) = P_N(n, t).P_N(0, dt) + P_N(n - 1, t).P_N(1, dt)$$

From the definition of the arrival rate and the third assumption,

$$P_N(1, dt) = \lambda.dt \qquad \text{and} \qquad P_N(0, dt) = 1 - \lambda.dt$$

Introducing this in the previous equation, upon dividing by dt, one gets

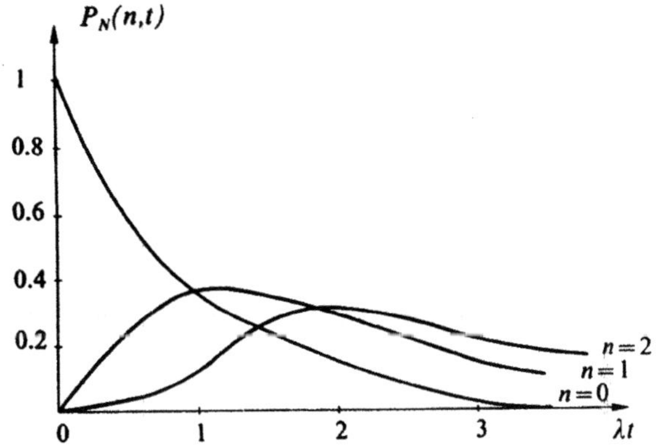

Figure 4.7: Probability function $P_N(n,t)$ of a uniform Poisson process.

$$\frac{P_N(n, t + dt) - P_N(n, t)}{dt} = \lambda[P_N(n-1, t) - P_N(n, t)]$$

and, taking the limit for $dt \to 0$,

$$\frac{d}{dt}P_N(n,t) + \lambda P_N(n,t) = \lambda P_N(n-1,t) \tag{4.24}$$

One can check that (4.23) is solution of this differential equation.

The variation of $P_N(n,t)$ with the non-dimensional parameter λt is represented in Fig.4.7 for various values of n. The average number of arrivals is

$$\mathrm{E}[N(t)] = \sum_{n=0}^{\infty} n P_N(n,t) = \lambda t \tag{4.25}$$

Alternatively, one may be interested in the *arrival time T_k of the kth event.* Unlike the counting process $N(t)$ which can take only integer values, T_k is a continuous random variable. Its probability density function can be obtained from the probability function of $N(t)$ by noting the identity between the following events:

$$\{T_k \leq t\} \equiv \{N(t) > k - 1\} \tag{4.26}$$

Both of these events express the fact that at least k arrivals have taken place in $[0, t]$. Since identical events have the same probability, the probability distribution function of the random variable T_k can be expressed in term of that of the counting process $N(t)$ by

$$F_{T_k}(t) = P[T_k \leq t] = P[N(t) > k - 1]$$

$$= 1 - P[N(t) \leq k - 1] = 1 - F_{N(t)}(k-1)$$

Since

$$F_{N(t)}(k-1) = \sum_{i=0}^{k-1} P[N(t)=i] = \sum_{i=0}^{k-1} e^{-\lambda t}\frac{(\lambda t)^i}{i!}$$

one gets

$$F_{T_k}(t) = 1 - e^{-\lambda t}\sum_{i=0}^{k-1}\frac{(\lambda t)^i}{i!} \qquad (t \geq 0) \tag{4.27}$$

Upon differentiating, one gets the probability density function

$$p_{T_k}(t) = \lambda e^{-\lambda t}\frac{(\lambda t)^{k-1}}{(k-1)!} \qquad (t \geq 0) \tag{4.28}$$

If, instead of T_k, one considers the non-dimensional random variable λT_k, its probability density function reads

$$p_{\lambda T_k}(\lambda t) = e^{-\lambda t}\frac{(\lambda t)^{k-1}}{(k-1)!} \qquad (t \geq 0) \tag{4.29}$$

It is identical to (4.23) if one substitutes $n = k-1$. It follows that the various curves in Fig.4.7 also represent the probability density function of the reduced arrival time λT_{n+1} (the curve corresponding to $n = 0$ is the *PDF* of λT_1; that for $n = 1$ is the *PDF* of λT_2, etc...). The mean of the distribution is

$$\mathrm{E}[\lambda T_k] = k \tag{4.30}$$

4.6.3 Non-uniform Poisson process

If the arrival rate is not constant, but varies with time, the Poisson process is said to be non-uniform. It still has independent increments, but it is nonstationary. Its probability function reads

$$P_N(n,t) = \frac{1}{n!}[\int_0^t \lambda(\tau)d\tau]^n \exp[-\int_0^t \lambda(\tau)d\tau] \tag{4.31}$$

where $\lambda(t)$ is the expected arrival rate at t.

More general counting processes can be constructed by removing the assumption of independent arrivals. In that case, the higher order probability functions do not factorize in terms of those of the first order any more, as in (4.22). The more general description (4.21) is required. More elaborate counting processes have been studied by Stratonovich (1963).

4.7 Random pulses

An interesting class of random processes can be constructed by superposition of random pulses arriving at random times. In this section, we restrict ourselves to the case where the pulses have a deterministic shape but random amplitudes:

$$X(t) = \sum_{k=1}^{N(t)} Y_k w(t, \tau_k) \tag{4.32}$$

where $N(t)$ is a counting process with arrival times τ_k, $w(t, \tau_k)$ is the deterministic, but possibly time-varying shape of the pulses, and Y_k is the random amplitude of the kth pulse. It is assumed that the random variables Y_k are mutually independent and identically distributed. It is also assumed that the shape function is such that $w(t, \tau) = 0$ for $t < \tau$. If we regard $w(t, \tau)$ as the impulse response of a linear system, Equ.(4.32) is the response to a train of Dirac delta functions with random intensity:

$$S(t) = \sum_{k=1}^{N(t)} Y_k \delta(t - \tau_k) \tag{4.33}$$

For a Poisson process with arrival rate $\lambda(t)$, it can be demonstrated that the mean and autocovariance function of $X(t)$ are given by

$$\mu_x(t) = \mu_y \int_0^t w(t, \tau)\lambda(\tau) d\tau$$

$$\kappa_{xx}(t_1, t_2) = \mathrm{E}[Y^2] \int_0^{\min(t_1, t_2)} w(t_1, \tau) w(t_2, \tau) \lambda(\tau) d\tau \tag{4.34}$$

More generally, the higher order cumulants read

$$\kappa_m[X(t_1), ..., X(t_m)] = \mathrm{E}[Y^m] \int_0^{\min(t_1, ..., t_m)} w(t_1, \tau) ... w(t_m, \tau) \lambda(\tau) d\tau \tag{4.35}$$

If the system is time-invariant, the impulse response depends only on the difference of its arguments, $w(t, \tau) = w(t - \tau)$. If the system is stable and dissipative, $w(u)$ is absolutely integrable. If, in addition to that, the arrival rate λ is uniform, $X(t)$ tends to a steady state for large values of t. Replacing the limits of the integrals by $-\infty$ and $+\infty$, one finds

$$\mu_x = \lambda \mu_y \int_{-\infty}^{+\infty} w(u) du$$

$$\kappa_{xx}(t_1, t_2) = \Gamma_{xx}(\tau) = \lambda \mathrm{E}[Y^2] \int_{-\infty}^{+\infty} w(u) w(\tau + u) du$$

$$\sigma_x^2 = \Gamma_{xx}(0) = \lambda \mathrm{E}[Y^2] \int_{-\infty}^{+\infty} w^2(u) du$$

These equations are known as *Campbell's theorem*.

For any value of t, the value of $X(t)$ arises from the superposition of the contributions of the random pulses. From the central limit theorem, one would expect that, if the arrival rate of the pulses becomes unbounded, the process $X(t)$ would tend to a Gaussian process. This can be established as follows. Consider the arrival rate $\lambda(t) = \lambda_0 \nu(t)$ where $\nu(t)$ is a given function of t and λ_0 is a scaling parameter which goes to infinity. In order to keep the variance finite, we also apply a scaling to the random intensity: $Y_k = Z_k \lambda_0^{-1/2}$, where Z_k are identically distributed random variables. From Equ.(4.35), the higher order cumulants read

$$\kappa_m = \lambda_0^{1-\frac{m}{2}} \mathrm{E}[Z^m] \int_0^{\min(t_1,\dots,t_m)} w(t_1,\tau)\dots w(t_m,\tau)\nu(\tau)d\tau \tag{4.36}$$

Clearly, since the exponent of λ_0 is $1 - m/2$, the cumulants of order m larger than 2 vanish at the limit, when $\lambda_0 \to \infty$. As we saw in section 4.6, this is a necessary and sufficient condition for $X(t)$ to be Gaussian.

4.8 Shot noise

A *shot noise* is the non-stationary counterpart of white noise; it is defined by

$$\mu_s(t) = 0$$

$$\kappa_{ss}(t_1, t_2) = I(t_1)\delta(t_2 - t_1) \tag{4.37}$$

$I(t)$ is the *Intensity function*. If the intensity is constant, this definition reduces to that of white noise. From the result of the previous section, *one* shot noise can be constructed from (4.33) with $\mathrm{E}[Y] = 0$ and $I(t) = \mathrm{E}[Y^2]\lambda(t)$.

4.9 References

W.B.DAVENPORT, *Probability and Random Processes*, McGraw-Hill, 1970.
Y.K.LIN, *Probabilistic Theory of Structural Dynamics*, McGraw-Hill, 1967.
A.PAPOULIS, *Probability, Random Variables and Stochastic Processes*, McGraw-Hill, 1965.
E.PARZEN, *Stochastic Processes*, Holden Day, 1962.
R.L.STRATONOVICH, *Topics in the Theory of Random Noise*, 1, Gordon & Breach, N-Y, 1963.

4.10 Problems

P.4.1 *Let X_i be independent random variables with uniform distribution between* -1 and *1:*

$$p_{x_i}(x) = 1/2 \qquad (-1 \le x \le 1)$$

Using the characteristic function, show that the random variable $Y = \sqrt{3/n}(X_1 + \ldots + X_n)$ *tends to be unity Gaussian as* $n \to \infty$.

P.4.2 *Let X_i be independent random variables of identical but arbitrary distribution with zero mean and standard deviation σ. Using the characteristic function, show that* $Y = (X_1 + \ldots + X_n)/(\sqrt{n}\sigma)$ *tends to a unity Gaussian distribution as* $n \to \infty$.

P.4.3 *Using the characteristic function (4.3), show that the moments of a Gaussian random variable satisfy the recursive formula (4.4).*

P.4.4 *Plot the binomial distribution (4.6) for $N = 8$ and $p = 1/2$.*

P.4.5 *Consider the random process*

$$X(t) = \sum_{k=1}^{N(t)} Y_k w(t - \tau_k)$$

where $N(t)$ is a uniform Poisson process, Y_k are independent random variables identically distributed with zero mean and standard deviation σ, and $w(t)$ is the rectangular approximation of the Dirac function $\delta(\tau)$:

$$w(\tau) = \frac{1}{\Delta} \qquad 0 \le \tau \le \Delta$$

Show that $X(t)$ approaches a white noise as $\Delta \to 0$. Show that

$$\Phi_{xx}(\omega) = \frac{\lambda\sigma^2}{2\pi} \cdot \frac{\sin^2(\frac{\omega\Delta}{2})}{(\frac{\omega\Delta}{2})^2}$$

P.4.6 *Determine the variance of the process*

$$X(t) = \sum_{k=1}^{N(t)} Y_k \mathbf{1}(t - \tau_k)$$

where $N(t)$ is a uniform Poisson process, $\mathbf{1}(t)$ is the Heaviside's step function and Y_k are mutually independant random variables identically distributed with zero mean and standard deviation σ [this process is the integral of (4.33)].

P.4.7 *Consider the Wiener process defined as*

$$\dot{X} = F(t) \qquad X(0) = 0$$

where $F(t)$ is a Gaussian white noise of intensity I. Show that $X(t)$ is non-stationary Gaussian with variance $\sigma_x^2(t) = It$.

P.4.8 *Let X be unity Gaussian. Show that the characteristic function of the random variable defined as $Y = X^2$ is*

$$M_Y(\theta) = (1 - j2\theta)^{-1/2}$$

(This is a chi-squared *probability distribution with one degree of freedom).*
P.4.9 *Let $N(t)$ be a uniform Poisson process with arrival rate λ. Show that, for a fixed t, $N(t)$ has the following characteristic function*

$$M_N(\theta) = \exp[\lambda t(e^{j\theta} - 1)]$$

Chapter 5

Random Response of a Single Degree of Freedom Oscillator

5.1 Response of a linear system

A single input single output (*SISO*) time invariant linear system is entirely characterized by its ***impulse response*** $h(t)$, which represents the response of the system, initially at rest, to a unit Dirac delta function $\delta(t)$. Equivalently, the system is completely characterized by its ***transfer function***, $H(s)$, which is the Laplace transform of $h(t)$:

$$H(s) = \int_0^\infty h(t)e^{-st}dt \tag{5.1}$$

Usually, $H(s)$ can be determined directly from the differential equation of the system. For an arbitrary excitation, the input-output relationship in the time

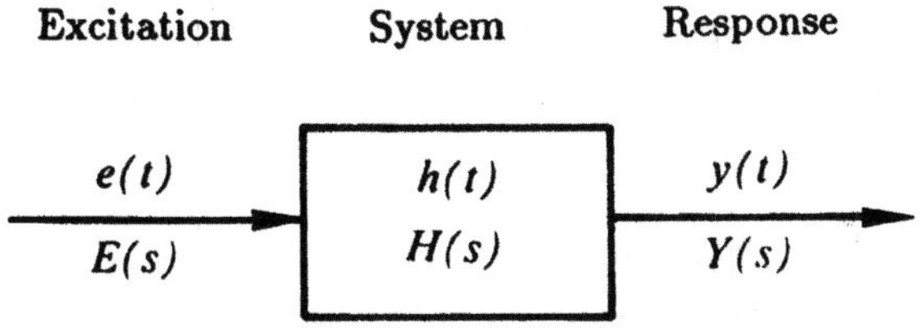

Figure 5.1: Time invariant single input single output linear system.

domain can be derived from the principle of superposition as the *convolution*

$$y(t) = \int_{-\infty}^{\infty} e(\tau)h(t-\tau)d\tau$$

$$= \int_{-\infty}^{\infty} h(\tau)e(t-\tau)d\tau = h \star e \quad (5.2)$$

Any physical system is *causal*: it cannot respond before being excited. In mathematical terms, this implies that $h(\tau) = 0, \tau < 0$. In that case, the bounds of the foregoing integrals can be reduced to

$$y(t) = \int_{-\infty}^{t} e(\tau)h(t-\tau)d\tau = \int_{0}^{\infty} h(\tau)e(t-\tau)d\tau \quad (5.3)$$

If the system is not initially at rest, the effect of the initial conditions must be added. Upon Laplace transforming the above equation, assuming zero initial conditions, one gets

$$Y(s) = H(s)E(s) \quad (5.4)$$

where $Y(s)$ and $E(s)$ are the Laplace transforms of the response and the excitation, repectively, and $H(s)$ is the transfer function of the system. From Equ.(5.3), the response to a harmonic excitation $e(t) = e^{j\omega t}$ is given by

$$y(t) = \int_{0}^{\infty} h(\tau)e^{j\omega(t-\tau)}d\tau = e^{j\omega t}H(j\omega) \quad (5.5)$$

where $H(j\omega)$ is the Fourier transform of the impulse response [we have used the fact that $h(t)$ is causal]. $H(j\omega)$ or $H(\omega)$ is called the *frequency response function*, but it is also frequently called the *transfer function* of the system. Equation (5.5) states that the response to any steady state harmonic excitation is also harmonic, with the same frequency; the complex magnification (amplitude and phase) is given by $H(\omega)$.

5.2 Single degree of freedom oscillator

In this section, we recall the equation of motion of the lightly damped single degree of freedom oscillator for the two modes of excitation represented in Fig.5.2. When subjected to an external force $f(t)$, the oscillator is governed by the differential equation

$$m\ddot{y} + c\dot{y} + ky = f(t) \quad (5.6)$$

where m is the mass, k the stiffness, and c the viscous damping coefficient. The natural frequency ω_n and the fraction of critical damping ξ are defined as

$$\omega_n^2 = \frac{k}{m} \qquad 2\xi\omega_n = \frac{c}{m} \quad (5.7)$$

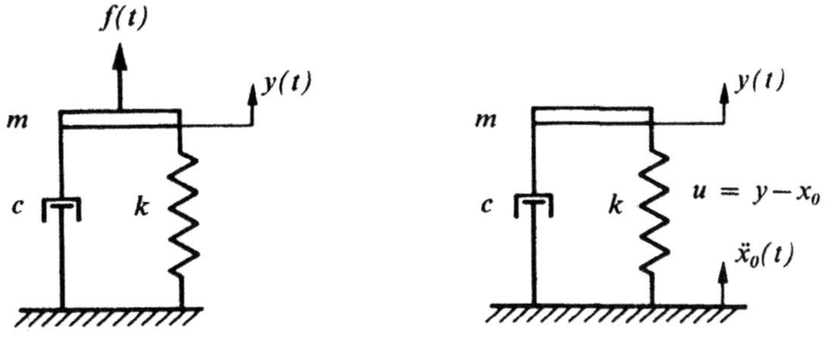

Figure 5.2: Single degree of freedom oscilator. (a)Excited by an external force. (b)Seismic excitation.

($\xi << 1$ if the oscillator is lightly damped). With these notations, the equation of motion can be rewritten as

$$\ddot{y} + 2\xi\omega_n\dot{y} + \omega_n^2 y = \frac{f(t)}{m} \tag{5.8}$$

The transfer function between the excitation and the absolute response is readily obtained from the differential equation as

$$H(s) = \frac{1}{m}\,\frac{1}{s^2 + 2\xi\omega_n s + \omega_n^2} \tag{5.9}$$

and the frequency response function

$$H(\omega) = \frac{Y(\omega)}{F(\omega)} = \frac{1}{m}\,\frac{1}{(\omega_n^2 - \omega^2) + 2j\xi\omega\omega_n} = \frac{1}{k}\,\frac{1}{(1 - \frac{\omega^2}{\omega_n^2}) + 2j\xi\frac{\omega}{\omega_n}} \tag{5.10}$$

It is represented in Fig.5.3. The impulse response is the solution of (5.8) for a unit impulse loading $f(t) = \delta(t)$. Integrating the equation between $t = 0^-$ and $t = 0^+$ and neglecting all the non impulsive loads, we obtain the impulse response as the free response to an intial velocity $1/m$. One easily gets

$$h(t) = \frac{e^{-\xi\omega_n t}}{m\omega_d}\sin\omega_d t \quad (t \geq 0); \qquad h(t) = 0 \quad (t < 0) \tag{5.11}$$

where $\omega_d = \omega_n\sqrt{1-\xi^2}$ is the damped resonance frequency. $h(t)$ is represented in Fig.5.4 for two values of the damping ratio. The exponential decay has a time constant $\tau = (\xi\omega_n)^{-1}$. This indicates that the *memory* of the system (i.e. the time during which a perturbation continues to affect the behaviour of the system after it has disappeared) increases when the damping decreases.

For a seismic excitation, defined by the acceleration of the support, the equation of motion is readily obtained from a free body diagram of the mass:

$$m\ddot{y} + c(\dot{y} - \dot{x}_0) + k(y - x_0) = 0 \tag{5.12}$$

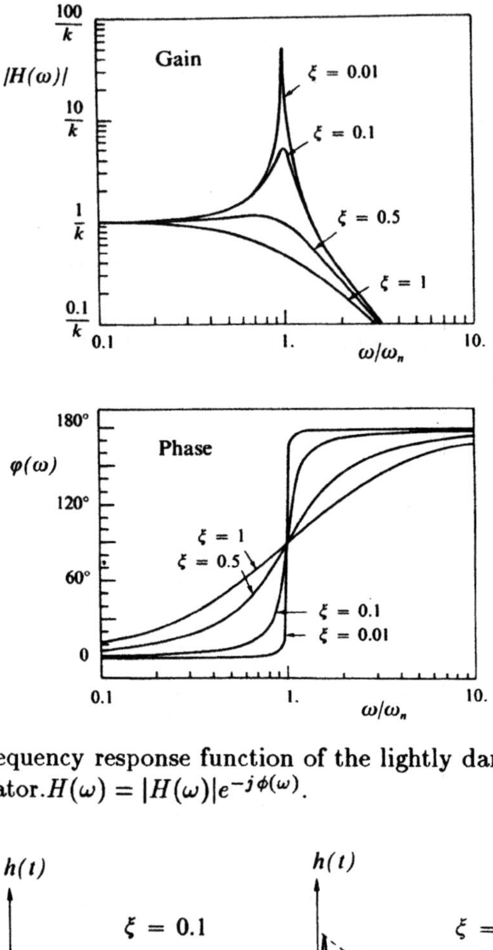

Figure 5.3: Frequency response function of the lightly damped single degree of freedom oscillator.$H(\omega) = |H(\omega)|e^{-j\phi(\omega)}$.

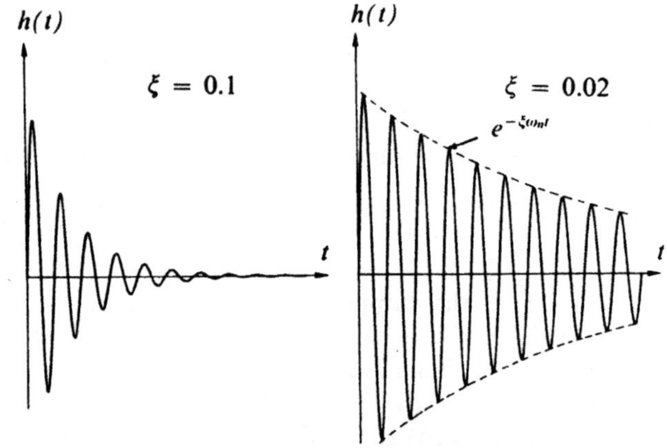

Figure 5.4: Impulse response of the linear oscillator.

Both the elastic restoring force and the viscous damping force depend on the relative motion $u = y - x_0$. Using that new variable and the definitions (5.7), we find that Equ.(5.12) becomes

$$\ddot{u} + 2\xi\omega_n\dot{u} + \omega_n^2 u = -\ddot{x}_0 \tag{5.13}$$

This equation tells us that the relative displacement, under seismic excitation, depends only on the natural frequency ω_n and the damping ratio ξ; it is therefore independent of the scale of the system. The transfer function between the excitation and the relative displacement is readily obtained:

$$H(s) = \frac{-1}{s^2 + 2\xi\omega_n s + \omega_n^2} \tag{5.14}$$

and the frequency response function

$$H(\omega) = \frac{U(\omega)}{\ddot{X}_0(\omega)} = \frac{-1}{(\omega_n^2 - \omega^2) + 2j\xi\omega\omega_n} \tag{5.15}$$

Except for a constant factor, it is identical to that represented in Fig.5.3. The impulse response reads

$$h(t) = -\frac{e^{-\xi\omega_n t}}{\omega_d}\sin\omega_d t \quad (t \geq 0); \qquad h(t) = 0 \quad (t < 0) \tag{5.16}$$

Except for the negative sign and the constant factor $1/m$, it is identical to that represented in Fig.5.4. The transfer function between the support acceleration and the *absolute* acceleration is obtained by noting that, from (5.13),

$$\ddot{y} = \ddot{u} + \ddot{x}_0 = -2\xi\omega_n\dot{u} - \omega_n^2 u$$

As a result,

$$H^1(s) = -(2\xi\omega_n s + \omega_n^2)H(s) \tag{5.17}$$

and

$$H^1(\omega) = \frac{\ddot{Y}(\omega)}{\ddot{X}_0(\omega)} = \frac{\omega_n^2 + 2j\xi\omega_n\omega}{(\omega_n^2 - \omega^2) + 2j\xi\omega\omega_n} \tag{5.18}$$

5.3 Stationary response of a linear system

Consider a linear, time invariant, *SISO* system subjected to a stationary random excitation $X(t)$. According to Equ.(5.3), the random response reads

$$Y(t) = \int_0^\infty h(\tau)X(t-\tau)d\tau \tag{5.19}$$

This integral exists in the mean square sense if the autocorrelation function ϕ_{yy} exists. Assume that this is so; if one considers two instants of time t and $t+\tau$ separated by some delay τ, one has

$$X(t)Y(t+\tau) = \int_0^\infty X(t)h(\xi)X(t+\tau-\xi)d\xi$$

$$Y(t)Y(t+\tau) = \int_0^\infty \int_0^\infty h(\xi)X(t-\xi)h(\eta)X(t+\tau-\eta)d\xi d\eta$$

Taking the mathematical expectation of these two expressions, one gets

$$R_{yx}(\tau) = \mathrm{E}[X(t)Y(t+\tau)] = \int_0^\infty h(\xi)R_{xx}(\tau-\xi)d\xi \tag{5.20}$$

$$R_{yy}(\tau) = \mathrm{E}[Y(t)Y(t+\tau)] = \int_0^\infty \int_0^\infty h(\xi)h(\eta)R_{xx}(\tau+\xi-\eta)d\xi d\eta \tag{5.21}$$

$$= \int_0^\infty h(\xi)R_{yx}(\tau+\xi)d\xi \tag{5.22}$$

Equation (5.20) states that the cross-correlation between the response and the excitation is expressed as the ***convolution*** of the autocorrelation of the excitation with the impulse response of the system, while Equ.(5.22) tells that the autocorrelation of the response is given by the ***correlation*** integral of $R_{yx}(\tau)$ with the impulse response of the system. The properties of the Fourier transform (section 1.3) give the relationship between the power spectral density functions as

$$\Phi_{yx}(\omega) = H(\omega)\Phi_{xx}(\omega) \tag{5.23}$$

$$\Phi_{yy} = H(\omega)^{\star}\Phi_{yx}(\omega) = |H(\omega)|^2\Phi_{xx}(\omega) \tag{5.24}$$

Note that Equ.(5.23) relates complex quantities; it therefore contains amplitude and phase information. By contrast, Equ.(5.24) relates positive real quantities and does not contain any phase information. Also, observe that the output *PSD* at one frequency depends only on the *PSD* of the input and the frequency response function for that same frequency; different frequencies can be treated completely independently.

5.4 Stationary response of the linear oscillator. White noise approximation.

Consider the oscillator of Fig.5.2 subject to a white noise seismic excitation of constant *PSD* S_0. Substituting Equ.(5.15) into (5.24), one gets the *PSD* of the relative displacement

$$\Phi_{yy}(\omega) = \frac{S_0}{(\omega_n^2-\omega^2)^2 + 4\xi^2\omega^2\omega_n^2} \tag{5.25}$$

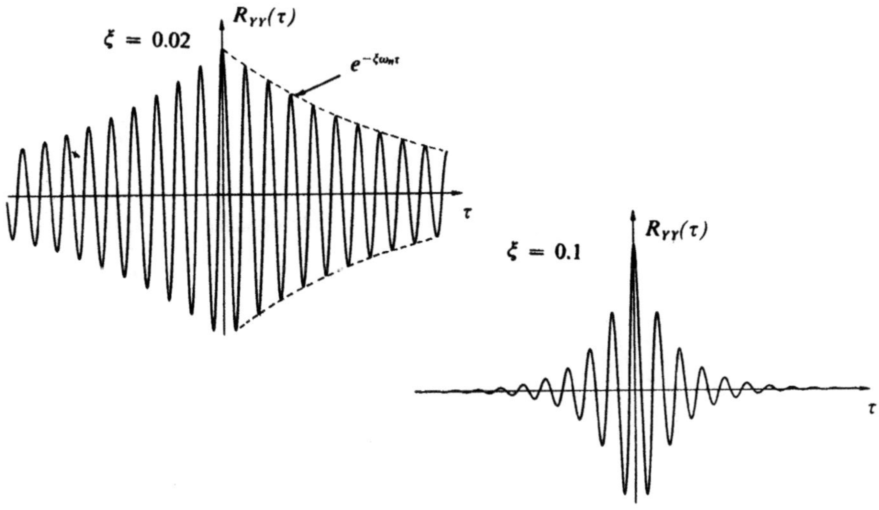

Figure 5.5: Autocorrelation function of the response of the linear oscillator to a white noise seismic excitation.

It is strongly peaked in the vicinity of the natural frequency of the oscillator, ω_n, where the oscillator, which acts as a filter, amplifies the excitation. The autocorrelation function reads

$$R_{yy}(\tau) = \sigma^2 e^{-\xi\omega_n|\tau|}[\cos\omega_d\tau + \frac{\xi}{\sqrt{1-\xi^2}}\text{sign}(\tau)\sin\omega_d\tau] \tag{5.26}$$

where

$$\sigma^2 = R_{yy}(0) = m_0 = \int_{-\infty}^{\infty}\Phi_{yy}(\omega)d\omega = \frac{\pi S_0}{2\xi\omega_n^3} \tag{5.27}$$

$R_{yy}(\tau)$ is represented in Fig.5.5 for two values of the damping ratio; just as for the impulse response, the decay rate (the memory) is controlled by the damping of the system.

Note that, although the variance of the excitation is unbounded, the variance of the response is finite provided that the system is damped. This is why modelling excitation processes by white noises is widely used in practice. Because $|H(\omega)|^2$ is strongly peaked near ω_n for lightly damped systems and decays like ω^{-4} at high frequency, the following approximation is often acceptable for an arbitrary excitation

$$R_{yy}(0) = \int_{-\infty}^{\infty}\Phi_{yy}(\omega)d\omega = \int_{-\infty}^{\infty}|H(\omega)|^2\Phi_{xx}(\omega)d\omega$$

$$\simeq \Phi_{xx}(\omega_n)\int_{-\infty}^{\infty}|H(\omega)|^2 d\omega = \frac{\pi\Phi_{xx}(\omega_n)}{2\xi\omega_n^3} \tag{5.28}$$

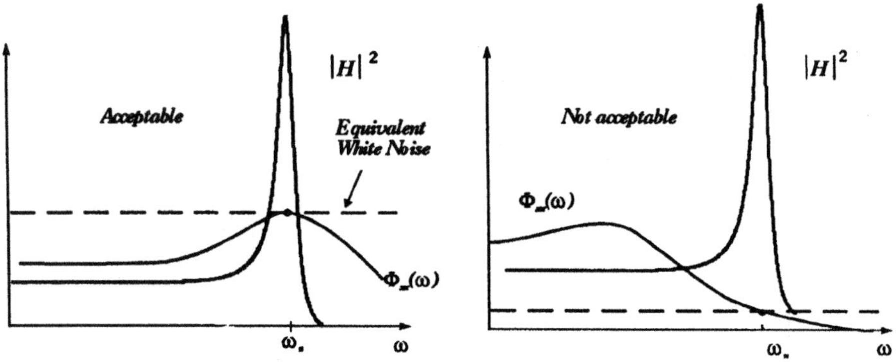

Figure 5.6: Condition of application of the white noise approximation.

This amounts to replacing the actual exitation by a fictituous white noise with a constant *PSD* equal to that of the actual excitation at the natural frequency of the system, $\Phi(\omega_n)$. The approximation is good if the dominant contribution to the integral comes from the vicinity of ω_n and if $\Phi_{xx}(\omega)$ varies slowly near ω_n (Fig.5.6).

5.5 Transient response

In this section, we shall restrict ourselves to the case where the transient character of the response comes from starting the system from rest. We pospone until chapter 8 the general treatment of nonstationary random processes.

5.5.1 Excitation applied from $t = 0$

Consider the random response $Y(t)$ to a random excitation $X(t)$ applied from $t = 0$; the system is assumed to be initially at rest. The input-output relationship in the time domain is

$$Y(t) = \int_0^t X(\tau)h(t-\tau)d\tau$$

It follows that the moments and cumulants of $Y(t)$ are given by

$$\mathrm{E}[Y(t_1)\cdots Y(t_2)] = \int_0^{t_1}\cdots\int_0^{t_n} \mathrm{E}[X(\tau_1)\cdots X(\tau_n)]h(t_1-\tau_1)\cdots h(t_n-\tau_n)d\tau_1 d\tau_n \tag{5.29}$$

$$\kappa_n[Y(t_1)\cdots Y(t_n)] = \int_0^{t_1}\cdots\int_0^{t_n} \kappa_n[X(\tau_1)\cdots X(\tau_n)]h(t_1-\tau_1)\cdots h(t_n-\tau_n)d\tau_1 d\tau_n \tag{5.30}$$

These two equations relate the correlation structure of the response to that of the excitation. If the excitation is Gaussian, all its cumulants of order larger than 2 vanish. Clearly, from (5.30), so do the cumulants of the response. This establishes that ***the response of a linear system to a Gaussian excitation is also Gaussian.*** In that case, it is completely specified by its mean and its covariance function

$$\kappa_1 = \mathrm{E}[Y(t)] = \int_0^t \mathrm{E}[X(\tau)]h(t-\tau)d\tau$$

$$\kappa_2[Y(t_1)Y(t_2)] = \int_0^{t_1}\int_0^{t_2} \kappa_2[X(\tau_1)X(\tau_2)]h(t_1-\tau_1)(t_2-\tau_2)d\tau_1 d\tau_2 \tag{5.31}$$

Note that even when the excitation is not Gaussian, the first two moments contain the most important information about the response and, often, allow us to judge the reliability of the system. Besides, the statistics of the excitation, which are obtained from tests, are known beyond the second moment only on very rare occasions.

5.5.2 Stationary excitation

If the excitation is weakly stationary, but is applied at $t = 0$ on the system at rest, the response is not stationary. The autocorrelation of the response can be expressed in terms of the *PSD* of the excitation, $\Phi_{xx}(\omega)$, by substituting

$$\mathrm{E}[X(\tau_1)X(\tau_2)] = R_{xx}(\tau_1-\tau_2) = \int_{-\infty}^{\infty} \Phi_{xx}(\omega)e^{j\omega(\tau_1-\tau_2)}d\omega$$

into Equ.(5.29):

$$\phi_{yy}(t_1,t_2) = \int_0^{t_1} d\tau_1 \int_0^{t_2} d\tau_2 \int_{-\infty}^{\infty} d\omega\, \Phi_{xx}(\omega)\, e^{j\omega(\tau_1-\tau_2)}\, h(t_1-\tau_1)h(t_2-\tau_2)$$

Integrating with respect to τ_1 and τ_2, one gets, after some algebraic manipulations,

$$\phi_{yy}(t_1,t_2) = \int_{-\infty}^{\infty} \Phi_{xx}(\omega)\mathcal{H}(\omega,t_1)\mathcal{H}^*(\omega,t_2)\, e^{j\omega(t_1-t_2)}\, d\omega \tag{5.32}$$

where

$$\mathcal{H}(\omega,t) = \int_0^t h(u)e^{-j\omega u}du \tag{5.33}$$

is the *truncated Fourier transform* of the impulse response (the lower bound of the integral can be changed to $-\infty$ if the system is causal). $\mathcal{H}(\omega,t)$ always exists for a damped system. For the linear oscillator,

$$\mathcal{H}(\omega,t) = H(\omega)[1-(\cos\omega_d t + \frac{\xi\omega_n + j\omega}{\omega_d}\sin\omega_d t)e^{-(\xi\omega_n+j\omega)t}] \tag{5.34}$$

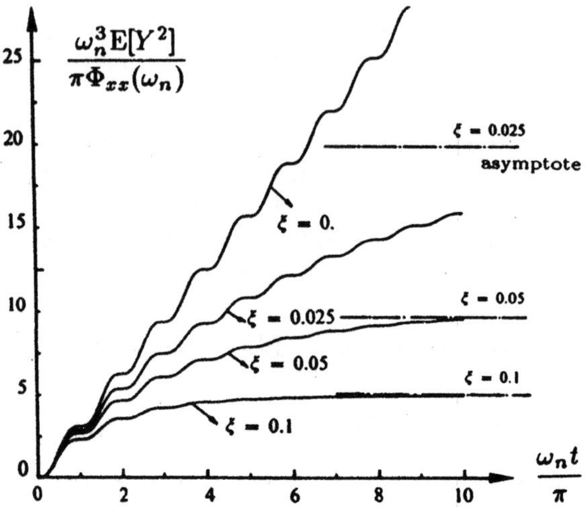

Figure 5.7: Transient response of a linear oscillator starting from rest and excited by a white noise.

It converges towards the frequency response function $H(\omega)$ after a duration which depends on the memory of the system. Subsequently, Equ.(5.32) shows that the autocorrelation function $\phi_{yy}(t_1, t_2)$ depends only on the difference $t_1 - t_2$, which indicates that the response becomes also weakly stationary. If the white noise approximation is applicable, the following approximate relationship holds for the mean square response

$$\phi_{yy}(t,t) = \mathrm{E}[Y^2(t)] \simeq \frac{\pi\Phi_{xx}(\omega_n)}{2\xi\omega_n^3}\{1 - \frac{e^{-2\xi\omega_n t}}{\omega_d^2}[\omega_d^2 + 2(\xi\omega_n \sin\omega_d t)^2 + \xi\omega_n\omega_d \sin 2\omega_d t]\} \quad (5.35)$$

It is illustrated in Fig.5.7 for various values of the damping. For each curve, a damped sinusoidal oscillation is superimposed on a monotonic rise. If one neglects the harmonic oscillations, (5.35) can be approximated by

$$\mathrm{E}[Y^2(t)] \simeq \frac{\pi\Phi_{xx}(\omega_n)}{2\xi\omega_n^3}(1 - e^{-2\xi\omega_n t}) \quad (5.36)$$

This form shows that the approach of the steady state response involves the time constant $\tau = (2\xi\omega_n)^{-1}$. The larger the damping of the system, the sooner the steady state is reached. If the time during which the system is exposed to the stationary excitation is large compared to this time constant, the response can be studied as if it were stationary. For very small values of its argument, the exponential can be expanded in power series; restricting to the first two terms, one gets

$$\mathrm{E}[Y^2(t)] = \frac{\pi\Phi_{xx}(\omega_n)}{\omega_n^2}t \quad (5.37)$$

This relationship is independent of the damping ratio ξ and applies for all t when $\xi = 0$; the mean square response grows indefinitely.

5.6 Spectral moments

5.6.1 Definition

The *spectral moment* m_i of order i of a stationary random process $X(t)$ is defined as

$$m_i = \int_{-\infty}^{\infty} |\omega|^i \Phi_{xx}(\omega) d\omega \tag{5.38}$$

Equation (3.32) and the Wiener-Khintchine theorem (3.50) show that, for a process with zero mean,

$$\sigma_x^2 = m_0 = \int_{-\infty}^{\infty} \Phi_{xx}(\omega) d\omega \tag{5.39}$$

$$\sigma_{\dot{x}}^2 = m_2 = \int_{-\infty}^{\infty} \omega^2 \Phi_{xx}(\omega) d\omega \tag{5.40}$$

$$\sigma_{\ddot{x}}^2 = m_4 = \int_{-\infty}^{\infty} \omega^4 \Phi_{xx}(\omega) d\omega \tag{5.41}$$

provided that these integrals exist. This depends on the shape of $\Phi_{xx}(\omega)$. For example, m_0 does not exist for a white noise; m_2 does not exist for a process with exponential correlation [Equ.(3.59)], and m_4 does not exist for the response of a linear oscillator to a white noise [Equ.(5.25)]; this process is not twice differentiable in the mean square sense.

5.6.2 Computation for the linear oscillator

The power spectral density function of the response of a lightly damped oscillator to a wide band excitation is strongly peaked in the vicinity of the natural frequency of the oscillator (Fig.5.8). This makes it difficult to evaluate the spectral moments using quadrature formulae, because they require a very large number of points in the vicinity of ω_n. The following method circumvents this difficulty.

The first step consists of calculating the spectral moments of the oscillator response to a band limited white noise $\Phi_0(\omega) = S_0, \quad \omega_1 \le |\omega| \le \omega_2$. For a seismic excitation, the power spectral density of the response reads (Fig.5.8)

$$\Phi_{xx}(\omega) = \frac{S_0}{(\omega_n^2 - \omega^2)^2 + 4\xi^2\omega^2\omega_n^2} \qquad \omega_1 \le |\omega| \le \omega_2 \tag{5.42}$$

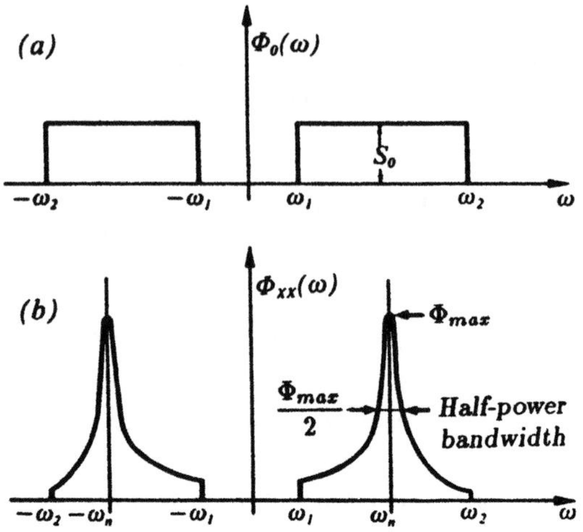

Figure 5.8: (a)Band limited white noise excitation. (b)Oscillator response.

Indefinite integrals for m_0, m_1, m_2 can be found in tables (e.g.see G.Petit Bois, 1961). For m_0, one gets

$$m_0 = \frac{\pi S_0}{2\xi\omega_n^3} \left[I(\frac{\omega_2}{\omega_n}, \xi) - I(\frac{\omega_1}{\omega_n}, \xi)\right] \tag{5.43}$$

where $I(\frac{\omega}{\omega_n}, \xi)$ is defined as

$$I(\frac{\omega}{\omega_n}, \xi) = \frac{1}{\pi} \arctan \frac{2\xi\frac{\omega}{\omega_n}}{1 - (\frac{\omega}{\omega_n})^2}$$

$$+ \frac{\xi}{2\pi\sqrt{1-\xi^2}} . \ln \frac{1 + (\frac{\omega}{\omega_n})^2 + 2\sqrt{1-\xi^2}(\frac{\omega}{\omega_n})}{1 + (\frac{\omega}{\omega_n})^2 - 2\sqrt{1-\xi^2}(\frac{\omega}{\omega_n})} \tag{5.44}$$

The first term in (5.43) is the variance of the response to a white noise excitation while the expression between brackets is a correcting term; this accounts for the fact that the white noise is band limited. $I(\frac{\omega}{\omega_n}, \xi)$ is a monotonically increasing function of ω/ω_n with values between 0 and 1. It is represented in Fig.5.9 where one notices that most of the variation takes place near the natural frequency of the system, in a frequency range equal to the bandwidth ($2\xi\omega_n$) of the oscillator. I goes to 1 as $\omega \to \infty$.

Similarly,

$$m_1 = \frac{\pi S_0}{2\xi\omega_n^2\sqrt{1-\xi^2}} \left[L(\frac{\omega_2}{\omega_n}, \xi) - L(\frac{\omega_1}{\omega_n}, \xi)\right] \tag{5.45}$$

where

$$L(\frac{\omega}{\omega_n}, \xi) = \frac{1}{\pi} \arctan \frac{(\frac{\omega}{\omega_n})^2 - 1 + 2\xi^2}{2\xi\sqrt{1-\xi^2}} \tag{5.46}$$

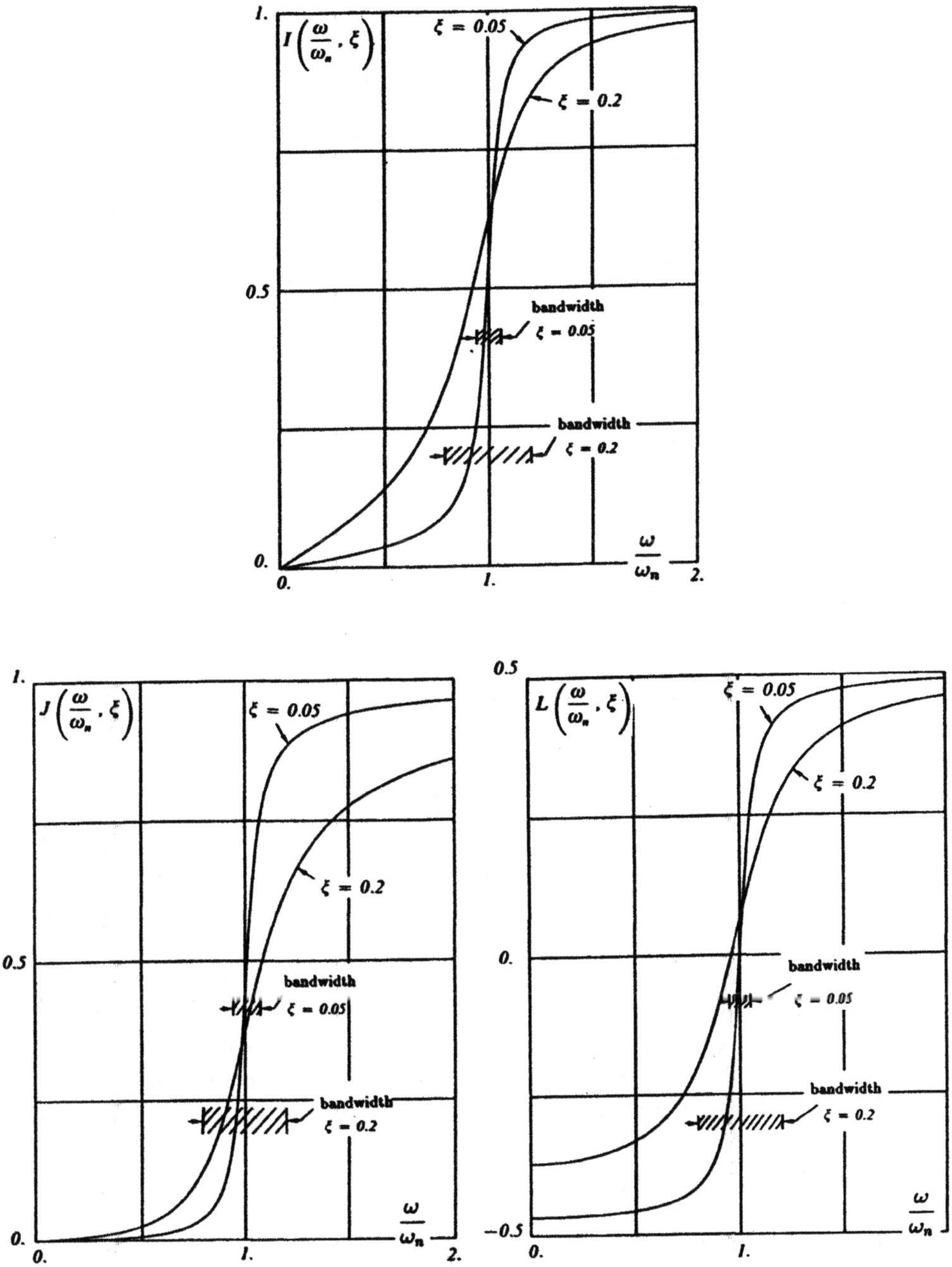

Figure 5.9: Functions I, J, L appearing in the calculation of the spectral moments.

and

$$m_2 = \frac{\pi S_0}{2\xi\omega_n}\,[J(\frac{\omega_2}{\omega_n},\xi) - J(\frac{\omega_1}{\omega_n},\xi)] \tag{5.47}$$

where

$$J(\frac{\omega}{\omega_n},\xi) = \frac{1}{\pi}\arctan\frac{2\xi\frac{\omega}{\omega_n}}{1-(\frac{\omega}{\omega_n})^2}$$

$$-\frac{\xi}{2\pi\sqrt{1-\xi^2}}.\ln\frac{1+(\frac{\omega}{\omega_n})^2+2\sqrt{1-\xi^2}(\frac{\omega}{\omega_n})}{1+(\frac{\omega}{\omega_n})^2-2\sqrt{1-\xi^2}(\frac{\omega}{\omega_n})} \tag{5.48}$$

In formulae (5.44) and (5.48), arctan( ) is assumed to belong to $[0,\pi]$. The behaviour of the functions $L(\frac{\omega}{\omega_n},\xi)$ and $J(\frac{\omega}{\omega_n},\xi)$ is also represented in Fig.5.9. Their general trend is the same as that of $I(\frac{\omega}{\omega_n},\xi)$, although L does not start from 0.

Formulae (5.43) to (5.48) are helpful in calculating numerically the spectral moments of the stationary response to an arbitrary *PSD* $\Phi_0(\omega)$: If the *PSD* of the excitation is decomposed into a set of elementary band limited white noises, each in a frequency band $[\omega_i,\omega_{i+1}]$, the contribution of each band to the spectral moments is

$$dm_0 = \frac{\pi S_i}{2\xi\omega_n^3}\,[I(\frac{\omega_{i+1}}{\omega_n},\xi) - I(\frac{\omega_i}{\omega_n},\xi)] \tag{5.49}$$

$$dm_1 = \frac{\pi S_i}{2\xi\omega_n^2\sqrt{1-\xi^2}}\,[L(\frac{\omega_{i+1}}{\omega_n},\xi) - L(\frac{\omega_i}{\omega_n},\xi)] \tag{5.50}$$

$$dm_2 = \frac{\pi S_i}{2\xi\omega_n}\,[J(\frac{\omega_{i+1}}{\omega_n},\xi) - J(\frac{\omega_i}{\omega_n},\xi)] \tag{5.51}$$

where

$$S_i = \frac{1}{2}[\Phi_0(\omega_i) + \Phi_0(\omega_{i+1})] \tag{5.52}$$

In this way, the spectral moments can be calculated numerically with a frequency discretization which provides only a good representation of the excitation, without regard to the bandwidth of $H(\omega)$.

$2\xi\omega_n$ is often called ***half-power bandwidth*** of the linear oscillator, because it is equal to the difference between the frequencies where the amplitude of the *PSD* of the response to a white noise is half its maximum (Fig.5.8).

5.6.3 Rice formulae

It will be established in chapter 10 that, for a zero-mean stationary Gaussian process, the ***average number of zero crossings with a positive slope per unit of time*** and the ***average number of maxima per unit of time*** (Fig.5.10) are given by the ***Rice formulae***:

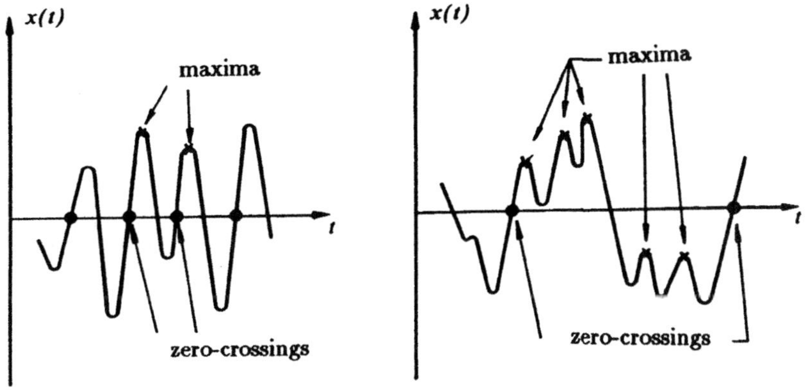

Figure 5.10: Zero crossings and maxima for (a) a narrow band process and (b) a wide band process.

- zero crossings:

$$\nu_0 = \frac{1}{2\pi}(\frac{m_2}{m_0})^{1/2} = \frac{1}{2\pi}\frac{\sigma_{\dot{x}}}{\sigma_x} \tag{5.53}$$

- maxima:

$$\nu_1 = \frac{1}{2\pi}(\frac{m_4}{m_2})^{1/2} = \frac{1}{2\pi}\frac{\sigma_{\ddot{x}}}{\sigma_{\dot{x}}} \tag{5.54}$$

provided that the spectral moments appearing in these expressions exist. ν_0 is called the *central frequency* of the process; it is a measure of the average frequency where the energy is concentrated in the signal. For example, consider the response of a linear oscillator to a white noise. From Equ.(5.43) and (5.47),

$$m_0 = \frac{\pi S_0}{2\xi\omega_n^3} \qquad m_2 = \frac{\pi S_0}{2\xi\omega_n} \tag{5.55}$$

one therefore gets that the central frequency is identical to the natural frequency of the oscillator:

$$\omega_0 = 2\pi\nu_0 = \omega_n \tag{5.56}$$

As illustrated in Fig.5.10, the number of maxima is always larger or equal to the number of zero crossings; ν_0 is closer to ν_1 for a narrow band process while $\nu_1 \gg \nu_0$ for a wide band process. The ratio

$$0 \le (\frac{\nu_0}{\nu_1})^2 = \frac{m_2^2}{m_0 m_4} \le 1 \tag{5.57}$$

can be considered as a measure of the bandwidth of the process; it is close to 1 for a narrow band process and decreases as the bandwidth increases. Note that this measure of the bandwidth is not often used in practice, because it

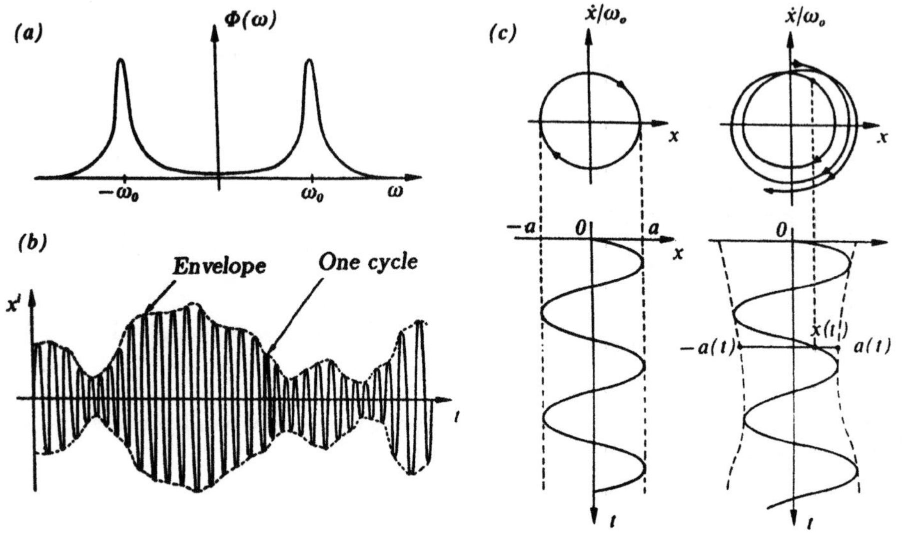

Figure 5.11: Definition of the envelope of a narrow band process. (a) Power spectral density. (b) Typical sample. (c) Trajectory in the phase plane.

requires that the spectral moments be defined up to the order 4 which is not always the case: For example, m_4 (which represents the variance of the relative acceleration) is unbounded for the response of a linear oscillator to a white noise. Another measure of the bandwidth, using the spectral moments of lower orders, will be defined in chapter 10.

5.7 Envelope of a narrow band process

5.7.1 Crandall & Mark's definition

A narrow band process has its power distribution concentrated in the vicinity of the central frequency ω_0 (Fig.5.11.a). A sample of such a process (Fig.5.11.b) looks like a sine function with slowly varying amplitude and frequency. The envelope can be visualized as the curve connecting the extrema of the sample. To define it formally, consider the trajectory of the sample in the phase plane $(x, \dot{x}/\omega_0)$ (Fig.5.11.c): A harmonic motion of constant amplitude, $x = a \sin \omega_0 t$, has a circular trajectory of radius a; the image point rotates at a constant angular velocity ω_0. The trajectory corresponding to a zero-mean narrow-band process consists of a smooth curve rotating clockwise at a frequency varying slowly about the central frequency ω_0; the radius is also a slowly varying function of time. Crandall and Mark define the envelope process $A(t)$ as the *radius of the image point of the process in the phase plane*:

$$A^2(t) = X^2(t) + (\frac{\dot{X}(t)}{\omega_0})^2 \tag{5.58}$$

For the sine function, this definition leads to a pair of straight lines at $x = \pm a$. For a sample of a narrow band process, the resulting curve is slowly varying and tangent to the sample $x(t)$ at the maxima. This definition of the envelope applies to narrow-band processes. Other definitions which apply to wide-band processes will be discussed in detail in chapter 10.

5.7.2 Joint distribution of X and $\dot{X}$

In Crandall and Mark's definition, the envelope process depends on $X(t)$ and $\dot{X}(t)$; the joint distribution of X and $\dot{X}$ is therefore required to establish the probability density function of the envelope. This distribution is also important in other problems, because $(X, \dot{X})^T$ is the state vector of the response of a single degree of freedom oscillator.

If $X(t)$ is stationary, Gaussian with zero mean, so is $\dot{X}(t)$, because the Gaussian character is preserved by differentiation, which is a linear transformation. According to Equ.(4.9), their joint distribution is completely defined by σ_x^2, $\sigma_{\dot{x}}^2$ and $\varrho_{x\dot{x}}$. The first two are related to the power spectral density by Equ.(5.39) and (5.40), while $\varrho_{x\dot{x}} = 0$, because a stationary random process is *orthogonal to its derivative* $[R_{\dot{x}x}(0) = 0]$. The joint distribution reads

$$p_{x\dot{x}}(x, t; \dot{x}, t) = \frac{1}{2\pi\sigma_x\sigma_{\dot{x}}} \exp[-\frac{1}{2}(\frac{x^2}{\sigma_x^2} + \frac{\dot{x}^2}{\sigma_{\dot{x}}^2})] \tag{5.59}$$

5.7.3 Probability distribution of the envelope

Introduce $Y(t) = \dot{X}(t)/\omega_0$, $A(t) = [X(t)^2 + Y(t)^2]^{1/2}$. In chapter 2, the probability distribution of this random variable was shown to be the *Rayleigh distribution*:

$$p_a(a) = \frac{a}{\sigma_x^2} \exp(-\frac{a^2}{2\sigma_x^2}) \qquad (a > 0) \tag{5.60}$$

It is illustrated in Fig.5.12. The mean square value of the envelope process is given by

$$\mathrm{E}[a^2] = \int_0^\infty a^2 p(a) da = 2\sigma_x^2 \tag{5.61}$$

5.8 References

J.BENDAT & A.PIERSOL, *Random Data: Analysis and Measurement Procedures*, Wiley-Interscience, 1971.

S.H.CRANDALL & W.D.MARK, *Random Vibration in Mechanical Systems*, Academic Press, 1963.

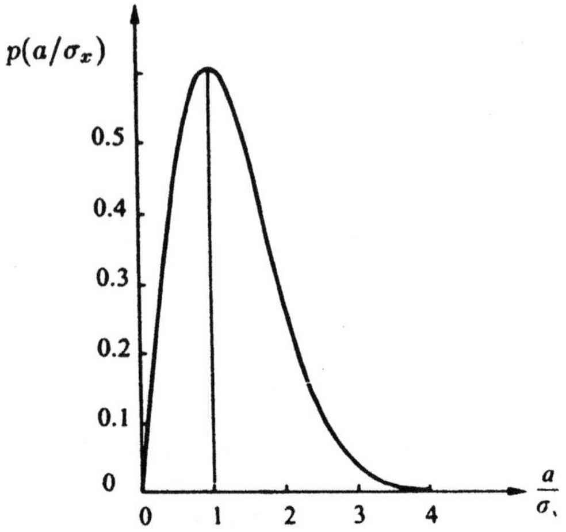

Figure 5.12: Rayleigh distribution.

Y.K.LIN, *Probabilistic Theory of Structural Dynamics*, McGraw-Hill, 1967.
G.PETIT BOIS, *Tables of Indefinite Integrals*, Dover, N-Y, 1961.
S.O.RICE, Mathematical analysis of random noise, *Bell System Tech. J.* 23, 282-332, 1944; 24,46-156, 1945. Reprinted in *Selected Papers on Noise and Stochastic Processes*, Wax ed., Dover, N-Y, 1954.

5.9 Problems

P.5.1 *Show that the impulse response of a single degree of freedom oscillator is given by Equ.(5.11).*
P.5.2 *Consider the first order differential system*

$$\dot{X} + aX = bF$$

excited by a white noise $\Phi_{ff}(\omega) = S_0$. *Calculate the PSD and the variance of the response. Does* m_2 *exist for this process?*
P.5.3 *In earthquake engineering, it is frequently assumed that the PSD of the ground acceleration at one point can be represented by the product of a finite duration shape function and a stationary process with PSD*

$$\Phi_0(\omega) = S_0 \frac{1 + 4\xi_g^2(\omega/\omega_g)^2}{[1 - (\omega/\omega_g)^2]^2 + 4\xi_g^2(\omega/\omega_g)^2}$$

where the parameters ω_g *and* ξ_g *characterize the local ground conditions (Kanai-Tajimi PSD). Show that this process can be viewed as the acceleration response of a single degree of freedom oscillator to a white noise support acceleration. Calculate the variance, using the results of section 5.6.*

P.5.4 *Consider a single degree of freedom oscillator seismically excited by a band-limited white noise*

$$\Phi_0(\omega) = S_0 \qquad (\omega_1 \leq |\omega| \leq \omega_2)$$

State the conditions under which the central frequency is independent of the bandwidth of the excitation. [Hint: base your analysis on the curves of Fig.5.9].

P.5.5 *Consider the transient response of a single degree of freedom oscillator starting from rest and excited by a stationary random excitation. Show that the variance approximately satisfies the differential equation*

$$\frac{1}{2\xi\omega_n}\frac{\mathrm{d}}{\mathrm{d}t}[\sigma(t)^2] + \sigma(t)^2 = \sigma_{st}^2$$

where σ_{st}^2 is the stationary variance.

P.5.6 *Consider the response of a lightly damped linear oscillator to a white noise. Show that the half-power bandwidth, that is the separation between the frequencies where the PSD is half its maximum value, is $2\xi\omega_n$.*

P.5.7 *Let $X(t)$ be a stationary random process with PSD $\Phi_{xx}(\omega)$ and consider the time averaging (smoothing) process*

$$Y(t) = \frac{1}{2\varepsilon}\int_{t-\varepsilon}^{t+\varepsilon} X(\tau)\,d\tau$$

Show that $Y(t)$ can be seen as the response of a linear system of impulse response

$$h(t) = \frac{1}{2\varepsilon} \qquad |t| < \varepsilon$$

[($h(t) = 0$ outside]. Show that

$$\Phi_{yy}(\omega) = \Phi_{xx}(\omega).\frac{\sin^2 \varepsilon\omega}{\varepsilon^2\omega^2}$$

$h(t)$ does not define a causal system. What change to the integral would lead to a causal system? What would be the corresponding output PSD?

P.5.8 *Consider the discrete averaging defined by the difference equation*

$$Y(t) = \frac{1}{4}X(t-T) + \frac{1}{2}X(t) + \frac{1}{4}X(t+T)$$

Show that

$$\Phi_{yy}(\omega) = \Phi_{xx}(\omega)(\cos\frac{\omega T}{2})^4$$

P.5.9 *Consider the response of a linear oscillator to an ideal low-pass process $\Phi(\omega) = S_0$, ($|\omega| < \omega_c$). What would be a reasonable approximation of the variance if $\omega_n \gg \omega_c$.*

Chapter 6

Random Response of Multi Degree of Freedom Systems

6.1 Some concepts of structural dynamics

6.1.1 Equation of motion

A discrete vibrating system with a finite number n of degrees of freedom (d.o.f.) is governed by the differential equation

$$M\ddot{x} + C\dot{x} + Kx = f \tag{6.1}$$

where x is the vector of [generalized] structural displacements. Equation (6.1) expresses the equilibrium between the external, elastic, inertia and damping forces. For lack of better knowledge, the damping is often assumed viscous. The matrices M, C and K are symmetric semi positive definite. M and K result from the discretization of the structure, most of the time using finite elements. A diagonal mass matrix is often sufficient to provide an acceptable representation of the inertia of the structure. The damping matrix C represents the various dissipation mechanisms, usually poorly known. To compensate for this lack of knowledge, it is customary to make assumptions on its form. One of the most popular hypotheses is the *Rayleigh damping*:

$$C = \alpha M + \beta K \tag{6.2}$$

The coefficients α and β are selected to fit the structure under consideration. Note that the Rayleigh damping tends to overestimate the damping of the high frequency modes.

6.1.2 Input-output relationship

The transfer matrix between the generalized structural displacements and the external forces is readily obtained by Fourier transform of Equ.(6.1):

$$X(\omega) = H(\omega)F(\omega) = [-\omega^2 M + j\omega C + K]^{-1}F(\omega) \tag{6.3}$$

$H(\omega)$ is often called the ***dynamic flexibility matrix***; it is non-singular for a stable dissipative system. It is related to the matrix of impulse responses $h(t)$ by the Fourier transform:

$$H(\omega) = \int_{-\infty}^{\infty} h(t)e^{-j\omega t}dt \tag{6.4}$$

$$h(t) = \frac{1}{2\pi}\int_{-\infty}^{\infty} H(\omega)e^{j\omega t}d\omega \tag{6.5}$$

The physical interpretation of $H(\omega)$ and $h(t)$ is as follows: The (k, j) component of $H(\omega)$ defines the amplitude of the coordinate x_k due to a harmonic excitation of unit amplitude and frequency ω, applied to coordinate x_j. Similarly, the (k, j) component of $h(t)$ provides the response of coordinate x_k to a unit impulse loading applied to coordinate x_j. For a causal system, $h(t) = 0$ for $t < 0$. If M, K and C are symmetric, so are H and h.

The input-output relationship in the time domain is the convolution:

$$x(t) = \int_{-\infty}^{\infty} h(t-\tau)f(\tau)d\tau = \int_{-\infty}^{\infty} h(\tau)f(t-\tau)d\tau = h \star f \tag{6.6}$$

It is formally the same as for a single d.o.f. oscillator, except that $x(t)$ and $f(t)$ are vectors with n components and $h(t)$ is a matrix n by n.

6.1.3 Modal decomposition

Let ϕ_i, $(i = 1, ..., n)$ be the mode shapes of the conservative (undamped) system defined by Equ(6.1); they are solutions of

$$(K - \omega_i^2 M)\phi_i = 0 \tag{6.7}$$

and satisfy the orthogonality conditions

$$\phi_i^T M \phi_j = \mu_i \delta_{ij} \tag{6.8}$$

$$\phi_i^T K \phi_j = \mu_i \omega_i^2 \delta_{ij} \tag{6.9}$$

where ω_i is the natural frequency and μ_i is the generalized mass of mode i (it is usual to normalize the modes so that $\mu_i = 1$). In modal coordinates

$$x = \Xi y \tag{6.10}$$

where $\Xi = (\phi_1, ..., \phi_n)$ is the matrix whose columns are the mode shapes and y is the vector of modal amplitudes, Equ.(6.1) becomes

$$M\Xi\ddot{y} + C\Xi\dot{y} + K\Xi y = f \tag{6.11}$$

Upon premultiplying by Ξ^T and using the orthogonality relationships (6.8) and (6.9), one gets

$$diag(\mu_i)\,\ddot{y} + \Xi^T C\,\Xi\dot{y} + diag(\mu_i\omega_i^2)\,y = \Xi^T f = p \tag{6.12}$$

where p is the vector of generalized modal forces, representing the work of the external forces on the various modes.

If the matrix $\Xi^T C\Xi$ is diagonal, the damping is said to be *classical, proportional*, or *normal*. The *modal fraction of critical damping*, ξ_i, is then defined by

$$\phi_i^T C\phi_j = 2\xi_i\omega_i\mu_i\delta_{ij} \tag{6.13}$$

It is readily checked that the Rayleigh damping (6.2) complies with this condition, with

$$\xi_i = \frac{1}{2}(\frac{\alpha}{\omega_i} + \beta\omega_i) \tag{6.14}$$

Under the condition (6.13), the modal equations are *decoupled* and Equ.(6.12) can be rewritten

$$\ddot{y} + 2\xi\Omega\dot{y} + \Omega^2 y = \mu^{-1}p \tag{6.15}$$

with the notations

$$\xi = diag(\xi_i)$$

$$\Omega = diag(\omega_i) \tag{6.16}$$

$$\mu = diag(\mu_i)$$

Apart from the classical damping assumption, the only difference between Equ.(6.1) and (6.15) lies in the change of coordinates (6.10). However, the structural response is usually dominated by the first few modes and it is possible to restrict the integration of (6.15) to these modes. This is essential, because the reduction of the number of coordinates may involve several orders of magnitude. For a seismic excitation, with a low frequency content ($< 30Hz$), it is common to restrict the analysis to less than 10 modes, while the structure may contain thousands of d.o.f.. For a wind loading, the reduction can be even more drastic, due to the very low frequency content of the wind spectrum; the first mode carries most of the dynamic response of the structure.

Equation (6.15) shows that the transfer matrix between the generalized modal forces p and the modal amplitudes reads

$$Y(\omega) = H(\omega)P(\omega) \tag{6.17}$$

where

$$H(\omega) = diag\{\frac{1}{\mu_i[(\omega_i^2 - \omega^2) + 2j\xi_i\omega_i]}\} \tag{6.18}$$

Using Equ.(6.10) and (6.12),we can readily obtain the ***spectral development*** of the dynamic flexibility matrix:

$$[-\omega^2 M + j\omega C + K]^{-1} = \sum_{i=1}^{n} \frac{\phi_i \phi_i^T}{\mu_i[(\omega_i^2 - \omega^2) + 2j\xi_i\omega_i]} \tag{6.19}$$

where the sum extends to all the n modes. For frequencies within a limited bandwidth, $\omega^2 < \omega_c^2 \ll \omega_m^2$, the development can be split into the contributions of the modes which respond dynamically (those within the bandwidth ω_c of the excitation) and the high frequency modes which respond statically:

$$= \sum_{i=1}^{m} \frac{\phi_i \phi_i^T}{\mu_i[(\omega_i^2 - \omega^2) + 2j\xi_i\omega_i]} + \sum_{i=m+1}^{n} \frac{\phi_i \phi_i^T}{\mu_i \omega_i^2} \tag{6.20}$$

Note that the computation of the second term of this expression does not require the knowledge of the high frequency modes, since, for $\omega = 0$, Equ.(6.19) gives

$$\sum_{i=m+1}^{n} \frac{\phi_i \phi_i^T}{\mu_i \omega_i^2} = K^{-1} - \sum_{i=1}^{m} \frac{\phi_i \phi_i^T}{\mu_i \omega_i^2} \tag{6.21}$$

This additional contribution to the flexibility is often called ***residual mode***.

The following values of the modal damping ratio can be regarded as typical: Satellites and space structures are generally very lightly damped ($\xi \simeq 0.001-0.005$), because of the extensive use of fiber reinforced composites, the absence of aerodynamic damping, and the low strain level. Mechanical engineering applications (steel structures, piping,...) are in the range of $\xi \simeq 0.01-0.02$; most dissipation takes place in the joints, and the damping increases with the strain level. For civil engineering applications, $\xi \simeq 0.05$ is typical and, when radiation damping through the ground is involved, it may reach $\xi \simeq 0.20$, depending on the local soil conditions. The assumption of classical damping is often justified for light damping, but it is questionable when the damping is large, as in problems involving soil-structure interaction.

6.1.4 State variable form

If the mass matrix is non-singular, Equ.(6.1) can be rewritten

$$\begin{pmatrix} \dot{x} \\ \ddot{x} \end{pmatrix} = \begin{pmatrix} 0 & I \\ -M^{-1}K & -M^{-1}C \end{pmatrix} \begin{pmatrix} x \\ \dot{x} \end{pmatrix} + \begin{pmatrix} 0 \\ M^{-1} \end{pmatrix} f \tag{6.22}$$

or

$$\dot{z} = Az + Bf \tag{6.23}$$

where the state vector has been defined as $z = (x^T, \dot{x}^T)^T$. Note that:

- if a lumped mass model is used, inverting M is straightforward.
- if some d.o.f. are devoid of inertia, they can be eliminated by static condensation (Guyan reduction) before inverting M.

A similar form can be obtained from the modal equation (6.15), using the state vector $z = (y^T, \dot{y}^T)^T$:

$$\dot{z} = \begin{pmatrix} 0 & I \\ -\Omega^2 & -2\xi\Omega \end{pmatrix} z + \begin{pmatrix} 0 \\ \mu^{-1}\Xi^T \end{pmatrix} f \tag{6.24}$$

It is worth insisting that if the structural response is dominated by the first few modes, the size of the state vector in modal coordinates is considerably smaller than that in structural coordinates.

Any response quantity r, linearly related to the modal amplitudes (e.g. a displacement or a stress component) can be expressed as

$$r = Dz \tag{6.25}$$

From (6.23), the transfer matrix between f and r reads

$$R(\omega) = D(j\omega I - A)^{-1} B\, F(\omega) \tag{6.26}$$

6.1.5 Structural and hereditary damping

There is a class of problems which cannot be described by an equation of motion of the form (6.1), because part of the forces acting on the system at t depend on its history at earlier times ($\tau < t$). Such situations occur in aeroelasticity, because of the unsteady aerodynamic forces, or in structures made of viscoelastic materials. The damping is said to be ***hereditary***. The equation of motion reads

$$M\ddot{x} + \Psi \star \dot{x} + Kx = f \tag{6.27}$$

where Ψ is a square matrix of heredity functions. Each term in $\Psi \star \dot{x}$ involves a convolution integral

$$\psi \star \dot{x}_1 = \int_{-\infty}^{t} \psi(t-\tau)\dot{x}_1(\tau)d\tau \tag{6.28}$$

Ψ may be non-symmetric, as for unsteady aerodynamic forces.

In order to illustrate this situation, consider the simplest case represented in Fig.6.1. It consists of a spring mounted in parallel with a Maxwell unit. The system is completely described by the two equations

$$m\ddot{x}_1 + k_1x_1 + c(\dot{x}_1 - \dot{x}_2) = f(t) \tag{6.29}$$

$$k_2x_2 - c(\dot{x}_1 - \dot{x}_2) = 0 \tag{6.30}$$

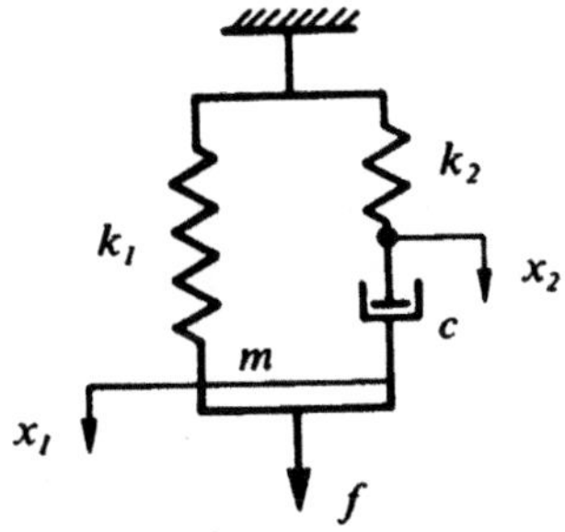

Figure 6.1: The simplest case of hereditary damping: Spring-mass system in parallel with a Maxwell unit.

The first equation expresses the equilibrium of the mass m while the second that of the point x_2. The complete description of the system requires three state variables: $x_1, \dot{x}_1$ and x_2. However, since no external force appears in Equ.(6.30), it can be used to eliminate the coordinate x_2 which becomes hidden in the system. Assuming the system at rest at $t = 0$, the following integro-differential equation is obtained:

$$m\ddot{x}_1 + k_1 x_1 + \int_0^t \psi(t-\tau)\dot{x}_1(\tau)d\tau = f(t) \tag{6.31}$$

where the heredity function is $\psi(t) = k_2 e^{-k_2 t/c}$. Notice that, because of a limiting form of the Dirac function, $\psi(t) \to c\delta(t)$ as $k_2 \to \infty$. Then, Equ.(6.31) is reduced to that of the viscous damping (this is obvious from Fig.6.1).

In the frequency domain, the transfer matrix associated to the linear system (6.27) is

$$H(\omega) = [-\omega^2 M + j\omega\Psi(\omega) + K]^{-1} \tag{6.32}$$

Equation (6.3) is the particular case for $\Psi(\omega) = C$. The transfer function associated with Equ.(6.31) is

$$H(\omega) = (-m\omega^2 + k_1 + \frac{j\omega k_2 c}{j\omega c + k_2})^{-1} \tag{6.33}$$

Even for non-classical damping, it may be convenient to express the equation of motion in the basis of the normal modes of the undamped structure, because, although the modal equations are coupled, the structural response remains dominated by the contribution of the first few modes. The reduction of the order of the system is achieved in a similar way.

In the aerospace industry, instead of assuming that the damping is viscous, it is customary to assume the following form:

$$j\omega\Psi(\omega) = jG\,\text{sign}(\omega) \tag{6.34}$$

This form of damping is called ***structural*** or ***hysteretic***. With the additional assumption $G = \gamma K$, the modal equations are again decoupled, because of the orthogonality condition (6.9).

6.1.6 Remarks

- The general form of C which satisfies the orthogonality condition (6.13) has been established by T.K.Caughey (1960).

- Although the damping mechanism defined by Equ.(6.34) is acceptable and widely used, *it cannot be associated with a causal system.* This can be established as follows: Consider a spring k coupled with an hysteretic damper of constant g. The transfer function of the system is

$$H(\omega) = [k + jg\,\text{sign}(\omega)]^{-1} \tag{6.35}$$

  The force $f(t)$ associated to a displacement $x(t)$ is therefore

$$f(t) = kx(t) + g\hat{x}(t) \tag{6.36}$$

  where $\hat{x}(t)$ is the *Hilbert transform* of $x(t)$. As we shall see in chapter 10, $\hat{x}(t)$ depends on the value of $x(t)$ over the whole range $-\infty < t < \infty$. As a result, $f(t)$ does not only depend on the past values of $x(t)$ but on the *future* ones as well (Fraejis de Veubeke, 1959; Crandall, 1970).

- In aeroelasticity, $\Psi(\omega)$ involves generalized *Theodorsen functions*; it is non-symmetric, because such are the aeroelastic operators.

- In general, the transfer matrix (6.32) is complex and non symmetric. It is non-singular for a stable dissipative system. When it becomes singular, the system is in condition of dynamic instability (e.g. flutter). The inverse matrix (6.32) must be computed for a set of control frequencies. When expressed in modal coordinates, the dimension of the transfer matrix is small and no particular numerical problems arise during the inversion process.

6.2 Seismic excitation

6.2.1 Equation of motion

Consider a multi-supported structure excited by the motion (possibly differential) of its supports. Partitioning the restrained and unrestrained d.o.f., we find the equation of motion

$$\begin{pmatrix} M_{11} & M_{10} \\ M_{01} & M_{00} \end{pmatrix} \begin{pmatrix} \ddot{x}_1 \\ \ddot{x}_0 \end{pmatrix} + \begin{pmatrix} C_{11} & C_{10} \\ C_{01} & C_{00} \end{pmatrix} \begin{pmatrix} \dot{x}_1 \\ \dot{x}_0 \end{pmatrix} + \begin{pmatrix} K_{11} & K_{10} \\ K_{01} & K_{00} \end{pmatrix} \begin{pmatrix} x_1 \\ x_0 \end{pmatrix} = \begin{pmatrix} 0 \\ f_0 \end{pmatrix} \tag{6.37}$$

where the subscript 1 refers to the unrestrained d.o.f., while the subscript 0 refers to those of the supports. f_0 represents the excitation force at the supports, that is, the support reactions.

The part of Equ.(6.37) relative to the unrestrained d.o.f. can be rewritten

$$M_{11}\ddot{x}_1 + C_{11}\dot{x}_1 + K_{11}x_1 = -M_{10}\ddot{x}_0 - C_{10}\dot{x}_0 - K_{10}x_0 \tag{6.38}$$

In this equation, x_1 represents the ***absolute*** displacements of the structure and x_0 the absolute displacements of the supports. To reduce the size of the system, it is appropriate to decompose the motion into the normal modes of the structure *fixed at its supports.* Before doing so, it is necessary to split x_1 in its dynamic and quasi-static contributions:

$$x_1 = y_1 + x_1^{qs} \tag{6.39}$$

where x_1^{qs} stands for the quasi-static response of the structure resulting from the support displacements, and y_1 is the dynamic response. x_1^{qs} can be obtained from Equ.(6.38) by cancelling out all the time derivatives:

$$K_{11}x_1^{qs} = -K_{10}x_0$$

or

$$x_1^{qs} = T_{qs}x_0 = -K_{11}^{-1}K_{10}x_0 \tag{6.40}$$

T_{qs} is the *quasi-static transmission matrix*; its i-th column contains the static displacements at the unrestrained d.o.f. resulting from a unit displacement at the i-th support d.o.f.. For a statically determinate structure, T_{qs} comes from rigid body kinematics; for statically over-determinate structures, its columns are obtained from static analyses. Since x_1^{qs} is linearly related to x_0, the following change of variables can be performed:

$$\begin{pmatrix} x_1 \\ x_0 \end{pmatrix} = \begin{pmatrix} I & T_{qs} \\ 0 & I \end{pmatrix} \begin{pmatrix} y_1 \\ x_0 \end{pmatrix} \tag{6.41}$$

where the dynamic displacements, y_1, satisfy homogeneous (zero) boundary conditions at the support, like the mode shapes of the fixed base structure. Combining Equ.(6.41) and (6.38), one gets

$$\begin{aligned} M_{11}\ddot{y}_1 + C_{11}\dot{y}_1 + K_{11} &= -(M_{11}T_{qs} + M_{10})\ddot{x}_0 \\ &\quad -(C_{11}T_{qs} + C_{10})\dot{x}_0 \\ &\quad -(K_{11}T_{qs} + K_{10})x_0 \end{aligned} \tag{6.42}$$

Substituting T_{qs} from Equ.(6.40), we see that the stiffness contribution to the excitation vanishes. So does the damping contribution if the damping matrix is proportional to the stiffness matrix; this term is usually small and is frequently neglected, as will be done here. Deleting the subscript 1, we can rewrite Equ.(6.42) as

$$M\ddot{y} + C\dot{y} + Ky = -(MT_{qs} + M_{10})\,\ddot{x}_0 \tag{6.43}$$

Because of the essentially low-pass character of most of the physical excitations, the major part of the response is usually concentrated in the first few modes. It is therefore appropriate to make a change of variables and decompose the dynamic displacements into their modal components z according to

$$y = \Xi z \tag{6.44}$$

where $\Xi = (\phi_1 \ldots, \phi_m)$ is the $n \times m$ matrix (n = number of d.o.f., m = number of modes considered in the analysis), whose columns are the normal modes of the fixed base structure, satisfying the orthogonality conditions

$$\Xi^T M \Xi = \mu = \text{diag}\,(\mu_i) \qquad \Xi^T K \Xi = \mu\Omega^2 = \text{diag}\,(\mu_i \omega_i^2)$$

(this change of coordinates is allowable since y and the mode shapes ϕ_i have the same boundary conditions). Substituting (6.44) in Equ.(6.43), multiplying both side of the equation by Ξ^T and taking into account the orthogonality conditions, we find

$$\begin{aligned} \mu\ddot{z} + \Xi^T C \Xi\,\dot{z} + \mu\Omega^2 z &= -\,\Xi^T\,(M T_{qs} + M_{10})\,\ddot{x}_0 \\ &= \Gamma \ddot{x}_0 \end{aligned} \tag{6.45}$$

where

$$\Gamma \stackrel{\text{def}}{=} -\,\Xi^T (M T_{qs} + M_{10}) \tag{6.46}$$

is the $m \times n_s$ *modal participation matrix*. A column of Γ gives the work done on each mode by the inertia forces associated with the quasi-static accelerations induced by a unit acceleration of the corresponding support d.o.f.. It is worth noting that in many cases, the term M_{10} is neglected in Equ.(6.43) and (6.46) (this term vanishes for a lumped mass matrix and is usually small).

6.2.2 Effective modal mass

When going from Equ.(6.43) to (6.45), a drastic reduction of the size of the system of equations is achieved ($m \ll n$). The question arises, then, of how many modes should be considered in the analysis (how large should m be). Obviously, the first criterion must be related to the frequency content of the excitation:

- *All the modes within the bandwidth of the excitation should be included in the analysis.*

Even so, this may not be enough to achieve a good accuracy for the support reactions, as first pointed out by G.H.Powell (1979). To understand this, one must think of the extreme case of a rigid structure excited at low frequency; none of the modes react dynamically and still there are support reactions associated with the *quasi-static inertia*.

The modal participation matrix provides a guide as to how well the structural mass is accounted for in the truncated modal basis. In fact, if one forms the matrix $\Gamma^T \mu^{-1} \Gamma$ *including all the modes* ($m = n$), one gets, after some algebra

$$\begin{aligned} \Gamma^T \mu^{-1} \Gamma &= T_{qs}^T M_{11} T_{qs} + T_{qs}^T M_{10} + M_{01} T_{qs} + M_{01} M_{11}^{-1} M_{10} \\ &= \hat{M}_{00} + M_{01} M_{11}^{-1} M_{10} - M_{00} \end{aligned} \tag{6.47}$$

where $\hat{M}_{00}$ is the so-called *Guyan mass matrix*, obtained by static condensation of the unrestrained d.o.f. according to

$$\begin{pmatrix} x_1 \\ x_0 \end{pmatrix} = \begin{pmatrix} T_{qs} \\ I \end{pmatrix} x_0 \tag{6.48}$$

$\hat{M}_{00}$ is obtained by expressing the conservation of the kinetic energy as

$$T = \frac{1}{2} \dot{x}_0^T \begin{pmatrix} T_{qs}^T & I \end{pmatrix} \begin{pmatrix} M_{11} & M_{10} \\ M_{01} & M_{00} \end{pmatrix} \begin{pmatrix} T_{qs} \\ I \end{pmatrix} \dot{x}_0 = \frac{1}{2} \dot{x}_0^T \hat{M}_{00} \dot{x}_0$$

which leads to

$$\hat{M}_{00} = T_{qs}^T M_{11} T_{qs} + T_{qs}^T M_{10} + M_{01} T_{qs} + M_{00} \tag{6.49}$$

$\hat{M}_{00}$ represents the inertia of the structure seen from the supports, when it responds statically. If one neglects $M_{01}\, M_{11}^{-1}\, M_{10}$ as second order term, Equ.(6.47) becomes

$$\Gamma^T \mu^{-1} \Gamma = \hat{M}_{00} - M_{00} \tag{6.50}$$

where all the modes are included in the left side. M_{00} is the mass matrix directly associated with the supports.

Now, if $\mathbf{1}_i$ stands for the unit rigid body translation of supports along a global axis i, a velocity v_0 along this axis corresponds to support velocities $v_0 \mathbf{1}_i$. The corresponding total kinetic energy is

$$T = \frac{1}{2} m_T v_0^2 = \frac{1}{2} v_0^2 \mathbf{1}_i^T \hat{M}_{00} \mathbf{1}_i$$

which means that the total mass of the structure, m_T, is related to $\hat{M}_{00}$ by

$$m_T = \mathbf{1}_i^T \hat{M}_{00} \mathbf{1}_i \tag{6.51}$$

for any direction i. From Equ.(6.50), we see that

$$\mathbf{1}_i^T \hat{M}_{00} \mathbf{1}_i - \mathbf{1}_i^T M_{00} \mathbf{1}_i - \mathbf{1}_i^T \Gamma^T \mu^{-1} \Gamma \mathbf{1}_i = 0$$

or

$$m_T - m_S - \mathbf{1}_i^T \Gamma^T \mu^{-1} \Gamma \mathbf{1}_i = 0 \tag{6.52}$$

where m_S is the total mass associated with the supports.

Equation (6.52) applies, provided that all the modes are included in the third term. If only m modes are included, it is not satisfied any longer and there is a residue, which represents the ***missing mass***, that is, the mass which is not accounted for by the truncated modal expansion (6.44). For a single support excitation, Equ.(6.52) is reduced to

$$m_T - m_S = \sum_{j=1}^{n} \frac{\Gamma_j^2}{\mu_j} \tag{6.53}$$

Γ_j^2/μ_j is called the ***effective modal mass*** of mode i. It represents the part of the total mass of the structure which is associated with mode i. In the multi-supported case, the component form of Equ.(6.52) is

$$m_T - m_S = \sum_{j=1}^{n} \frac{(\Gamma \, \mathbf{1}_i)_j^2}{\mu_j} \tag{6.54}$$

It is apparent that, in the truncated case, the missing mass depends on the direction i of the excitation. From the foregoing discussion, the second criterion for mode selection is

- *Any mode whose effective mass is a significant part of the total mass should be included in the analysis.*

6.2.3 Input-Output relationships in the frequency domain

- *Modal amplitudes*

Upon Fourier transforming Equ.(6.45), one gets the transfer matrix between the modal amplitudes $Z(\omega)$ and the excitations $\ddot{X}_0(\omega)$:

$$Z(\omega) = H(\omega)\Gamma\ddot{X}_0(\omega) \tag{6.55}$$

where

$$H(\omega) = \{\text{diag}\,[\mu_i(\omega_i^2 - \omega^2)] + j\omega\Xi^T C\Xi\}^{-1} \tag{6.56}$$

- *Absolute accelerations*

Let Ξ_p be the $n_p \times m$ matrix containing the modal amplitudes at the n_p d.o.f. where the *PSD* matrix of accelerations is to be calculated. One wishes to determinee the $n_p \times n_s$ transfer matrix between the acceleration at these nodes and the excitaticn. It consists of a dynamic and a quasi-static contribution, which are respectively

$$-\omega^2\Xi_p Z(\omega) \qquad \text{and} \qquad T_{qs}\ddot{X}_0(\omega)$$

Combining with Equ (6.55), one gets

$$\ddot{X}_p(\omega) = H_p(\omega)\ddot{X}_0(\omega) \tag{6.57}$$

$$H_p(\omega) = -\omega^2 \Xi_p H \Gamma + T_{qs} \tag{6.58}$$

- ***Support reactions***

If one neglects the contribution of the damping to the support reactions, the part of Equ.(6.37) relative to the restrained d.o.f. provides

$$f_0 = M_{01}\ddot{x}_1 + M_{00}\ddot{x}_0 + K_{01}x_1 + K_{00}x_0 \tag{6.59}$$

Substituting the modal expansion of the displacements,

$$x_1 = T_{qs}x_0 + \Xi z$$

one gets

$$f_0 = M_{00}\ddot{x}_0 + M_{01}(T_{qs}\ddot{x}_0 + \Xi\ddot{z}) + (K_{00} + K_{01}T_{qs})x_0 + K_{01}\Xi z \tag{6.60}$$

Apart from the second term, which represents the coupling inertia between the unrestrained and the restrained d.o.f., the meaning of the other terms is obvious: the first represents the support inertia, the third is related to the differential displacements and the fourth contains the dynamic modal reactions (the columns of $K_{01}\Xi$ are the modal reaction vectors). The above formulation is subject to the missing mass problem mentioned earlier. Equation (6.52) suggests that the results can be improved by applying a quasi-static correction (missing mass correction) of the form

$$(\hat{M}_{00} - M_{00} - \Gamma^T\mu^{-1}\Gamma)\ddot{x}_0 \tag{6.61}$$

It represents the quasi-static inertial effect of those modes which have not been included in the analysis. If $M_{10} = 0$, the corrected result is

$$f_0 = (K_{00} + K_{01}T_{qs})x_0 + K_{01}\Xi z + (\hat{M}_{00} - \Gamma^T\mu^{-1}\Gamma)\ddot{x}_0 \tag{6.62}$$

The three terms represent respectively the differential displacements, the dynamic response and the quasi-static inertial contribution (inertia of the supports plus missing mass).

An alternative (and more attractive) form can be obtained as follows. First, $K_{01}x_1$ is eliminated from Equ.(6.59) by using the first part of Equ.(6.37) (the damping is omitted for the sake of simplification):

$$x_1 = -K_{11}^{-1}(K_{10}x_0 + M_{11}\ddot{x}_1 + M_{10}\ddot{x}_0)$$

$$K_{01}x_1 = T_{qs}^T(K_{10}x_0 + M_{11}\ddot{x}_1 + M_{10}\ddot{x}_0)$$

Combining with Equ.(6.59), one gets

$$f_0 = (M_{01} + T_{qs}^T M_{11})\ddot{x}_1 + (M_{00} + T_{qs}^T M_{10})\ddot{x}_0 + (K_{00} + K_{01}T_{qs})x_0$$

Upon substituting the modal expansion of the accelerations

$$\ddot{x}_1 = T_{qs}\ddot{x}_0 + \Xi\ddot{z} \tag{6.63}$$

(this is often called the ***modal acceleration method***, as opposed to the modal displacement method discussed above), one gets, after some straightforward algebra

$$f_0 = \hat{M}_{00}\ddot{x}_0 - \Gamma^T\ddot{z} + (K_{00} + K_{01}T_{qs})x_0 \tag{6.64}$$

Upon Fourier transforming and using Equ.(6.55), one gets

$$F_0(\omega) = \hat{M}_{00}\ddot{X}_0(\omega) + \omega^2\Gamma^T H\Gamma\ddot{X}_0(\omega) + (K_{00} + K_{01}T_{qs})X_0(\omega) \tag{6.65}$$

This form is totally equivalent to (6.62); the first term represents the quasi-static inertia, the second is the dynamic response and the third refers to the differential displacements. The transfer matrix between the support reactions and the excitation is

$$F_0(\omega) = H_R(\omega)\ddot{X}_0(\omega) \tag{6.66}$$

$$H_R(\omega) = \hat{M}_{00} + \omega^2\Gamma^T H\Gamma - \frac{1}{\omega^2}(K_{00} + K_{01}T_{qs}) \tag{6.67}$$

This formulation is ***statically correct***, which means that the reaction forces will be computed accurately, even if a mode with a significant effective mass, but with a natural frequency above the bandwidth of the excitation, has been omitted. The contribution of this mode to the support reactions is included in the first component ($\hat{M}_{00}$). The third term in Equ.(6.67) vanishes for a single support excitation.

$$\mathbf{1}_i^T(\hat{M}_{00} + \omega^2\Gamma^T H\Gamma)\mathbf{1}_i \tag{6.68}$$

is often called the ***dynamic mass*** in the direction i.

For a unidirectional, single support excitation, if the damping is classical, the dynamic mass can be expanded into its modal components as (Problem P.6.1)

$$F_0(\omega) = \{m_S + \sum_i \frac{\Gamma_i^2}{\mu_i}[\frac{\omega_i^2 + 2j\xi_i\omega_i\omega}{\omega_i^2 - \omega^2 + 2j\xi_i\omega_i\omega}]\}\ddot{X}_0(\omega) \tag{6.69}$$

where m_S is the mass of the support; within a limited bandwidth ($\omega \ll \omega_m$), this can be approximated by

$$F_0(\omega) = \{m_S + \sum_{i=1}^{m} \frac{\Gamma_i^2}{\mu_i}[\frac{\omega_i^2 + 2j\xi_i\omega_i\omega}{\omega_i^2 - \omega^2 + 2j\xi_i\omega_i\omega}] + \sum_{i=m+1}^{n} \frac{\Gamma_i^2}{\mu_i}\}\ddot{X}_0(\omega)$$

We see that the contribution of the modes beyond the bandwidth of the excitation is simply the effective modal mass, Γ_i^2/μ_i. The total effective modal mass of the high frequency modes can be evaluated from Equ.(6.53). Equ.(6.69) provides an easy way to determine experimentally the effective modal masses, if both the reaction force and the acceleration are measured.

- *Generalized stresses*

Any stress or generalized stress component has a dynamic contribution and a static one arising from the differential displacements of the supports. If b stands for the modal components of the dynamic contribution and c for the static part, the general form of the response r is

$$r = b^T z + c^T x_0 \tag{6.70}$$

From Equ.(6.55), the $(1 \times n_s)$ transfer matrix between r and $\ddot{x}_0$ is

$$H_r(\omega) = b^T H(\omega)\Gamma - \frac{1}{\omega^2} c^T \tag{6.71}$$

For a multi-support excitation, it is a row vector; it becomes a scalar and the second term vanishes for a single support excitation (in this case, there are no differential displacements).

6.3 Response to a stationary excitation

The input-output relationship in the time domain is given by Equ.(6.6). If the forcing function $F(t)$ is random, so is the response $X(t)$:

$$X(t) = \int_{-\infty}^{\infty} h(t-\tau)F(\tau)d\tau \tag{6.72}$$

Taking the mathematical expectation of this equation, we find

$$E[X(t)] = \int_{-\infty}^{\infty} h(t-\tau)E[F(\tau)]d\tau \tag{6.73}$$

Similarly, the correlation matrix reads

$$E[X(t_1)X^T(t_2)] = \int\int_{-\infty}^{\infty} h(t_1-\tau_1)E[F(\tau_1)F^T(\tau_2)]h^T(t_2-\tau_2)d\tau_1 d\tau_2 \tag{6.74}$$

If the excitation is weakly stationary, its correlation matrix depends only on the time difference $\tau_1 - \tau_2$.

$$E[F(\tau_1)F^T(\tau_2)] = R_F(\tau_1 - \tau_2) \tag{6.75}$$

If each component of the correlation matrix can be Fourier transformed, we define the *Power Spectral Density Matrix* as

$$\Phi_F(\omega) = \frac{1}{2\pi}\int_{-\infty}^{\infty} R_F(\tau)e^{-j\omega\tau}d\tau \tag{6.76}$$

$$R_F(\tau) = \int_{-\infty}^{\infty} \Phi_F(\omega) e^{j\omega\tau} d\omega \tag{6.77}$$

Equation (3.17) implies that the elements of $R_F(\tau)$ satisfy the symmetry relationship

$$R_{F_iF_j}(\tau) = R_{F_jF_i}(-\tau) \tag{6.78}$$

so that $\Phi_F(\omega)$ is *Hermitian*:

$$\Phi_{F_iF_j}(\omega) = \Phi^{\star}_{F_jF_i}(\omega) \tag{6.79}$$

Following the same development as for a s.d.o.f. system (section 5.3), we may readily establish that Equ.(6.74) can be transformed in the frequency domain into

$$\Phi_X(\omega) = H(\omega)\Phi_F(\omega)H^{\star}(\omega) \tag{6.80}$$

where $\star$ stands for the conjugate transpose. This relationship is the vector extension of (5.24); it applies for any transfer matrix $H(\omega)$, provided that the system is linear and stable.

If r is a response quantity (e.g. a stress component or a displacement) linearly related to the modal amplitudes of a structure exposed to a field of random forces, its *PSD* function can be computed according to the following steps:

- Compute the modal excitation *PSD* matrix

$$p = \Xi^T f \Rightarrow \Phi_P(\omega) = \Xi^T \Phi_F(\omega) \Xi \tag{6.81}$$

- Compute the modal response *PSD* matrix

$$Y(\omega) = H(\omega)P(\omega) \Rightarrow \Phi_y(\omega) = H(\omega)\Phi_P(\omega)H^{\star}(\omega) \tag{6.82}$$

- Compute the response *PSD* function

$$r = b^T y \Rightarrow \Phi_r(\omega) = b^T \Phi_Y(\omega) b \tag{6.83}$$

Both $\Phi_P(\omega)$ and $\Phi_Y(\omega)$ are square and Hermitian, of dimension m. As we can see, the procedure is computationally simple as soon as the transfer matrices are available and the excitation has been defined. This latter point turns out to be the most difficult one for most physical problems. Several examples will be treated in detail later in this chapter.

6.4 Role of the cross-correlation

If the damping is classical, $H(\omega) = \text{diag}[H_i(\omega)]$ and Equ.(6.83) can be expanded as

$$\Phi_r(\omega) = \sum_i \sum_k b_i b_k \Phi_{P_iP_k}(\omega) H_i(\omega) H_k^{\star}(\omega) \tag{6.84}$$

$$= \sum_i b_i^2 \Phi_{P_iP_i}(\omega)|H_i(\omega)|^2 + \sum_i \sum_{k\neq i} b_i b_k \Phi_{P_iP_k}(\omega) H_i(\omega) H_k^\star(\omega) \tag{6.85}$$

where the diagonal and off-diagonal terms have been separated. The mean-square value is obtained by integrating over ω

$$\sigma_r^2 = \sum_i b_i^2 \beta_{ii} + \sum_i \sum_{k\neq i} b_i b_k \beta_{ik} \tag{6.86}$$

with the notation

$$\beta_{ik} = E[Y_i Y_k] = \int_{-\infty}^{\infty} \Phi_{P_iP_k}(\omega) H_i(\omega) H_k^\star(\omega) d\omega \tag{6.87}$$

β_{ii} is called *modal autocorrelation* and β_{ik} *modal cross-correlation* (between mode i and mode k). Note that, since $\beta_{ik} = \beta_{ki}^\star$, only the real part of β_{ik} contributes to the second sum of (6.86). This term represents the *interaction* between the modes. In the remainder of this section, we shall discuss the relative importance of the two contributions in (6.86) and establish the condition under which the cross-correlations can be neglected. Before doing that, consider the *mass-averaged mean square displacement* in the structure [compare to Equ.(6.8)]

$$E[X^T M X] = \sum_{ij} M_{ij} E[X_i X_j] \tag{6.88}$$

After transformation in modal coordinates according to (6.10), we find

$$E[X^T M X] = \sum_{ij} M_{ij} \sum_{kl} \Xi_{ik} \beta_{kl} \Xi_{jl}$$

where the definition (6.87) has been used. Upon reversing the order of summation and using the orthogonality condition $\Xi_{ik} M_{ij} \Xi_{jl} = \mu_k \delta_{kl}$, one gets

$$E[X^T M X] = \sum_k \mu_k \beta_{kk} \tag{6.89}$$

where all the modal cross-correlations have disappeared. This result indicates that *the modal cross-correlations contribute to the local variations of the MS response, but not to its mass-average over the entire structure.* If a pair of cross-correlations contribute positively to the *MS* response in one excited part of the structure, they will provide a negative contribution in another part, so that the total contribution to the mass-average is zero.

Consider the example of Fig.6.2, taken from Elishakoff (1982). The mass, stiffness, and damping matrices are easily obtained

$$M = \begin{pmatrix} m & 0 \\ 0 & m \end{pmatrix} K = \begin{pmatrix} k(1+\varepsilon) & -k\varepsilon \\ -k\varepsilon & k(1+\varepsilon) \end{pmatrix} C = \begin{pmatrix} c(1+\gamma) & -c\gamma \\ -c\gamma & c(1+\gamma) \end{pmatrix}$$

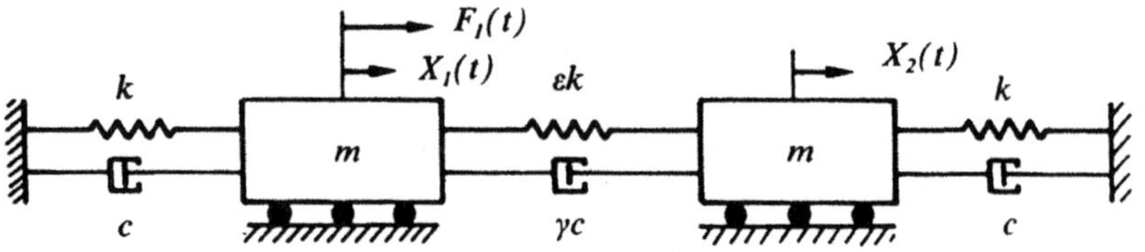

Figure 6.2: 2 d.o.f. oscillator used to illustrate the effect of cross-correlations.

The natural frequencies are

$$\Omega = (\frac{k}{m})^{1/2} \begin{pmatrix} 1 & 0 \\ 0 & (1+2\varepsilon)^{1/2} \end{pmatrix}$$

One observes that the parameter ε controls the spacing of the two modes. For $\varepsilon = \gamma = 0$, the system degenerates into two decoupled s.d.o.f. oscillators with the same natural frequency. The mode shapes (normalized so that $\mu_i = 1$) are

$$\Xi = \frac{1}{\sqrt{2m}} \begin{pmatrix} 1 & 1 \\ 1 & -1 \end{pmatrix}$$

The modal damping ratios are

$$\xi = \frac{c}{2\sqrt{km}} \begin{pmatrix} 1 & 0 \\ 0 & \frac{1+2\gamma}{\sqrt{1+2\varepsilon}} \end{pmatrix}$$

They become identical for $1 + 2\gamma = \sqrt{1+2\varepsilon}$. We shall assume that in what follows, in order to simplify the algebra. The system is excited by a white noise point force applied to d.o.f. 1:

$$\Phi_F = \begin{pmatrix} \Phi_0 & 0 \\ 0 & 0 \end{pmatrix}$$

$$\Phi_P = \Xi^T \Phi_F \Xi = \frac{\Phi_0}{2m} \begin{pmatrix} 1 & 1 \\ 1 & 1 \end{pmatrix}$$

Carrying on according to the scheme of section 6.3, we get

$$\Phi_{X_1X_1}(\omega) = \frac{\Phi_0}{4m^2}\{|H_1(\omega)|^2 + |H_2(\omega)|^2 + 2\mathrm{Re}[H_1(\omega)H_2^\star(\omega)]\} \qquad (6.90)$$

$$\Phi_{X_2X_2}(\omega) = \frac{\Phi_0}{4m^2}\{|H_1(\omega)|^2 + |H_2(\omega)|^2 - 2\mathrm{Re}[H_1(\omega)H_2^\star(\omega)]\} \qquad (6.91)$$

where we have used the identity

$$H_1(\omega)H_2^\star(\omega) + H_1^\star(\omega)H_2(\omega) = 2\mathrm{Re}[H_1(\omega)H_2^\star(\omega)] \qquad (6.92)$$

After some algebra, the corresponding *MS* displacements can be written as

$$E[X_1^2] = \frac{\pi\Phi_0}{4m^2}\{\frac{1}{2\xi\omega_1^3} + \frac{1}{2\xi\omega_2^3} + \frac{8\xi}{(\omega_1+\omega_2)[(\omega_1-\omega_2)^2+4\xi^2\omega_1\omega_2]}\} \tag{6.93}$$

$$E[X_2^2] = \frac{\pi\Phi_0}{4m^2}\{\frac{1}{2\xi\omega_1^3} + \frac{1}{2\xi\omega_2^3} - \frac{8\xi}{(\omega_1+\omega_2)[(\omega_1-\omega_2)^2+4\xi^2\omega_1\omega_2]}\} \tag{6.94}$$

The last terms of these expressions represent the effect of the cross-correlations. ***They contribute identically, but with opposite sign.*** They produce an ***increase of the MS response at the d.o.f. where the load is applied, and a decrease at the other d.o.f. in such a way that*** $E[X_1^2] + E[X_2^2]$ ***is unchanged.*** In order to examine the relative importance of the cross-correlations as compared to the auto-correlations, let us introduce

- the mean frequency

$$\omega_n = \frac{\omega_1+\omega_2}{2} \tag{6.95}$$

- the reduced frequency spacing

$$\alpha = \frac{\omega_2-\omega_1}{2\xi\omega_n} \tag{6.96}$$

α expresses the spacing between the natural frequencies in term of the bandwidth of the system. With these notations, we find

$$\frac{\text{crosscorrelation}}{\text{autocorrelation}} = 2\xi\omega_n^3\frac{8\xi}{(\omega_1+\omega_2)[(\omega_1-\omega_2)^2+4\xi^2\omega_1\omega_2]}$$

$$= \frac{1}{1+\alpha^2(1-\xi^2)} \simeq \frac{1}{1+\alpha^2} \tag{6.97}$$

This ratio is small if α is large, that is, if ***the spacing between the natural frequencies is much larger than the bandwidth of the oscillators.***

If the contribution of the cross-correlations is neglected, Equ.(6.93) and (6.94) tell us that the overall mean square response is the sum of the modal mean square responses. This amounts to considering the modal responses as ***statistically independent.*** As we have just seen, this is acceptable if the modes are ***well separated*** ($\alpha \gg 1$). This assumption is at the origin of the method known as the *SRSS* rule (*Square Root of the Sum of the Squares*) which is widely used in seismic analysis. According to that rule, the maximum response of each mode, z_i, is computed separately using response spectra; next, they are combined according to

$$z_{\max}^2 = \sum_{i=1}^{m} z_i^2 \tag{6.98}$$

It is well known that this rule may lead to serious errors for closely spaced modes [e.g. see (Der Kiureghian, 1981)].

6.5 Response to a stationary seismic excitation

The discussion of section 6.3 can be extended to a seismic input by using the proper transfer functions as developed in section 6.2.

- *Absolute accelerations*

In order to analyse a *secondary* structure, one needs to calculate the *PSD* matrix of the absolute acceleration of the anchor points on the primary structure. The corresponding transfer matrix is given by Equ.(6.58). From the fundamental input-output relationship for the stationary random response, one gets

$$\Phi_{\ddot{x}}(\omega) = H_p(\omega)\Phi_0(\omega)H_p^\star(\omega) \tag{6.99}$$

Both $\Phi_{\ddot{x}}(\omega)$ and $\Phi_0(\omega)$ are complex and hermitian; only one half of $\Phi_{\ddot{x}}(\omega)$ must be computed. The diagonal terms characterize the spectral content of each component of the acceleration, while the off-diagonal terms define their cross-correlation. For example, the white noise excitations

$$\begin{pmatrix} \Phi_0 & 0 \\ 0 & \Phi_0 \end{pmatrix} \text{and} \begin{pmatrix} \Phi_0 & \Phi_0 \\ \Phi_0 & \Phi_0 \end{pmatrix} \tag{6.100}$$

refer respectively to uncorrelated and fully correlated components; in the first case, the excitations are statistically independent, while in the second case, the same excitation is being applied at the two points. The cross-correlation has, of course, a major effect on the dynamic response as well as on the quasi-static one (differential displacements).

- *support reactions*

As for the absolute accelerations, the *PSD* matrix of the support reactions reads

$$\Phi_R(\omega) = H_R(\omega)\Phi_0(\omega)H_R^\star(\omega) \tag{6.101}$$

where $H_R(\omega)$ is given by Equ.(6.67). However, unlike the acceleration which is used in the analysis of the secondary structure, for which the off-diagonal terms are important, only the diagonal terms of $\Phi_R(\omega)$ are used in the subsequent strength analysis.

- *stresses*

Similarly, the *PSD* function of a stress component can be written as

$$\Phi_r(\omega) = H_r(\omega)\Phi_0(\omega)H_r^\star(\omega) \tag{6.102}$$

where $H_r(\omega)$ is the transfer matrix defined by Equ.(6.71). For a single support excitation, there are no differential displacements and the static contribution vanishes in (6.70); Equ.(6.102) can be reduced to

$$\Phi_r(\omega) = b^T H(\omega)\Gamma\, \Phi_0(\omega)\, \Gamma^T H^\star(\omega)\, b \tag{6.103}$$

All the above calculations must be performed for a set of control frequencies whose spacing allows a smooth representation of $\Phi_0(\omega)$ and $H(\omega)$. As a result, the spacing will be closely related to the natural frequency and the bandwidth of the various modes.

6.6 Continuous structures

Before addressing the definition of the excitation process and its discretization, this section will consider the continuous structures. The discussion follows Y.K.Lin (1967) closely.

6.6.1 Input-Output relationship

Just as the matrix of impulse responses for a discrete system, the *impulse influence function*, $h(r,\varrho;t)$, represents the displacement at r resulting from a unit impulse loading applied at ϱ, at $t = 0$, assuming that the structure is at rest. Similarly, the *frequency response function*, $H(r,\varrho;\omega)$ is the amplitude of the response at r to a unit harmonic excitation applied at ϱ. As in the discrete case, they are related by the Fourier transform, according to Equ.(6.4) and (6.5).

For a continuous structure, both the excitation $P(r,t)$ and the response $W(r,t)$ are functions of the space variable r and the time variable t. They constitute *random fields*. For a linear structure, the principle of superposition applies and the input-output relationship in the time domain consists of the convolution

$$W(r,t) = \int_{-\infty}^{t} \int_{R} h(r,\varrho;t-\tau) P(\varrho,\tau) d\tau d\varrho \tag{6.104}$$

where the spatial integral is extended to the complete physical domain. Limiting the time integral at t implies that the system is causal. Just as we did in section 6.3 for discrete systems, we can readily establish that

$$E[W(r,t)] = \int_{-\infty}^{t} \int_{R} h(r,\varrho;t-\tau) E[P(\varrho,\tau)] d\tau d\varrho \tag{6.105}$$

$$E[W(r_1,t_1)W(r_2,t_2)] = \int_{-\infty}^{t_1} \int_{-\infty}^{t_2} \int_{R} \int_{R} h(r_1,\varrho_1;t_1-\tau_1) h(r_2,\varrho_2;t_2-\tau_2)$$

$$E[P(\varrho_1,\tau_1)P(\varrho_2,\tau_2)] d\tau_1 d\tau_2 d\varrho_1 d\varrho_2 \tag{6.106}$$

$E[P(\varrho_1,\tau_1)P(\varrho_2,\tau_2)]$ is the cross-correlation between the excitation functions at the locations ϱ_1 and ϱ_2.

If the excitation is stationary, its mean does not depend on t, and the correlation function $E[P(\varrho_1,\tau_1)P(\varrho_2,\tau_2)] = R_{PP}(\varrho_1,\varrho_2;\tau_1-\tau_2)$ depends only on the time difference $\tau_1 - \tau_2$. Upon introducing this into Equ.(6.106) and Fourier transforming, we can readily establish that

$$\Phi_{WW}(r_1, r_2; \omega) = \int_R \int_R \Phi_{PP}(\varrho_1, \varrho_2; \omega) H(r_1, \varrho_1; \omega) H^\star(r_2, \varrho_2; \omega) d\varrho_1 d\varrho_2 \tag{6.107}$$

where $\Phi_{PP}(\varrho_1, \varrho_2; \omega)$ is the [cross] power spectral density function of the stationary excitation, related to the correlation function by the Fourier transform

$$R_{PP}(\varrho_1, \varrho_2; \tau) = \int_{-\infty}^{\infty} \Phi_{PP}(\varrho_1, \varrho_2; \omega) e^{j\omega\tau} d\omega \tag{6.108}$$

$$\Phi_{PP}(\varrho_1, \varrho_2; \omega) = \frac{1}{2\pi} \int_{-\infty}^{\infty} R_{PP}(\varrho_1, \varrho_2; \tau) e^{-j\omega\tau} d\tau \tag{6.109}$$

$\Phi_{PP}(\varrho_1, \varrho_2; \omega)$ depends on the frequency ω and on the spatial coordinates ϱ_1 and ϱ_2. If the excitation field is ***spatially homogeneous***, its correlation structure depends only on the difference $\varrho_1 - \varrho_2$ and $\Phi_{PP}(\varrho_1, \varrho_2; \omega) = \Phi_{PP}(\varrho_1 - \varrho_2; \omega)$. In that case, the *PSD* function can be Fourier transformed with respect to the spatial coordinates

$$\Phi_{PP}(\varrho_1 - \varrho_2; \omega) = \int \Psi_{PP}(k, \omega) e^{jk^T(\varrho_1 - \varrho_2)} dk \tag{6.110}$$

$$\Psi_{PP}(k, \omega) = \frac{1}{(2\pi)^\alpha} \int_R \Phi_{PP}(\varrho; \omega) e^{-jk^T \varrho} d\varrho \tag{6.111}$$

where α is the dimension of the vectors k and ϱ, and k is the vector of ***wave numbers***. Just as a *PSD* function consists of a frequency decomposition of the power in a random process, $\Psi_{PP}(k, \omega)$ provides a decomposition of the power of a weakly stationary, weakly spatially homogeneous random field in the space (k, ω); k is in general tridimensional, like the spatial coordinate. A specific value of k corresponds to an harmonic variation in the direction k, with a wavelength $\lambda = 2\pi/|k|$. It is clear from Equ.(6.110) that a given wave number leads to the same component of the *PSD* everywhere in a plane perpendicular to k.

Combining (6.107) and (6.110) one gets

$$\Phi_{WW}(r_1, r_2; \omega) = \int_k dk\, \Psi_{PP}(k; \omega) \int_R \int_R H(r_1, \varrho_1; \omega) H^\star(r_2, \varrho_2; \omega) e^{jk^T(\varrho_1 - \varrho_2)} d\varrho_1 d\varrho_2 \tag{6.112}$$

or

$$\Phi_{WW}(r_1, r_2; \omega) = \int_k dk\, \Psi_{PP}(k; \omega) G(r_1, k; \omega) G^\star(r_2, k; \omega) \tag{6.113}$$

with

$$G(r, k; \omega) = \int_R H(r, \varrho; \omega) e^{jk^T \varrho} d\varrho \tag{6.114}$$

$$= \int_{-\infty}^{\infty} \int_R h(r, \varrho; t) e^{j(k^T \varrho - \omega t)} d\varrho \, dt$$

$G(r, k; \omega)$ is called the ***sensitivity function***. It is a characteristic of the structure: it represents the sensitivity of the structural response at r to a harmonic excitation of frequency ω and of spacewise variation defined by the wave number k. The impulse influence function, $h(r, \varrho; t)$, the frequency response function, $H(r, \varrho; \omega)$, and the sensitivity function, $G(r, k; \omega)$, can be expressed elegantly in terms of the normal modes if they exist.

6.6.2 Structure with normal modes

Consider a structure with an equation of motion of the form

$$m\ddot{w} + c\dot{w} + \mathcal{L}(w) = p \qquad (6.115)$$

where the first two terms represent respectively the inertia and viscous damping forces (in general, m and c depend on the space coordinate), and $\mathcal{L}(w)$ is a differential operator, linear in the space variables, representing the elastic restoring forces. As examples of such operators,

- uniform beam ($EI \equiv$ bending stiffness): $\mathcal{L}() = EI \frac{\partial^4}{\partial x^4}$
- taut string ($T \equiv$ string tensile force): $\mathcal{L}() = -T \frac{\partial^2}{\partial x^2}$
- uniform flat plate: $\mathcal{L}() = \frac{Eh^3}{12(1-\nu^2)} \left(\frac{\partial^4}{\partial x^4} + 2 \frac{\partial^4}{\partial x^2 \partial y^2} + \frac{\partial^4}{\partial y^4} \right)$

The normal modes of the undamped system, $f_i(r)$, are solutions of the eigen value problem

$$m\omega_i^2 f_i(r) = \mathcal{L}[f_i(r)] \qquad (6.116)$$

and satisfy the orthogonality relationship

$$\int_R m(r) f_i(r) f_j(r) \, dr = \mu_i \delta_{ij} \qquad (6.117)$$

where μ_i is the generalized mass of mode i. The impulse influence function, $h(r, \varrho; t)$, is, by definition, the response to a unit impulse load, $\delta(t)$, applied at $r = \varrho$; it is solution of

$$m\ddot{h} + c\dot{h} + \mathcal{L}(h) = \delta(r - \varrho)\delta(t) \qquad (6.118)$$

We seek a solution of the form

$$h(r, \varrho; t) = \sum_{j=1}^{\infty} a_j(\varrho; t) f_j(r) \qquad (6.119)$$

where $f_j(r)$ are the normal modes defined by Equ.(6.116). Introducing this form into Equ.(6.118), one gets

$$m\sum_{j=1}^{\infty}\ddot{a}_j(\varrho;t)f_j(r)+c\sum_{j=1}^{\infty}\dot{a}_j(\varrho;t)f_j(r)+m\sum_{j=1}^{\infty}a_j(\varrho;t)\omega_j^2 f_j(r)=\delta(r-\varrho)\delta(t) \tag{6.120}$$

where Equ.(6.116) has been used. Premultiplying by $f_i(r)$, integrating over the space coordinate r and using the orthogonality condition (6.117), one gets

$$\mu_i\ddot{a}_i(\varrho;t)+\sum_{j=1}^{\infty}\dot{a}_j(\varrho;t)\int_R cf_i(r)f_j(r)dr+\mu_i\omega_i^2 a_i(\varrho;t)=f_i(\varrho)\delta(t) \tag{6.121}$$

One notices that, if the following orthogonality condition is fulfilled,

$$\int_R cf_i(r)f_j(r)dr=2\mu_i\omega_i\xi_i\delta_{ij} \tag{6.122}$$

Equ.(6.121) is reduced to a set of uncoupled equations

$$\ddot{a}_i(\varrho;t)+2\xi_i\omega_i\dot{a}_i(\varrho;t)+\omega_i^2 a_i(\varrho;t)=\frac{f_i(\varrho)\delta(t)}{\mu_i} \tag{6.123}$$

The orthogonality condition (6.122) is quite similar to (6.13). Since Equ.(6.123) is identical to that of a single degree of freedom oscillator, its solution reads

$$a_i(\varrho;t)=f_i(\varrho)h_i(t) \tag{6.124}$$

where $h_i(t)$ is the impulse response of a s.d.o.f. oscillator of natural frequency ω_i, mass μ_i and damping ratio ξ_i. Substituting in Equ.(6.119), one gets

$$h(r,\varrho;t)=\sum_{i=1}^{\infty}f_i(r)f_i(\varrho)h_i(t) \tag{6.125}$$

Thus, under the assumption of classical damping, the impulse influence function can be expanded in terms of the mode shapes as above. Note that it is symmetric with respect to the coordinates r and ϱ. Upon successive Fourier transforming with respect to the variables t and ϱ, one gets

$$H(r,\varrho;\omega)=\sum_{i=1}^{\infty}f_i(r)f_i(\varrho)H_i(\omega) \tag{6.126}$$

$$G(r,k;\omega)=\sum_{i=1}^{\infty}f_i(r)S_i(k)H_i(\omega) \tag{6.127}$$

where $H_i(\omega)$ is the transfer function of the s.d.o.f. oscillator and $S_i(k)$ is a spatial Fourier decomposition of the mode shape $f_i(\varrho)$

$$S_i(k) = \int_R f_i(\varrho) e^{jk^T\varrho} d\varrho \tag{6.128}$$

Upon introducing Equ.(6.126) into (6.107), we find

$$\Phi_{WW}(r_1, r_2; \omega) = \int_R \int_R \Phi_{PP}(\varrho_1, \varrho_2; \omega) \sum_{i=1}^{\infty} \sum_{j=1}^{\infty} f_i(r_1) f_i(\varrho_1) H_i(\omega)$$

$$f_j(r_2) f_j(\varrho_2) H_j^*(\omega) d\varrho_1 d\varrho_2$$

$$= \sum_{i=1}^{\infty} \sum_{j=1}^{\infty} f_i(r_1) f_j(r_2) H_i(\omega) H_j^*(\omega) I_{ij}(\omega) \tag{6.129}$$

where I_{ij} is the cross *PSD* of the generalized forces in the modes i and j:

$$I_{ij}(\omega) = \int_R \int_R \Phi_{PP}(\varrho_1, \varrho_2; \omega) f_i(\varrho_1) f_j(\varrho_2) d\varrho_1 d\varrho_2 \tag{6.130}$$

It is the continuous counterpart of modal excitation *PSD* matrix defined by Equ.(6.81). In principle the double sum in Equ.(6.129) must be extended to all the modes; in practice, it can be truncated after a limited number of modes, because

- $\Phi_{PP}(\varrho_1, \varrho_2; \omega)$ decays at high frequency for all physical excitations;
- the correlation length of the excitation process becomes large with respect to the wavelength of the high frequency modes; this reduces the corresponding contributions $I_{ij}(\omega)$.

If the excitation is spatially homogeneous, Φ_{PP} depends only on the difference $\varrho_1 - \varrho_2$. In that case, it is customary to introduce the *co-spectrum*

$$C_{PP}(\varrho_1 - \varrho_2; \omega) = \frac{\Phi_{PP}(\varrho_1 - \varrho_2; \omega)}{\Phi_{PP}(\omega)} \tag{6.131}$$

where $\Phi_{PP}(\omega) = \Phi_{PP}(0; \omega)$ is the *PSD* at any point in the field. The co-spectrum is a measure of the *coherence* (see section 7.2) between the components of the excitation at points located $\varrho_1 - \varrho_2$ apart. Substituting into Equ.(6.130), we find

$$I_{ij}(\omega) = \Phi_{PP}(\omega) A_{ij}(\omega) \tag{6.132}$$

where A_{ij} is the *joint acceptance function*

$$A_{ij} = \int_R \int_R C_{PP}(\varrho_1 - \varrho_2; \omega) f_i(\varrho_1) f_j(\varrho_2) d\varrho_1 d\varrho_2 \tag{6.133}$$

If the excitation is fully correlated, $C_{PP}(\varrho_1, \varrho_2; \omega) = 1$; if it is completely spatially uncorrelated, $C_{PP}(\varrho_1 - \varrho_2; \omega) = \delta(\varrho_1 - \varrho_2)$. In that latter case, the joint acceptance functions read

$$A_{ij} = \int_R f_i(\varrho_1) f_j(\varrho_1) d\varrho_1 \tag{6.134}$$

If one compares this expression to the orthogonality condition (6.117), one notices that the off-diagonal terms vanish for a uniform mass distribution and the double sum (6.129) is reduced to the single sum

$$\Phi_{WW}(r_1, r_2; \omega) = \sum_{i=1}^{\infty} f_i(r_1) f_i(r_2) |H_i(\omega)|^2 \Phi_{PP}(\omega) A_{ii} \tag{6.135}$$

In practice, as we discussed in section 6.4, for a lightly damped structure with well separated modes, the double sum (6.129) is always dominated by the diagonal terms. Just as in the case of a discrete system, the mass-averaged mean square response does not depend on the off-diagonal contributions; in fact,

$$\begin{aligned} \int_R m E[W^2] dr &= \int_R \int_{-\infty}^{\infty} m\, \Phi_{WW}(\omega) d\omega\, dr \\ &= \sum_{i=1}^{\infty} \sum_{j=1}^{\infty} \beta_{ij} \int_R m f_i(r) f_j(r) dr \end{aligned} \tag{6.136}$$

with the notation

$$\beta_{ij} = \int_{-\infty}^{\infty} H_i(\omega) H_j^*(\omega) I_{ij}(\omega) d\omega \tag{6.137}$$

Owing to the orthogonality relationship (6.117), Equ.(6.136) is reduced to a single sum

$$\int_R m E[W^2] dr = \sum_{i=1}^{\infty} \beta_{ii} \mu_i \tag{6.138}$$

This relationship is identical to (6.89). Here again, it implies that the off-diagonal terms contribute to the local variations of the *MS* response, the mass-average over the entire structure being dependent on the diagonal terms alone. This result is due to A.Powell (1958).

6.7 Co-spectrum

As explained in the previous section, the co-spectrum defines the spatial coherence of the excitation. For a spatially homogeneous excitation, the *PSD* is the same at every point and the co-spectrum depends on the difference of coordinates

$$C(r_1 - r_2; \omega) = \frac{\Phi_{PP}(r_1 - r_2; \omega)}{\Phi_P(\omega)} \tag{6.139}$$

In general, if the excitation is non-homogeneous, the *PSD* varies with the spatial coordinate; the co-spectrum is defined as

$$\mathcal{C}(r_1, r_2; \omega) = \frac{\Phi_{PP}(r_1, r_2; \omega)}{\Phi_P^{1/2}(r_1, \omega)\Phi_P^{1/2}(r_2, \omega)} \tag{6.140}$$

In the extreme cases, the random field can be

- Fully correlated: $\mathcal{C}(r_1, r_2; \omega) = 1$
- Completely uncorrelated: $\mathcal{C}(r_1, r_2; \omega) = \delta(r_1 - r_2)$. The excitation forces acting at different points are statistically independent.

In practice, the physical processes are such that the coherence decreases with the distance, as for example

$$\mathcal{C}(r_1 - r_2; \omega) = e^{-a|r_1 - r_2|} \cos b|r_1 - r_2| \tag{6.141}$$

Often, the physics of the process is such that $\mathcal{C}$ depends on the frequency ω. This is the case, for example, for turbulent flows, where the pressure fluctuation is related to the transport of eddies by the flow. The co-spectrum is, in this case, a decreasing function of the ratio

$$\frac{|r_1 - r_2|}{\lambda} \tag{6.142}$$

where λ is the characteristic size of the eddies. If the eddies are transported with a convection velocity U_c, the frequency of the pressure fluctuation is related to λ and U_c according to

$$\omega \sim \frac{U_c}{\lambda} \quad \text{which implies} \quad \lambda \sim \frac{U_c}{\omega} \tag{6.143}$$

If one substitutes this into (6.142), one finds the co-spectrum to be a function of $|r_1 - r_2|\omega/U_c$ and, upon introducing the reduced coordinate $\zeta = |r_1 - r_2|/L$, where L is a characteristic length of the system, one finds the following general form

$$\mathcal{C}[|\zeta|.\frac{\omega L}{U_c}] \tag{6.144}$$

This generic form applies to many physical problems involving the transport of eddies. The dimensionless ratio

$$S = \frac{\omega L}{U_c} \tag{6.145}$$

is known as the *Strouhal number*; it appears in most harmonic problems in unsteady fluid dynamics.

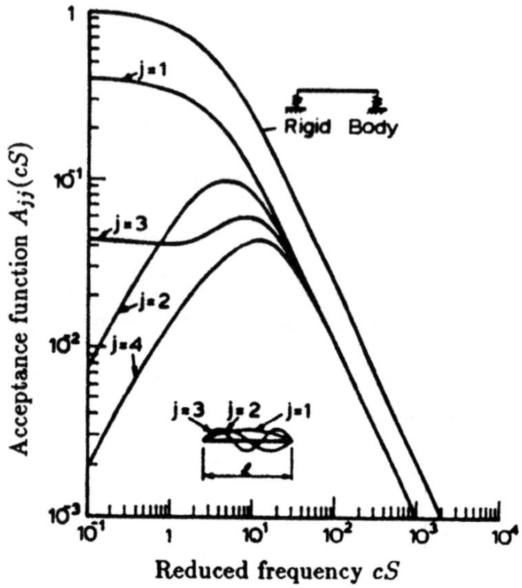

Figure 6.3: Acceptance functions of a simply supported beam.

The importance of a good representation of the spatial coherence of the excitation can be assessed from Fig.6.3, taken from (Novak, 1983). The acceptance functions are represented for a simply supported beam subject to a co-spectrum with exponential form

$$\mathcal{C}(x_1' - x_2'; \omega) = \exp(-cS|x_1' - x_2'|) \tag{6.146}$$

where $x_i' = x_i/L$ and $S = \omega L/V$. According to Equ.(6.133), the diagonal components are

$$A_{ii} = \int_0^1 \int_0^1 e^{-cS|x_1'-x_2'|} \phi_i(x_1') \phi_i(x_2') dx_1' dx_2' \tag{6.147}$$

In Fig.6.3, one observes that, for $c = 0$ (perfect coherence), only the modes with odd numbers are excited. The maximum value of the acceptance function is obtained for a value of cS which increases with the order of the mode, and the magnitude of the maximum decreases rapidly with the order of the mode.

6.8 Example: Boundary layer noise

Figure 6.4 shows typical observations of the correlation in the pressure field along the x axis which coincides with the convection velocity (Tack et al.,1961). The

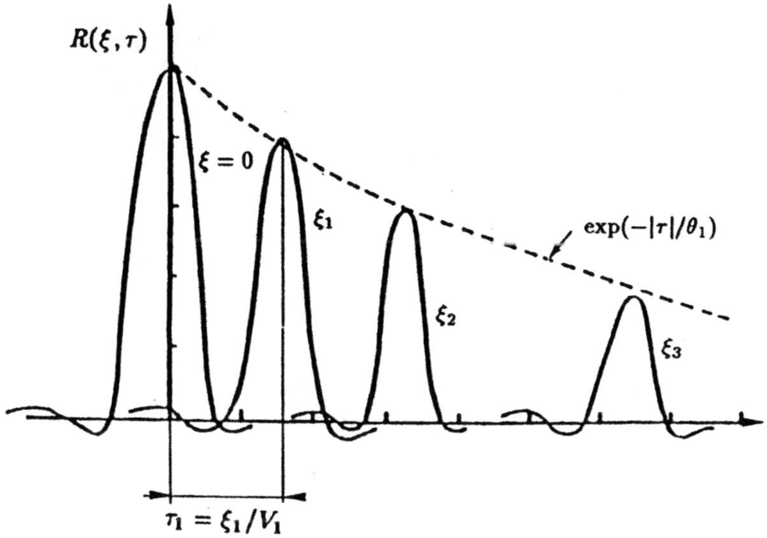

Figure 6.4: Boudary layer noise: cross-correlation function for various streamwise distances.

cross-correlation between points separated by a distance x exhibits a maximum after a delay $\tau = x/U_c$, which corresponds to the time taken to transport the perturbation a distance x streamwise, at the convection velocity U_c. Also, the maximum in the cross-correlation closely follows a decaying exponential. The same decreasing behaviour (but with a different decay rate) is observed along y. This suggests the following form for the correlation function

$$R(x, y; \tau) = R_0(\tau - x/U_c)e^{-\alpha_1|x|}e^{-\alpha_2|y|} \tag{6.148}$$

where $R_0(\tau)$ is the autocorrelation function of the pressure field at any point, and x and y are the distances between the points in the streamwise and the transverse directions, respectively. The argument $\tau - x/U_c$ takes care of the transport velocity; the streamwise decay rate is related to the lifetime of the eddies. Upon Fourier transforming Equ.(6.148), we find the *PSD* function

$$\Phi(x, y; \omega) = \Phi_0(\omega)C(x, y; \omega) = \Phi_0(\omega)\exp(-\frac{j\omega x}{U_c})e^{-\alpha_1|x|}e^{-\alpha_2|y|} \tag{6.149}$$

where $\Phi_0(\omega)$ is the *PSD* of the pressure field at any point (which is easy to measure) and the complex exponential arises from the translation theorem of the Fourier transform. Note that the foregoing form of the co-spectrum has very few

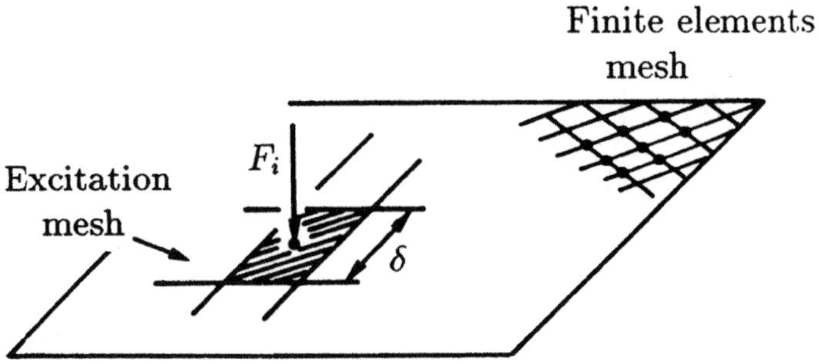

Figure 6.5: Finite elements mesh and excitation mesh.

free parameters: they can be determined experimentally relatively easily. Equation (6.149) constitutes a nice generic form which can fit numerous situations (e.g. acoustic fatigue).

6.9 Discretization of the excitation

In the finite element discretization of continuous structures, the mesh is essentially related to the representation of the stiffness of the structure. It would not be economical, nor practical, to define the excitation *PSD* matrix at all the structural nodes, in order to compute the modal excitation *PSD* matrix according to Equ.(6.81). In practice, the excitation is defined according to a regular mesh, the nodes of which are a subset of the structural nodes (Fig.6.5). The condition for achieving convergence [i.e. Equ.(6.81) being a good approximation of Equ.(6.130)] is that the typical size of the excitation mesh, δ, satisfies the following conditions

- $\delta \ll$ *wavelength of the modes contributing significantly to the response.*
- $\delta \ll$ *correlation length of the excitation process.*

To illustrate this, Fig.6.6 shows the first three modes of a simply supported flat plate subject to a fully coherent acoustic random pressure field. The plate is discretized by 100 finite elements. Two pressure meshes have been used as represented on the first mode. In the cruder one, there are 25 pressure nodes, each one having a tributary area corresponding to 4 finite elements; the finer one has 81 pressure nodes, each one with a tributary area equal to that of a single finite element. The convergence of the *RMS* value of the displacement (expressed in $10^{-4}m$) at the center of the plate is as follows: Crude mesh (25 pressure nodes): 3.895; fine mesh (81 pressure nodes): 3.700; exact (analytic): 3.658.

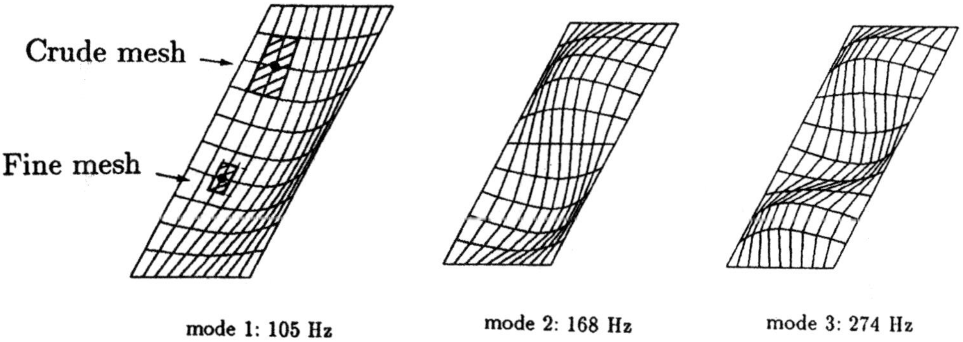

Figure 6.6: Convergence study for the pressure mesh on a flat plate.

6.10 Along-wind response of a tall building

6.10.1 Along-wind aerodynamic forces

The flow around a massive structure is very complicated and not yet fully understood. Although the structure changes the local flow conditions, the current practice assumes that the drag forces can be expressed in terms of the unperturbed flow. Accordingly, the drag force at a point j reads

$$F_j(t) = \frac{1}{2}\varrho A_j C_D V_j^2(t) \tag{6.150}$$

where ϱ is the air density, A_j is the area associated with point j, C_D is the drag coefficient and V_j is the relative velocity between the *unperturbed* wind and the building, at j. The velocity of the building is usually small and V_j can be taken as the velocity of the unperturbed wind. Besides, because the turbulent component is small as compared to the mean wind $\bar{u}_j$, V_j can be written as

$$V_j(t) = \bar{u}_j[1 + \xi_i(t)] \tag{6.151}$$

where $\xi_i(t)$ is the non-dimensional turbulent velocity. It is a stationary random process with zero-mean and satisfies the condition $|\xi_i(t)| \ll 1$. As a result, the drag force acting on the building can be *linearized* as

$$F_j(t) \cong \frac{1}{2}\varrho A_j C_D \bar{u}_j^2[1 + 2\xi_j(t)] \tag{6.152}$$

where the second order term in ξ_i has been neglected. The first term is a constant which corresponds to the static loading of the mean wind; it can be dealt with

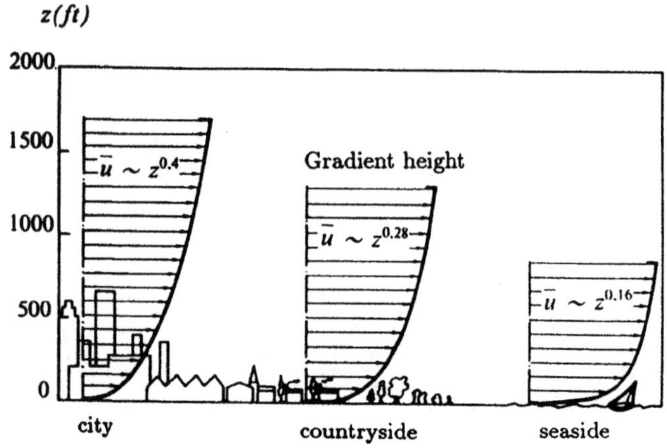

Figure 6.7: Average wind profiles for various terrain conditions.

separately. The cross *PSD* of the fluctuating part of the drag forces at the nodes i and j can be expressed as

$$\Phi_{F_iF_j}(\omega) = 4\nu_i\nu_j\Phi_{\xi_i\xi_j}(\omega) \tag{6.153}$$

with

$$\nu_i = \frac{1}{2}\varrho A_i C_D \bar{u}_i^2 \tag{6.154}$$

Its characterization requires the knowledge of the mean wind $\bar{u}_i$ and the cross *PSD* of the reduced turbulent velocity $\xi_j(t)$.

6.10.2 Mean wind

The direction of the mean velocity does not change appreciably with the altitude; the velocity profile is essentially function of the ground roughness as illustated in Fig.6.7. In the boundary layer, the mean wind profile can be represented by the power law

$$\bar{u}_i = u_g(\frac{z_i}{z_g})^\alpha \qquad\qquad 0 \le z_i \le z_g \tag{6.155}$$

where z_g is called the ***gradient height***: it is the altitude above which the velocity becomes constant, and equal to the ***gradient velocity***, u_g; α and z_g are constants which depend on the ground roughness.

6.10.3 Spectrum at a point

The *RMS* value of the turbulent component, σ_u, varies little with the altitude; it follows the approximate relationship

$$\sigma_u = K u_{10} \tag{6.156}$$

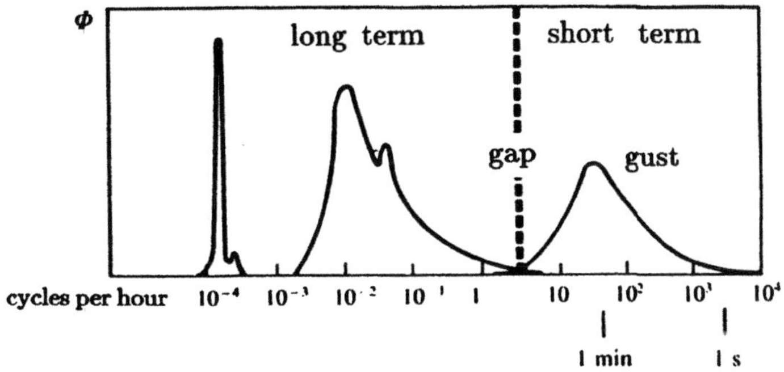

Figure 6.8: Spectral distribution of the wind velocity (Van der Hoven curve).

where K is a constant depending on the site, and u_{10} is the reference mean wind velocity at 10 m above the ground.

The frequency distribution of the wind power is represented in Fig.6.8 for a wide range of frequencies. The part of the figure corresponding to periods larger than 1 hour represents seasonal and daily variations; it has nothing to do with the dynamic response of the structure. The effect of the long term variations is included in the choice of the reference mean velocity (u_g or u_{10}). Thanks to the gap in the wind spectrum for periods close to 1 hour, the short term variations (gust) can be treated as stationary. Note that the first natural frequency of all the existing tall buildings (of the order of 0.2 Hz or more) is always in the tail of the gust spectrum. In the frequency range of interest for the dynamic response of buildings, the *PSD* of the turbulent component in the direction of the mean wind can be represented approximately by

$$\Phi_{uu}(\omega) = \frac{2\kappa u_{10}^2}{|\omega|} \frac{(\frac{600\omega}{\pi u_{10}})^2}{[1+(\frac{600\omega}{\pi u_{10}})^2]^{4/3}} \tag{6.157}$$

where κ is a constant depending on the roughness of the terrain and u_{10} is the mean wind reference velocity, 10 m above the ground.

6.10.4 Davenport spectrum

The co-spectrum of the wind for the vertical direction agrees with the following form

$$C(z, z+\Delta z;\omega) = \frac{\Phi_{u_i u_j}(z, z+\Delta z;\omega)}{\Phi_i^{1/2}(z,\omega)\Phi_j^{1/2}(z+\Delta z,\omega)} = \exp(-\frac{|\omega|}{2\pi u_{10}} C|\Delta z|) \tag{6.158}$$

where C is a correlation constant, $C \cong 7$. Note that this expression agrees with the general form (6.144). Equations (6.157) and (6.158) can be combined to

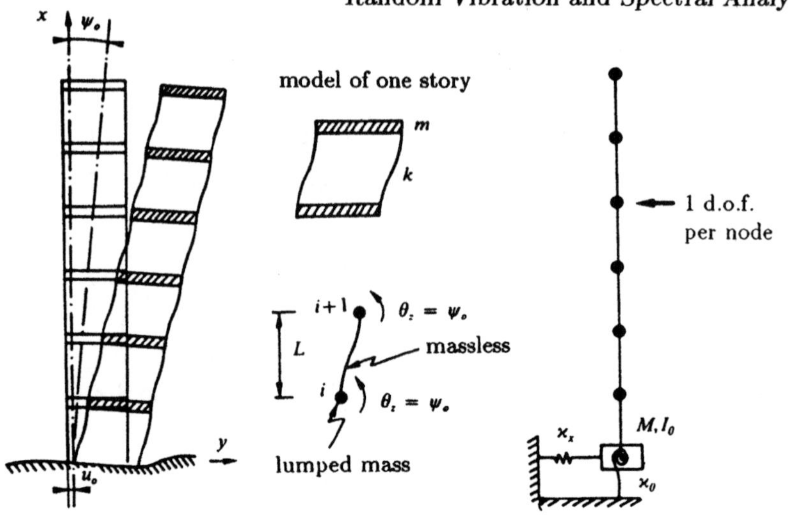

Figure 6.9: Structural model of the building.

provide the cross *PSD* of the non-dimensional field ξ_i along a vertical axis:

$$\Phi_{\xi_i\xi_j}(z_i, z_j; \omega) = \frac{2\kappa u_{10}^2}{\bar{u}_i \bar{u}_j |\omega|} \frac{(\frac{600\omega}{\pi u_{10}})^2}{[1 + (\frac{600\omega}{\pi u_{10}})^2]^{4/3}} \exp(-\frac{|\omega|}{2\pi u_{10}} C|z_i - z_j|) \tag{6.159}$$

This form was first proposed by Davenport (1961).

6.10.5 Example

A simplified model of a building can be obtained by assuming that it consists of rigid floors connected by massless columns. An accurate finite element model of such a structure can be easily obtained with very few d.o.f. as indicated on Fig.6.9: Each floor is represented by a single beam element with a bending stiffness EI such as to match the rigidity k of the floor (no improvement would result from using several elements since they are assumed massless). The rigidity of the floors is enforced by requesting that the rotation of all the nodes be the same as that of the base ($\theta_z = \psi_0$). The compliance of the ground can be represented by two d.o.f. as indicated on the figure. Using Equ.(6.153) and the Davenport spectrum, we can generate the *PSD* matrix of the excitation at the nodes of the structural model and find the response following the procedure of section 6.3.

Figure 6.10 shows the *PSD* of the displacement at the top of a representative 10 story building. Two damping mechanisms are considered: the so-called ***absolute*** damping, where the damping force acting on a floor is assumed proportional to the absolute velocity of that floor, and the ***inter-story*** damping, where the damping force is proportional to the relative velocity between the floors; both

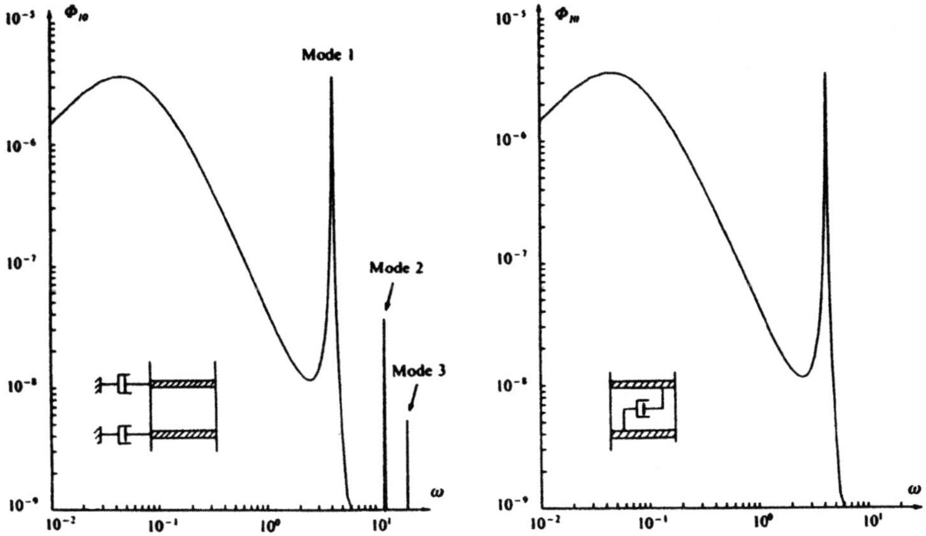

Figure 6.10: Top displacement PSD for (a)absolute damping; (b)inter-story damping.

have been normalized to provide $\xi_1 = 0.01$ in the first mode, which is reasonable for a tall building (Davenport, 1967; Paquet, 1979).

The general shape of the *PSD* is typical of the wind response of tall structures: it consists of a large quasi-static contribution which duplicates more or less the excitation, and a dynamic contribution from the flexible modes, which amplify the tail of the excitation (note that the logarithmic scales give a misleading idea of the power content of the signal).

Comparing the two damping mechanisms, one notices that three flexible modes can be identified in the response for the absolute damping, while only the first mode appears for the inter-story damping. The reason is that the absolute damping is essentially proportional to the mass matrix, while the inter-story damping is proportional to the stiffness matrix. Both are particular cases of the Rayleigh damping, but the former leads to modal damping ratios decreasing with the order of the modes, while the latter leads to increasing ones. In all cases, because of the fast decay of the Davenport spectrum at high frequency (like $\omega^{-5/3}$), only the first mode contributes significantly to the top displacement. Finally it should be mentioned that for high rise buildings, limiting the amplitude of the response is often associated with the comfort of the occupants, rather than with the risk of structural damage.

6.11 Earthquake

6.11.1 Response spectrum

The methodology for evaluating the random response of a structure subjected to a stationary excitation from the supports has been developed in sections 6.2 and 6.5. The situation may be that of a satellite during the launch, or a nuclear plant during an earthquake. In this section, we discuss some specific aspects of earthquake engineering.

Typical time-histories of the ground acceleration during earthquakes are shown in Fig.8.9. The expected characteristics of the transient signal (maximum acceleration, duration, frequency content) depend on the seismicity of the site and the local geological conditions. For lack of better information about future earthquakes, seismologists have defined standard shapes of ***response spectra***, corresponding to various local soil conditions (e.g. bedrock, alluvium). These spectra are normalized with respect to the local seismicity, expressed by the maximum acceleration for the site. A typical set of response spectra is shown in Fig.6.11. Each curve represents the maximum response versus the natural frequency of the oscillator, for a specific damping ratio ξ. The ***displacement spectrum***, $S_d(\omega_n, \xi)$, is defined as the maximum relative displacement of a single d.o.f. oscillator of natural frequency ω_n and damping ration ξ to the ground acceleration $\ddot{x}_0$. The ***pseudo-velocity*** spectrum is defined as

$$S_v(\omega_n, \xi) = \omega_n S_d(\omega_n, \xi) \tag{6.160}$$

and the ***pseudo-acceleration*** spectrum

$$S_a(\omega_n, \xi) = \omega_n S_v(\omega_n, \xi) = \omega_n^2 S_d(\omega_n, \xi) \tag{6.161}$$

S_v is different from the maximum velocity of the response, but usually, S_a is very close to the maximum absolute acceleration of the oscillator. In the log-log representation of the pseudo-velocity spectrum, constant values of the relative displacement (S_d) and of the acceleration (S_a) appear as straight lines (Fig.6.11). For natural frequencies much larger than the cut-off frequency of the excitation, the structure behaves like a rigid body: its motion tends to follow closely that of the support. As a result, the high frequency asymptotic value of S_a is independent of the damping and equal to the maximum acceleration for the site. The response spectra are normalized to specific sites by moving the diagram vertically until the high frequency asymptote matches the expected maximum acceleration for the site (a_{max} can be from $0.15g$ in areas of moderate seismicity to $0.3g$ in areas of high seismic activity).

The ***design*** response spectra define the envelope of all the possible ground motions for a given site, over some period of time. They depend very much on the return period considered. In practice, for the design of nuclear power plants, two sets of response spectra are defined at a site: the ***Operating Basis***

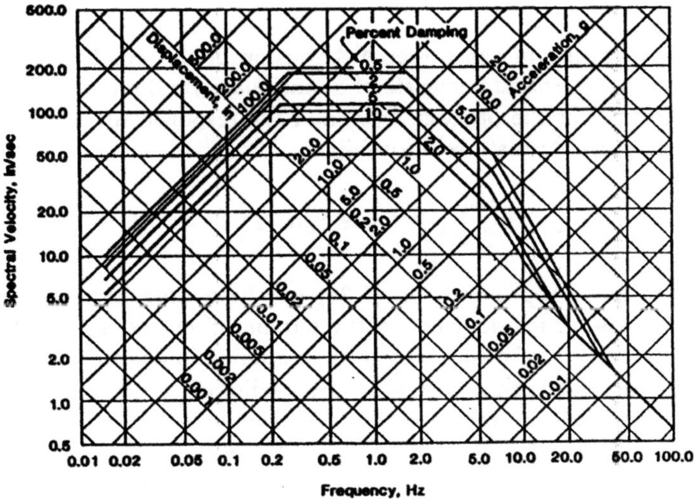

Figure 6.11: Newmark design reponse spectra for alluvium soil, normalized to $a_{max} = 1g$.

Earthquake (OBE) is the earthquake that the plant is likely to experience once during its lifetime (say over 20 years); the plant is supposed to be able to restart after minor repairs. On the contrary, the *Safe Shut-down Earthquake (SSE)* is the maximum possible earthquake for the site, which can destroy the plant, providing it is safely shut down and no release of fission products occurs.

The reason why the response spectrum has been used, historically, is that if the structural response is dominated by a single mode, the maximum response can be directly evaluated by

$$\phi_i \frac{\Gamma_i}{\mu_i} S_d(\omega_i, \xi)$$

When several modes are involved, the maximum response of each mode can be evaluated in the same way, but it is not clear how to combine the various modal responses. For well separated modes, the *SRSS* rule can clearly be applied, as we discussed in section 6.4, but for closely spaced modes, it may lead to very inaccurate results. One alternative is to use a more accurate combination rule like the *CQC* (Der Kiureghian, 1981), or to rely on an equivalent stationary random vibration analysis, which requires the knowledge of the *PSD* of the excitation. That alternative is especially attractive for multi-supported structures where the information about the correlation between the various excitations is lost in the response spectra.

If the acceleration time-history, $\ddot{x}_0(t)$ is known, the response spectrum $S_d(\omega_n, \xi)$ is entirely determined. The reverse is not true but, because the response of the s.d.o.f. oscillator is strongly influenced by the energy contained in the signal in the vicinity of the natural frequency ω_n, there is a strong relationship between the response spectrum and the power distribution of the acceleration. If one assumes that the accelerogram consists of a stationary random process of finite duration T, the PSD of the process, $\Phi(\omega)$ and the response spectrum $S_d(\omega,\xi)$ are related by the approximate relationship

$$[\frac{\pi\Phi(\omega)}{2\xi\omega^3}]^{1/2}.\eta \cong S_d(\omega,\xi) \tag{6.162}$$

where the white noise approximation (5.28) has been used and η is the *peak factor* of the process, which depends on the number of cycles, $\omega T/2\pi$, over the duration T. The concept of peak factor will be studied in chapter 10. The foregoing approximate relationship is good in the medium frequency range; it does not apply very well at high frequency because the white noise approximation is no longer true there, and not very well at very low frequency either, because the stationary assumption is not applicable. More refined models for converting $S_d(\omega,\xi)$ into $\Phi(\omega)$ are available (e.g. Mertens, 1993).

In general, different response spectra are used for the vertical and the two horizontal components of the acceleration at a point. The three components can be regarded as independent.

6.11.2 Cascade analysis

Figure 6.12 shows a simplified stick model as often used in the design of nuclear power plants. The primary structure consists of two vertical beams excited along the axis y; the secondary structure consists of a pipe supported at three points, two of them on one beam and the other on the second beam. This makes the differential displacements particularly important. If the secondary structure is considerably lighter than the primary structure, they are decoupled and the dynamic analysis can be performed in two steps: *(i)* Analysis of the primary structure to obtain the acceleration *PSD* matrix of the support points of the secondary structure. In this step, the interaction of the secondary structure is neglected. *(ii)* Analysis of the secondary structure subjected to the multidimensional excitation computed at step *(i)*. Since the motions of the various supports are out of phase, the information about their correlation (i.e. the off-diagonal terms of the *PSD* matrix) is particularly important.

A nice feature of the random vibration approach, as compared to the time-history analysis, is that, once the *PSD* of the response has been computed, the theory of extreme values can be applied to draw a ***probability of exceedance curve***, whose ordinate gives the probability that a specific amplitude be exceeded. Establishing these curves would otherwise require a large number of time-history analyses. In Fig.6.12, such curves are shown for the reaction forces on the primary structures. They are compared to similar curves obtained from 10 time-histories. Needless to say, the numerical effort involved in the time-history analyses is much greater.

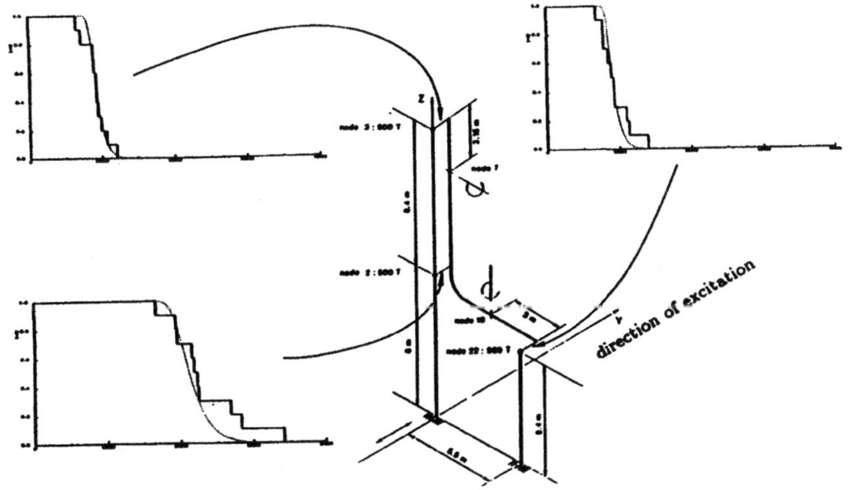

Figure 6.12: Cascade analysis of a pipe: probability of exceedance curves of the support reactions; comparison between random vibration and time-histories.

6.12 Remark on sound pressure level

According to our definition of the *PSD*, the pressure spectrum at a point is expressed in $Pa^2.s/rad$ or in Pa^2/Hz for the unilateral spectrum $G(f)$ (section 3.7.2). In acoustics, the current practice consists of defining the sound environment by ***Sound Pressure Level Spectra (SPL)***. On can define the 1/3 ***octave SPL*** and the ***octave SPL***, noted respectively $SPL_{1/3}$ and SPL_1, depending on the bandwidth of the narrow-band filter involved in its definition.

The *SPL* at the central frequency f_c is defined as the level in *dB* of the mean square pressure after narrow-band filtering centered on f_c:

$$SPL(dB) = 10.\log(\frac{p_{RMS}^2}{p_0^2}) \tag{6.163}$$

where p_{RMS}^2 is the mean square pressure after narrow-band filtering and $p_0 = 2.10^{-5}Pa$ is the reference pressure. Two kinds of narrow-band filters can be used:

$$1/3 \; octave: \qquad [0.89\, f_c, 1.12\, f_c]$$

$$1 \; octave: \qquad [0.707\, f_c, 1.404\, f_c]$$

The first one correspond to $SPL_{1/3}$ and the second one to SPL_1. Since p_{RMS}^2 represents the amount of power in the signal within the bandwidth of the filter, it is related to the unilateral spectrum by

$$p_{RMS}^2 = G(f_c)\,\Delta f \tag{6.164}$$

The conversion formulae from SPL to $G(f)$ are therefore

$$p_{RMS}^2 = p_0^2.10^{SPL/10} \tag{6.165}$$

$$G(f_c) = \frac{p_{RMS}^2}{0.232\, f_c} \qquad (1/3\ octave)$$

$$G(f_c) = \frac{p_{RMS}^2}{0.707\, f_c} \qquad (octave)$$

6.13 References

R.D.BLEVINS, *Flow-Induced Vibration*, Van Nostrand Reinhold Co, 1977.

T.K.CAUGHEY, Classical normal modes in damped linear dynamic systems, *Trans. ASME, J. of Applied Mechanics, Vol.27, No 2, pp.269-271, 1960.*

R.W.CLOUGH & J.PENZIEN, *Dynamics of Structures*, McGraw-Hill, 1975.

S.H.CRANDALL, The role of damping in vibration theory, *Journal of Sound and Vibration 11(1), pp.3-18, 1970.*

A.G.DAVENPORT, The application of statistical concepts to the wind loading of structures, *Proc. Inst. Civ. Eng., Vol.19, pp.449-471, August 1961.*

A.G.DAVENPORT, The treatment of wind loading on tall buildings, *Proceedings of the Symposium on Tall Buildings*, University of Southampton, Pergamon Press, London, 1966.

A.DER KIUREGHIAN, A response spectrum method for random vibration analysis of MDF systems, *Earthquake Engineering and Structural Dynamics, Vol.9, pp.419-435, 1981.*

I.ELISHAKOFF, *Probabilistic Methods in the Theory of Structures*, Wiley, 1982.

B.FRAEJIS de VEUBEKE, Influence of internal damping on aircraft resonance, AGARD report, November 1959.

Y.K.LIN, *Probabilistic Theory of Structural Dynamics*, McGraw-Hill, 1967.

P.G.MERTENS & A.PREUMONT, Improved generation of *PSD* functions, artificial accelerograms and spectra, fully compatible with a design response spectrum, *SMIRT-12*, paper K 13/3, Stuttgart 1993.

M.NOVAK, Random vibrations of structures, Proceedings of *ICASP-4*, Florence, pp.539-550, June 1983.

J.PAQUET, Etude experimentale in situ de l'effet du vent sur la tour Maine-Montparnasse, *Annales de l'Institut Technique du Batiment et des Travaux Publics, No 376, October 1979.*

A.POWELL, On the fatigue failure due to vibrations excited by random pressure fields. *The Journal of the Acoustical Society of America, Vol.30, No 12, pp.1130-1135, December 1958.*

G.H.POWELL, *Missing mass* correction in modal analysis of systems, *SMIRT-5*, paper K10/3, Berlin, 1979.

D.H.TACK, M.W.SMITH & R.F.LAMBERT, Wall pressure correlations in turbulent airflow. *The Journal of the Acoustical Society of America, Vol.33, No 4, pp.410-418, April 1961.*
E.H.VANMARCKE, Structural response to earthquakes, Ch.8 of *Seismic Risk and Engineering Decisions*, C.LOMNITZ & E.ROSENBLUETH, Eds., Elsevier, 1976.
J.N.YANG & Y.K.LIN, Along-wind motion of multistory building, *Proc. ASCE, J. of the Engineering Mechanics Division, Vol 107, EM2, pp.295-307, April 1981.*

6.14 Problems

P.6.1 *Show that for a unidirectional, single support seismic excitation, if the damping is assumed classical, the dynamic mass can be expanded into its modal components as*

$$F_0(\omega) = \{m_S + \sum_i \frac{\Gamma_i^2}{\mu_i}[\frac{\omega_i^2 + 2j\xi_i\omega_i\omega}{\omega_i^2 - \omega^2 + 2j\xi_i\omega_i\omega}]\}\ddot{X}_0(\omega)$$

P.6.2 *Consider the structure of Fig.6.2, excited by the support accelerations. Compute the response for the following excitations:*
(a) in-phase white noise excitations (the two supports move in phase, with a white noise acceleration);
(b) out of phase white noise excitations;
(c) uncorrelated white noise excitations.
For each case, write the excitation PSD matrix and calculate the PSD function of the effort in the spring and dash-pot connecting the two masses.
P.6.3 *Show that the 1/3 octave SPL and the octave SPL are related by*

$$SPL_1 = 10.\log[\sum_{i=1}^{3} 10^{\frac{(SPL_{1/3})_i}{10}}]$$

P.6.4 *The overall SPL is defined by*

$$SPL_{overall} = 10\log(\frac{p_{RMS}^2}{p_0^2})$$

where p_{RMS}^2 is the overall mean square pressure (without narrow band filtering). Show that

$$SPL_{overall} = 10.\log[\sum_i 10^{\frac{(SPL)_i}{10}}]$$

P.6.5 *In a diagram [$G(f)$ vs. $\log f$], the area below the curve $G(f)$ does not provide a fair idea of the fraction of the power in the signal between given* *frequencies. Show that this information can be obtained from a diagram [$f.G(f)$ vs. $\log f$].*

P.6.6 *The road profile $W(t)$ seen by a vehicle travelling at a speed v can be approximated by the response of a first order system to a white noise excitation. The corresponding autocorrelation and PSD functions are respectively*

$$R_{ww}(\tau) = \sigma^2 e^{-av|\tau|} \qquad \Phi_{ww}(\omega) = \frac{\sigma^2}{\pi}\frac{av}{\omega^2 + a^2v^2}$$

where σ is the RMS value, v is the vehicle speed and a is a parameter depending of the roughness of the road. If the rear wheels are a distance l behind the front wheels, compute the correlation and the PSD matrices (2 by 2) of the excitation at the front and rear wheels. [Hint: The rear wheel sees the same excitation as the front wheel, with a delay l/v.]

Chapter 7

Input-Output Relationship for Physical Systems

7.1 Estimation of frequency response functions

Consider the linear time-invariant single-input single-output system of Fig. 7.1. According to section 5.3, the stationary input-output relationships in the frequency domain are

$$\Phi_{yx}(\omega) = H(\omega)\Phi_{xx}(\omega) \tag{7.1}$$

$$\Phi_{xy}(\omega) = \Phi^{\star}_{yx}(\omega) = H^{\star}(\omega)\Phi_{xx}(\omega) \tag{7.2}$$

$$\Phi_{yy}(\omega) = H^{\star}(\omega)\Phi_{yx}(\omega) = H(\omega)\Phi_{xy}(\omega) \tag{7.3}$$

$$\Phi_{yy}(\omega) = |H(\omega)|^2\Phi_{xx}(\omega) \tag{7.4}$$

where $\Phi_{xx}(\omega)$ and $\Phi_{yy}(\omega)$ are the power spectral density functions of the input and the output, respectively, and $\Phi_{yx}(\omega)$ is the cross power spectral density of

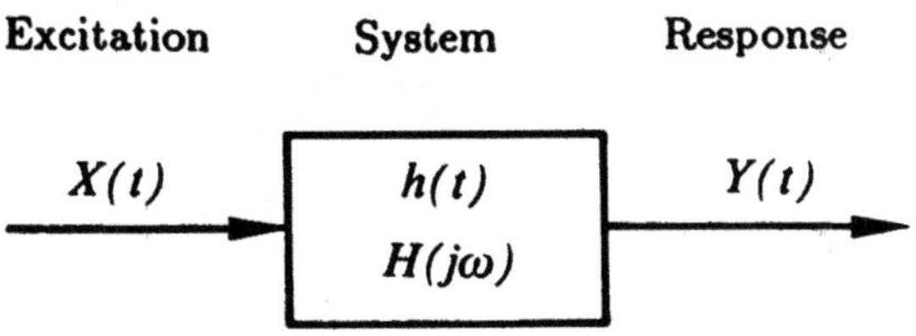

Figure 7.1: Time-invariant single-input single-output linear system.

$Y(t)$ and $X(t)$. $\Phi_{xx}(\omega)$ and $\Phi_{yy}(\omega)$ are real quantities, while $\Phi_{yx}(\omega)$ is complex. Equations (7.1) to (7.3) contain *amplitude* as well as *phase* information, while Equ.(7.4) relates only the amplitude of the *PSD*'s. Introducing

$$H(\omega) = |H(\omega)|e^{j\theta(\omega)} \tag{7.5}$$

one gets

$$\frac{\Phi_{yx}(\omega)}{\Phi_{xy}(\omega)} = \frac{H(\omega)}{H^\star(\omega)} = e^{2j\theta(\omega)} \tag{7.6}$$

Equations (7.1) to (7.3) provide the following relationships for the frequency response function

$$H_1(\omega) = \frac{\Phi_{yx}(\omega)}{\Phi_{xx}(\omega)} \tag{7.7}$$

$$H_2(\omega) = \frac{\Phi_{yy}(\omega)}{\Phi_{xy}(\omega)} \tag{7.8}$$

These formulae allow the estimation of system frequency response functions from measured input-output data. Note that they allow the determination of both the amplitude and the phase of $H(\omega)$. On the contrary,

$$|H_3(\omega)|^2 = \frac{\Phi_{yy}(\omega)}{\Phi_{xx}(\omega)} \tag{7.9}$$

supplies only the amplitude of $H(\omega)$. In the ideal case of a *linear* system where the measurements are not contaminated by *noise*, all three estimators are identical. This condition is rarely met in practice, and the estimators are different. One way to evaluate how far from ideal the conditions are is provided by the *coherence function*.

7.2 Coherence function

The coherence function between the input $X(t)$ and the output $Y(t)$ is a real valued quantity defined by

$$\gamma^2_{xy}(\omega) = \frac{|\Phi_{xy}(\omega)|^2}{\Phi_{xx}(\omega)\Phi_{yy}(\omega)} \tag{7.10}$$

[Compare with the co-spectrum, Equ.(6.140)]. From the inequality (3.69),

$$0 \leq \gamma^2_{xy}(\omega) \leq 1 \tag{7.11}$$

For a linear time-invariant system, substituting Equ.(7.1) to (7.4) leads to $\gamma^2_{xy}(\omega) = 1$. This is the ideal situation. If X and Y are uncorrelated, $\gamma^2_{xy}(\omega) = 0$. If $\gamma^2_{xy}(\omega)$ is between 0 and 1, then one or more of the following conditions exist:

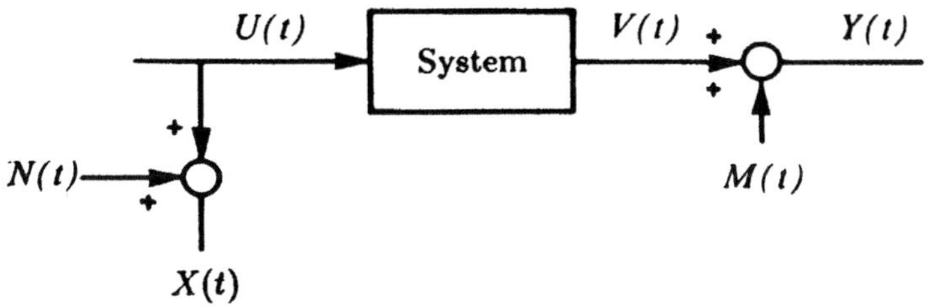

Figure 7.2: System with measurement noise.

- Extraneous noise is present in the measurements.
- The system is not linear.
- Besides $X(t)$, there are other inputs affecting the output $Y(t)$.

When the spectral estimates have a finite frequency resolution, one should add the ***resolution bias error*** which may be substantial for narrow band processes. The digital estimation of *PSD* functions will be addressed in chapter 12. As we are going to see, for linear systems, the coherence function $\gamma_{xy}^2(\omega)$ can be interpreted as the ***fraction of the mean square output*** $Y(t)$ ***which can be attributed to the input*** $X(t)$, ***for every frequency*** ω. It is therefore a measure of the ***causality*** between the excitation and the response.

Combining Eq. (7.7) to (7.9), we observe that

$$\gamma_{xy}^2(\omega) = \frac{H_1}{H_2} = \frac{|H_1|^2}{|H_3|^2} = \frac{|H_3|^2}{|H_2|^2} \tag{7.12}$$

7.3 Effect of measurement noise

Consider the system of Fig. 7.2. The actual input and output are respectively $U(t)$ and $V(t)$, but the measured values are

$$X(t) = U(t) + N(t)$$

$$Y(t) = V(t) + M(t) \tag{7.13}$$

where $N(t)$ and $M(t)$ are respectively the input and output measurement noise, assumed statistically independent of each other and of the processes $U(t)$ and $V(t)$:

$$\Phi_{un}(\omega) = \Phi_{vm}(\omega) = \Phi_{nm}(\omega) = 0$$

It follows that

$$\Phi_{xx}(\omega) = \Phi_{uu}(\omega) + \Phi_{nn}(\omega)$$

$$\Phi_{yy}(\omega) = \Phi_{vv}(\omega) + \Phi_{mm}(\omega) \tag{7.14}$$
$$\Phi_{xy}(\omega) = \Phi_{uv}(\omega)$$

The coherence function between the input and the output of the system is

$$\gamma_{uv}^2(\omega) = \frac{|\Phi_{uv}(\omega)|^2}{\Phi_{uu}(\omega)\Phi_{vv}(\omega)} \tag{7.15}$$

and that between the measured signals

$$\gamma_{xy}^2(\omega) = \frac{|\Phi_{xy}(\omega)|^2}{\Phi_{xx}(\omega)\Phi_{yy}(\omega)} = \frac{|\Phi_{uv}(\omega)|^2}{[\Phi_{uu}(\omega) + \Phi_{nn}(\omega)][\Phi_{vv}(\omega) + \Phi_{mm}(\omega)]} \tag{7.16}$$

or

$$\gamma_{xy}^2(\omega) = \frac{\gamma_{uv}^2(\omega)}{[1 + \frac{\Phi_{nn}(\omega)}{\Phi_{uu}(\omega)}][1 + \frac{\Phi_{mm}(\omega)}{\Phi_{vv}(\omega)}]} < \gamma_{uv}^2(\omega) \tag{7.17}$$

Thus, *the presence of uncorrelated noise in the measurements will always result in lower values of the coherence function between the measured signals.* If there are other inputs to the system, uncorrelated to $X(t)$, their contribution to the response appears as an uncorrelated output noise, which produces a reduction of the coherence function.

Introducing Eq.(7.14) into Eq. (7.7) to (7.9), one gets

$$H_1(\omega) = \frac{\Phi_{vu}(\omega)}{\Phi_{uu}(\omega) + \Phi_{nn}(\omega)} \tag{7.18}$$

$$H_2(\omega) = \frac{\Phi_{vv}(\omega) + \Phi_{mm}(\omega)}{\Phi_{uv}(\omega)} \tag{7.19}$$

$$|H_3(\omega)|^2 = \frac{\Phi_{vv}(\omega) + \Phi_{mm}(\omega)}{\Phi_{uu}(\omega) + \Phi_{nn}(\omega)} \tag{7.20}$$

One observes that the estimator $H_3(\omega)$ is always biased, unless both $\Phi_{nn}(\omega)$ and $\Phi_{mm}(\omega)$ are zero, that is if $\gamma_{xy}^2(\omega) = 1$. On the contrary, $H_1(\omega)$ is insensitive to an uncorrelated noise at the output, while $H_2(\omega)$ is insensitive to an uncorrelated noise at the input. In particular, H_1 is unbiased in the case of multiple uncorrelated inputs, which appear as an output noise.

In practice, the frequency distribution of the input power, $\Phi_{uu}(\omega)$, is controllable to a large extent, while that of the response, $\Phi_{vv}(\omega)$, depends on the system to be identified. This means that the estimator H_1 is often superior to H_2. H_2 may be superior in the vicinity of the resonances, where the power level of the excitation drops to a lower value, close to that of the noise Φ_{nn}. Conversely, H_1 will be preferred near the anti-resonances (imaginary zeros), because the response level becomes very small, possibly lower than the measurement noise Φ_{mm}. From Eq. (7.17), one may anticipate that the coherence function can be substantially lower than 1 at the resonances and the anti-resonances of the frequency response function, even if the test is performed properly and if the structure is linear. In the former case this can be attributed to large values of Φ_{nn}/Φ_{uu} while in the latter case, this is due to large Φ_{mm}/Φ_{vv}.

For the linear system of Fig. 7.2, consider the coherence function $\gamma_{uy}^2(\omega)$ between the actual input U and the measured output Y. Since the system is linear, $\Phi_{uy}(\omega) = \Phi_{uv}(\omega) = H(\omega)\Phi_{uu}(\omega)$. The uncorrelated output noise $M(t)$ represents the measurement noise as well as the non-linearity and the other sources of excitation. One finds easily that

$$\gamma_{uy}^2(\omega) = \frac{|H(\omega)|^2\,\Phi_{uu}(\omega)}{|H(\omega)|^2\,\Phi_{uu}(\omega) + \Phi_{mm}(\omega)} \tag{7.21}$$

This relationship shows that, *for every frequency ω, the coherence function $\gamma_{uy}^2(\omega)$ represents the fraction of the output PSD resulting from the input $U(t)$.*

The ratio between the useful output signal and the output noise is called the *signal to noise ratio* it is related to the coherence function by

$$\frac{\text{signal}}{\text{noise}} = \frac{|H(\omega)|^2\Phi_{uu}(\omega)}{\Phi_{mm}(\omega)} = \frac{\gamma_{uy}^2(\omega)}{1-\gamma_{uy}^2(\omega)} \tag{7.22}$$

7.4 Example

The following example is taken from (Bendat & Piersol, 1971), pp.143-146. It illustrates the effect of noise and of multiple uncorrelated inputs on the coherence function.

Consider a plane flying through atmospheric turbulence (Fig. 7.3.a). The vertical gust wind velocity is taken as input to the system, $X(t)$, while the output $Y(t)$ is the vertical acceleration of the center of mass. Recorded *PSD*'s of $X(t)$ and $Y(t)$ are shown in Fig. 7.3.b. Their coherence function is displayed in Fig. 7.3.c where it can be observed that the coherence between the input and the output is large ($0.8 < \gamma_{xy}^2 < 0.9$) for the frequency range [0.3 *Hz*, 2 *Hz*] while it is considerably lower outside that interval. The origin of the lower coherence values is different at low and at high frequency, as explained below.

- At low frequency, a significant part of the vertical acceleration is due to the action of the pilot, rather than the atmospheric turbulence. The loss of coherence at low frequency can therefore be attributed to multiple inputs to the system.

- At frequencies above 2 *Hz*, one observes that there is little power in the input signal; this is still amplified in the output signal, because of the low-pass filter behaviour of the airplane (Fig.7.3.b). Since the measurement noise can be regarded as more or less uniform over the frequency range of interest, the signal to noise ratio is smaller at high frequency, which is responsible for the loss of coherence.

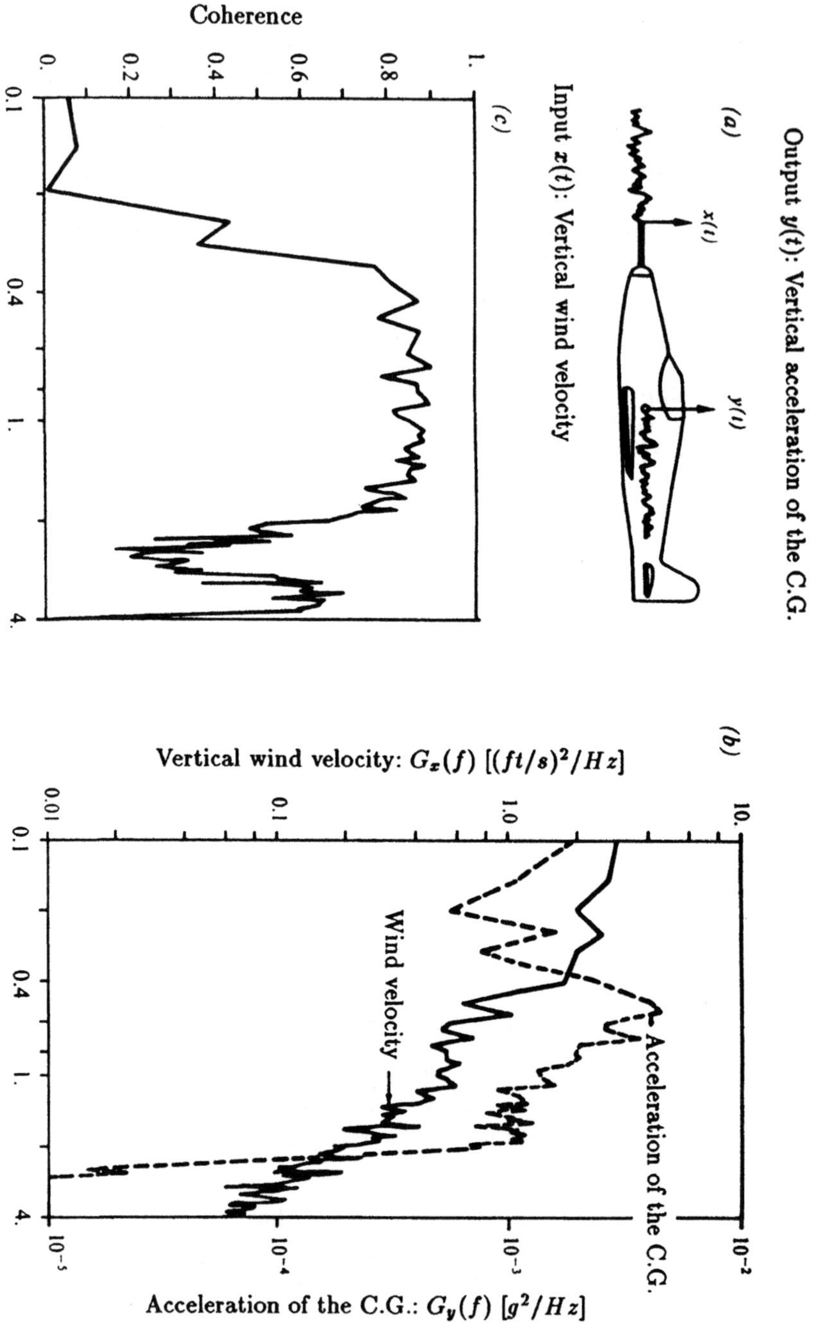

Figure 7.3: Coherence between the vertical wind velocity and the vertical acceleration of an airplane (Bendat & Piersol, 1971).

7.5 Remark

The foregoing discussion also applies to a transient excitation $X(t)$, if one replaces the power spectral densities $\Phi_{xx}(\omega)$ and $\Phi_{xy}(\omega)$ by the *energy* spectral density functions:

$$S_{xx}(\omega) = \frac{1}{2\pi}\mathrm{E}[\mathcal{X}(\omega)\mathcal{X}^{\star}(\omega)] \qquad \text{and} \qquad S_{xy}(\omega) = \frac{1}{2\pi}\mathrm{E}[\mathcal{X}(\omega)\mathcal{Y}^{\star}(\omega)]$$

respectively, where $\mathcal{X}(\omega)$ is the Fourier transform of $X(t)$.

7.6 References

J.BENDAT & A.PIERSOL, *Random Data: Analysis and Measurement Procedures*, Wiley-Interscience, 1971.
J.BENDAT & A.PIERSOL, *Engineering Applications of Correlation and Spectral Analysis*, Wiley-Interscience, 1980.
D.J.EWINS, *Modal Testing: Theory and practice*, Wiley, 1984.
L.D.MITCHELL, Improved methods for the Fast Fourier Transform (FFT) calculation of the frequency response function, *ASME J. Mech. Design, Vol.104, pp.277-279, April 1982.*

Chapter 8

Spectral Description of Non-stationary Random Processes

8.1 Introduction

8.1.1 Stationary random process

The spectral description of a weakly stationary random process is given by the power spectral density function $\Phi_{xx}(\omega)$. The *local* character of the frequency decomposition can be seen as follows: Consider an ideal narrow-band filter as in Fig.1, where $H_i(\omega) = 1$ within the bandwidth and 0 outside. The corresponding impulse response is (Problem 1.8)

$$h_i(\tau) = \frac{2}{\pi\tau} \sin \frac{\Delta\omega\, \tau}{2} \cos \omega\tau \tag{8.1}$$

(This system is not causal and cannot be realized exactly). From Equ.(5.24), the mean square value of the filter output is related to the *PSD* of the input by

$$\mathrm{E}[X_i^2(t, \omega, \Delta\omega)] = 2 \int_{\omega-\Delta\omega/2}^{\omega+\Delta\omega/2} \Phi_{xx}(\nu)\, d\nu \tag{8.2}$$

This relation tells us that *the average power within any frequency interval is given by the area under the PSD function for that interval.* If the bandwidth is small this result can be approximated by

$$\mathrm{E}[X_i^2(t, \omega, \Delta\omega)] \simeq 2\, \Delta\omega\, \Phi_{xx}(\omega) \tag{8.3}$$

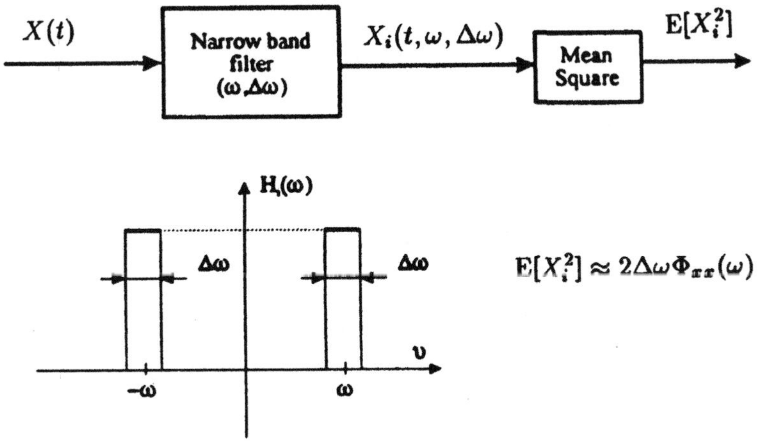

Figure 8.1: Estimation of the *PSD* from narrow band measurements.

If the process is ergodic, the ensemble average can be replaced by a time average

$$z(t) = \frac{1}{T}\int_{t-T}^{t} x_i^2(\tau)\,d\tau \tag{8.4}$$

Obviously, the measured quantity varies with the sample $x_i(t)$, the filter parameters ω and $\Delta\omega$ and the duration T. From the ergodicity theorem, we know that

$$z(t) \rightarrow \mathrm{E}[X_i^2(t,\omega,\Delta\omega)] \qquad \text{as} \qquad T \rightarrow \infty$$

For samples of finite duration, one can show that the relative fluctuation in the measured quantity is

$$\frac{\sigma_z}{\mathrm{E}[Z]} = \sqrt{\frac{2\pi}{\Delta\omega T}} \tag{8.5}$$

It can be reduced by increasing either $\Delta\omega$ or T. Of course widening $\Delta\omega$ decreases the resolution of the measurement, which is the integral (8.2) rather than a pointwise estimate at the central frequency of the filter, ω.

8.1.2 Non-stationary random process

Consider the transient process $X(t)$ and its Fourier transform (assuming it exists)

$$\mathcal{X}(\omega) = \int_{-\infty}^{\infty} X(t)e^{-j\omega t}dt \tag{8.6}$$

Using Parseval's theorem (1.7) and following the same development as in the previous section, we can readily establish that the *energy spectral density* function

$$S_{xx}(\omega) = \frac{1}{2\pi}\mathrm{E}[|\mathcal{X}(\omega)|^2] \tag{8.7}$$

is a *local* decomposition of the *energy* in the transient process, exactly in the same sense as the *PSD* function for the power in a stationary process. However, the energy spectral density function does not provide any information as to the *time evolution of the frequency distribution of the power* in the process. It is this type of energy mapping in the time-frequency plane that we shall now consider.

8.1.3 Objectives of a spectral description

For the past 20 years, the spectral analysis of nonstationary oscillatory processes has attracted a great deal of attention and several characterizations have been proposed. To be useful, such a representation should enjoy the following properties

- A clear *local* interpretation in the frequency-time plane.
- An estimate can be generated from a single sample record.
- A simple input-output relationship should exist for linear time-invariant systems.
- It should coincide with the power spectral density function when the process is stationary.

As we shall see below, a strictly local mapping does not exist, because a good resolution in one domain (e.g. frequency) can only be achieved at the expense of a poor resolution in the dual domain (time). This is known as the *uncertainty principle.* In fact, this can be understood easily by considering the narrow-band filter of Fig.8.1. The stationary relationship (8.3) can be extended to nonstationary processes as

$$\Psi_{xx}(t,\omega) = \frac{\mathrm{E}[X_i^2(t,\omega,\Delta\omega)]}{2\,\Delta\omega} \tag{8.8}$$

where X_i is the output of the narrow-band filter

$$X_i(t,\omega,\Delta\omega) = \int_{-\infty}^{\infty} h_i(\tau)X(t-\tau)\,d\tau \tag{8.9}$$

One sees that $X_i(t,\omega,\Delta\omega)$ consists of the *weighted average* of the values of the process $X(\tau)$ in the vicinity of t. The weighting function is the impulse response of the filter, defined by (8.1). Obviously, some smoothing occurs in the time domain. If, to increase the resolution in the frequency domain, we reduce the bandwith $\Delta\omega$ of the filter, the effective duration of the impulse response increases and the smoothing in the time domain involves an even longer period.

8.2 Instantaneous power spectrum

Let $\phi_{xx}(t_1, t_2)$ be the autocorrelation function of a non-stationary random process. With the following transformations

$$t = \frac{t_1 + t_2}{2} \qquad \tau = t_1 - t_2$$

$$t_1 = t + \frac{\tau}{2} \qquad t_2 = t - \frac{\tau}{2} \tag{8.10}$$

the autocorrelation function can be rewritten

$$\mathcal{R}_{xx}(t,\tau) = \phi_{xx}(t + \frac{\tau}{2}, t - \frac{\tau}{2}) = \mathrm{E}[X(t + \frac{\tau}{2})X(t - \frac{\tau}{2})] \tag{8.11}$$

$\mathcal{R}_{xx}(t,\tau)$ can be considered as an instantaneous autocorrelation function, at the average time t; it is an even function of the separation time τ. By analogy with stationary processes, the ***instantaneous*** power spectral density function is defined as

$$\Phi_{xx}(t,\omega) = \frac{1}{2\pi}\int_{-\infty}^{\infty} \mathcal{R}_{xx}(t,\tau)e^{-j\omega\tau}d\tau \tag{8.12}$$

It is a real, even function of the variable ω; the inverse relationship is

$$\mathcal{R}_{xx}(t,\tau) = \int_{-\infty}^{\infty} \Phi_{xx}(t,\omega)e^{j\omega\tau}d\omega \tag{8.13}$$

and, at $\tau = 0$, one gets the mean square value

$$\mathrm{E}[X^2(t)] = \mathcal{R}_{xx}(t,0) = \int_{-\infty}^{\infty} \Phi_{xx}(t,\omega)d\omega \tag{8.14}$$

These relationships are seemingly completely similar to the Wiener-Khintchine theorem for stationary processes. Equation (8.14) shows that $\Phi_{xx}(t,\omega)$ is indeed a frequency decomposition of the instantaneous mean square value at t. However,

- the global frequency decomposition cannot be particularized to any frequency interval as in the case of a stationary process;
- $\Phi(t,\omega)$ can even become ***negative***;
- $\Phi(t,\omega)$ cannot be measured directly in practice.

Thus, the instantaneous power spectrum does not fulfil the criteria stated in the foregoing section and one should not expect that it provides, except in particular situations, a correct mapping of the energy in the frequency-time plane. One particular case where the instantaneous spectrum is physically meaningful is that of a ***locally stationary*** process, whose autocorrelation function has the form

$$\mathcal{R}_{xx}(t,\tau) = R_1(t)R_2(\tau) \tag{8.15}$$

where $R_1(t)$ is a non negative function and $R_2(\tau)$ is the autocorrelation function of a weakly stationary process. Upon Fourier transforming, one gets

$$\Phi_{xx}(t,\omega) = R_1(t)\Phi_2(\omega) \tag{8.16}$$

where $\Phi_2(\omega)$ is the *PSD* associated to $R_2(\tau)$. A *shot noise* constitutes an example of a locally stationary process.

A *separable* process is defined as the product of a weakly stationary process $X(t)$ and a slowly varying function $a(t)$

$$Y(t) = a(t)X(t) \tag{8.17}$$

Its autocorrelation function is

$$\phi_{yy}(t_1,t_2) = \mathrm{E}[Y(t_1)Y(t_2)] = a(t_1)a(t_2)\mathrm{E}[X(t_1)X(t_2)]$$

$$\phi_{yy}(t_1,t_2) = a(t_1)a(t_2)R_{xx}(t_1 - t_2) \tag{8.18}$$

If the fluctuations of $a(t)$ are slow, compared to the correlation time of $X(t)$,

$$\mathcal{R}_{xx}(t,\tau) \simeq a^2(t)R_{xx}(\tau)$$

$$\Phi_{xx}(t,\omega) \simeq a^2(t)\Phi_{xx}(\omega) \tag{8.19}$$

The instantaneous *PSD* of a locally stationary process cannot become negative.

8.3 Mark's Physical Spectrum

8.3.1 Definition and properties

Consider a non-stationary oscillatory real process $X(u)$. One isolates the vicinity of $u = t$ by multiplying $X(u)$ by a *window* function $w(t-u)$ such that

- $w(t)$ is positive in the vicinity of $t = 0$,
- $|w(t)|$ is small except near $t = 0$,
- it is normalized according to

$$\int_{-\infty}^{\infty} w^2(t)\,dt = 1 \tag{8.20}$$

Examples of such windows are shown in Fig. 8.3.

According to Parseval's theorem, the energy spectrum is a frequency decomposition of the total energy in the process. Therefore, a frequency decomposition of the energy in the vicinity of $u = t$ is provided by the energy spectrum of $w(t-u)X(u)$.

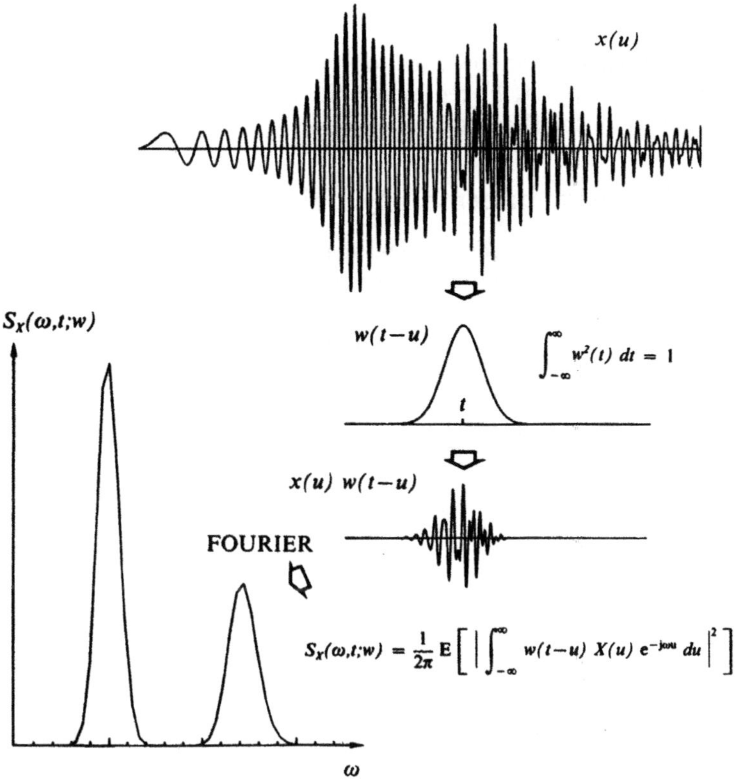

Figure 8.2: Definition of the physical spectrum.

The ***physical spectrum*** is defined as the ensemble average of the energy spectrum of $w(t-u)X(u)$:

$$S_x(\omega,t;w) = \frac{1}{2\pi}\mathrm{E}[|\int_{-\infty}^{\infty} w(t-u)X(u)e^{-j\omega u}du|^2] \tag{8.21}$$

It depends on the choice of the window $w(t)$. The operations associated with the definition of the physical spectrum are illustrated in Fig.8.2. For every $w(t)$, $S_x(\omega,t;w)$ supplies a non negative mapping, in the domain (ω,t), of the energy in the process. It is an even function of ω. According to Parseval's theorem [applied to the process $w(t-u)X(u)$],

$$\int_{-\infty}^{\infty} w^2(t-u)\mathrm{E}[X^2(u)]\,du = \int_{-\infty}^{\infty} S_x(\omega,t;w)\,d\omega \tag{8.22}$$

and, thanks to the normalizing condition (8.20), integrating with respect to time gives

$$E_x = \int_{-\infty}^{\infty} \mathrm{E}[X^2(t)]\,dt = \int_{-\infty}^{\infty}\int_{-\infty}^{\infty} S_x(\omega,t;w)\,d\omega\,dt \tag{8.23}$$

Thus, providing the window function is properly normalized, the volume under the surface $S_x(\omega, t; w)$ represents the average total energy in the process, independently of the shape of the window.

8.3.2 Duality, uncertainty principle

According to Equ.(8.22), the physical spectrum at t is a frequency decomposition of the local weighted average of the expected mean square in the vicinity of t. Similarly, the following dual interpretation can be established (Mark, 1970): The physical spectrum at ω is a time decomposition of the local weighted average of the energy spectrum in the vicinity of ω

$$\frac{1}{2\pi}\int_{-\infty}^{\infty} |\mathcal{W}(\omega-\nu)|^2 S_{xx}(\nu)\, d\nu = \int_{-\infty}^{\infty} S_x(\omega, t; w)\, dt \tag{8.24}$$

where the energy spectrum $S_{xx}(\nu)$ is defined by Equ.(8.7) and $\mathcal{W}(\omega)$ is the Fourier transform of the window function. Note that from Parseval's theorem, the normalization condition implies that

$$\int_{-\infty}^{\infty} |\mathcal{W}(\omega)|^2\, d\omega = 2\pi \tag{8.25}$$

The ***nominal duration*** T and the ***nominal width*** β of the window are defined respectively as

$$T = \frac{1}{w(0)}\int_{-\infty}^{\infty} |w(t)|\, dt \qquad \beta = \frac{1}{\mathcal{W}(0)}\int_{-\infty}^{\infty} |\mathcal{W}(\omega)|\, d\omega \tag{8.26}$$

According to the foregoing discussion, T is a measure of the resolution of the physical spectrum in the time domain and β in the frequency domain. It is easily shown that

$$T\beta = \frac{1}{w(0)}\int_{-\infty}^{\infty} |w(t)| dt \; \frac{1}{\mathcal{W}(0)}\int_{-\infty}^{\infty} |\mathcal{W}(\omega)| d\omega$$

$$\geq \frac{1}{w(0)}\int_{-\infty}^{\infty} w(t) dt \; \frac{1}{\mathcal{W}(0)}\int_{-\infty}^{\infty} \mathcal{W}(\omega) d\omega = 2\pi$$

Thus T and β must satify the following inequality

$$\beta T \geq 2\pi \tag{8.27}$$

which shows that the resolutions in the time and frequency domains are not independent. Once the resolution has been fixed in one domain, the resolution in the dual domain is fixed automatically. ***A good resolution in one domain can only be achieved at the expense of a poor resolution in the dual domain (uncertainty principle).***

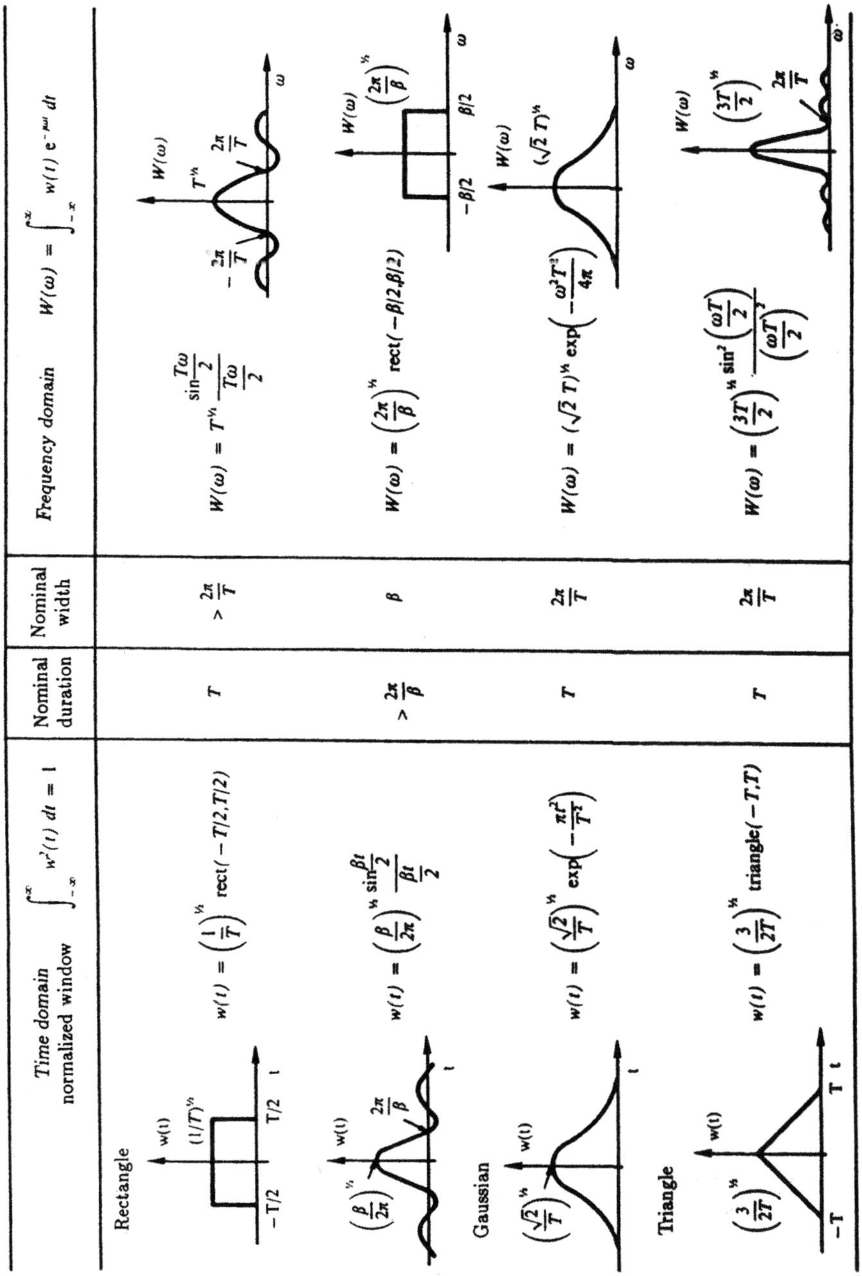

Figure 8.3: Examples of window functions.

Note that different window functions with the same nominal duration lead to different physical spectra. For example, it can be seen in Fig.8.3 that a Gaussian window decreases rapidly both in the time and frequency domains, as compared to a rectangular window; it will be more appropriate to distinguish small pulses close to larger ones in the plane (ω, t). It can be shown (Papoulis, 1962) that the Gaussian window is that minimizing the product of the second moments of $w^2(t)$ and $|\mathcal{W}(\omega)|^2$.

Unlike the instantaneous spectrum, the global energy decomposition (8.23) applies also *locally*:

$$\int_{\mathcal{D}} S_x(\omega, t; w)\, d\omega\, dt$$

represents the energy contribution to the signal from the domain $\mathcal{D}$ of the plane (ω, t). However, there is an ambiguity at the limits, because of the influence of the values outside the domain. The size of the ambiguity is of order β along the frequency axis and of order T along the time axis.

8.3.3 Relation to the *PSD* of a stationary process

By definition,

$$S_x(\omega_0, t; w) = \frac{1}{2\pi} \int\int_{-\infty}^{\infty} w(t-u_1)w(t-u_2)\mathrm{E}[X(u_1)X(u_2)]e^{-j\omega_0(u_1-u_2)}du_1\, du_2$$

Performing the change of variables $t - u_1 = \tau_1$ and $t - u_2 = \tau_2$ and assuming the process stationary, we can transform this equation into

$$S_x(\omega_0, t; w) = \frac{1}{2\pi} \int\int_{-\infty}^{\infty} w(\tau_1)w(\tau_2)R_{xx}(\tau_2 - \tau_1)e^{-j\omega_0(\tau_2-\tau_1)}d\tau_1\, d\tau_2$$

$$= \frac{1}{2\pi} \int\int_{-\infty}^{\infty} w(\tau_1)w(\tau_2) \int_{-\infty}^{\infty} \Phi_{xx}(\omega)e^{j\omega(\tau_2-\tau_1)}d\omega\, e^{-j\omega_0(\tau_2-\tau_1)}d\tau_1\, d\tau_2$$

$$= \frac{1}{2\pi} \int_{-\infty}^{\infty} \Phi_{xx}(\omega)\, d\omega \int_{-\infty}^{\infty} w(\tau_1)e^{-j(\omega-\omega_0)\tau_1}d\tau_1 \int_{-\infty}^{\infty} w(\tau_2)e^{j(\omega-\omega_0)\tau_2}d\tau_2$$

and finally,

$$S_x(\omega_0, t; w) = \frac{1}{2\pi} \int_{-\infty}^{\infty} \Phi_{xx}(\omega)\, |\mathcal{W}(\omega - \omega_0)|^2\, d\omega \tag{8.28}$$

Thus, the physical spectrum is a *weighted average of the PSD in the vicinity of* ω_0. The frequency resolution is of the order $1/T$; as the duration of the window increases, $\mathcal{W}(\omega)$ tends towards a Dirac delta function in the frequency domain and the physical spectrum tends to the *local* value of $\Phi_{xx}(\omega)$. The digital estimation of the *PSD* from sample records of finite duration is often based on Equ.(8.21). The window function is chosen to minimize the *leakage* introduced in the estimator by the convolution (8.28).

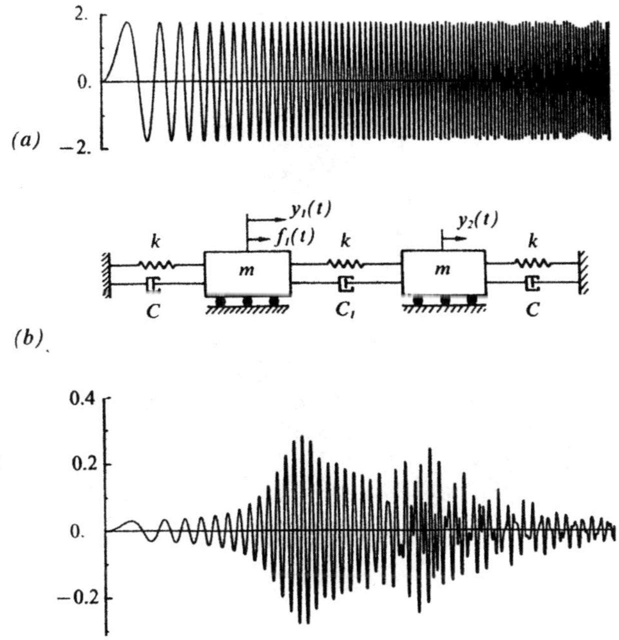

Figure 8.4: Response of a 2 d.o.f. oscillator to a sweep sine.

8.3.4 Example: Structural response to a sweep sine

Consider the 2 d.o.f. oscillator of Fig.8.4 ($\omega_1 = 2\pi$, $\omega_2 = 2\pi\sqrt{3}$; $\xi_1 = \xi_2 = 0.01$) excited by a ***sweep sine***

$$f_1(t) = \sin \frac{\varrho t^2}{2} \tag{8.29}$$

with a sweep rate $\varrho = \pi/10$, so that the instantaneous frequency (time derivative of the argument) varies linearly from 0 to 3 Hz over a period of 60 s.

Figure 8.5 shows the physical spectra computed from the time-history of the excitation and the response with a Gaussian window of effective duration equal to 5 s. The physical spectrum of the excitation appears as a surface with a large constant amplitude following a straight line in the (ω, t) plane; which is exactly what one would expect from an energy map of a sweep sine with constant sweep rate. Cross sections parallel to one of the axes have a Gaussian shape. The physical spectrum of the response shows clearly that initally the response is fairly small and occurs at the instantaneous frequency of the excitation; large amplitude oscillations are excited when the first natural frequency is reached. As the excitation moves away from ω_1, there is an exponential decay of the first mode and large oscillations of the second mode occur when the instantaneous frequency becomes in tune with ω_2. Thus, the physical spectrum appears to provide a meaningful mapping of the energy in the signal. Although the sweep sine is not a random process, it can be seen as the limit of the response of a time-

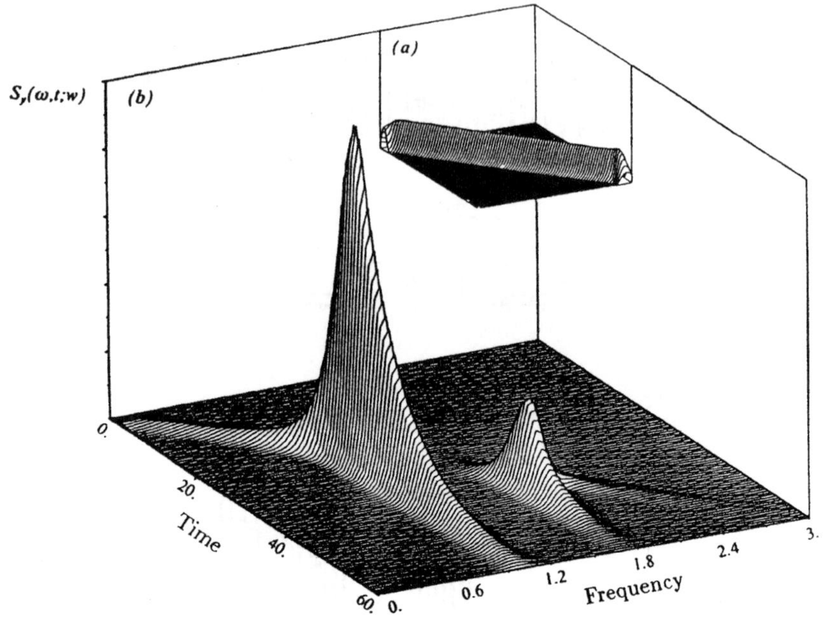

Figure 8.5: Mark's physical spectrum of the response of a 2 d.o.f. oscillator to a sweep sine. (a) Excitation. (b) Response.

varying lightly damped oscillator to a white noise excitation, or as a realization of a random process $F(t) = \sin(\varrho t^2/2 + \theta)$ where θ is a random phase.

The physical spectrum is a very convenient tool for signal analysis, but it does not provide a simple input-output relationship for linear systems, and it is fairly difficult to handle analytically. For these purposes, it is more convenient to use ***Priestley's evolutionary spectrum.***

8.4 Priestley's Evolutionary Spectrum

8.4.1 Generalized harmonic analysis

From the discussion in the previous chapters, we know that

- The Fourier transform is defined in the strict sense for signals which ***vanish at infinity.***

- By extension, the Fourier transform of a ***periodic signal*** consists of Dirac delta functions at frequencies equal to the harmonics of the signal, with amplitudes equal to the corresponding Fourier series coefficients.

- A ***stationary*** random process does not have a Fourier transform.

We now introduce a more general harmonic representation which can apply to the three types of signals:

$$X(t) = \int_{-\infty}^{\infty} e^{j\omega t} dZ(\omega) \tag{8.30}$$

where $Z(\omega)$ is a function uniquely determined by the form of $X(t)$, but which is ***not necessarily differentiable.***

- If $Z(\omega)$ is differentiable, $dZ(\omega) = \mathcal{X}(\omega)\, d\omega/2\pi$ and the harmonic representation (8.30) is identical to the Fourier transform.
- If the signal is periodic, since $dZ(\omega)$ consists of a set of Dirac delta functions, $Z(\omega)$ is a *staircase*, the steps being located at the various harmonics of the signal, with amplitudes equal to the Fourier series coeficients.
- A stationary random signal can also be represented according to (8.30), with $dZ(\omega) = O(\sqrt{d\omega})$, so that the power per unit bandwidth, i.e. the power spectral density, $|dZ(\omega)|^2/d\omega = \Phi_{xx}(\omega)$ is *finite*.

In fact, if the process $X(t)$ is *stationary*, the process $Z(\omega)$ has *orthogonal increments*, and reciprocally:

$$\mathrm{E}[dZ(\omega) dZ^\star(\omega')] = \Phi_{xx}(\omega)\, \delta(\omega - \omega')\, d\omega d\omega' \tag{8.31}$$

This can be shown as follows:

$$\mathrm{E}[X^2(t)] = \mathrm{E}[X^2(0)] = \int\int_{-\infty}^{\infty} \mathrm{E}[dZ(\omega)\, dZ^\star(\omega')]$$

$$= \int_{-\infty}^{\infty} \Phi_{xx}(\omega)\, d\omega = \int\int_{-\infty}^{\infty} \Phi_{xx}(\omega)\, \delta(\omega - \omega')\, d\omega d\omega'$$

Thus, Equ.(8.30) expresses a stationary process as a sum of harmonic components with uncorrelated amplitudes. This property is the origin of the local interpretation of the *PSD* for stationary processes: If $Z(\omega)$ and $W(\omega)$ are the generalized harmonic representations of respectively the input $X(t)$ and the output $Y(t)$ of a linear system with frequency response function $H(\omega)$, each harmonic component is amplified according to $dW(\omega) = H(\omega) dZ(\omega)$, so that

$$\mathrm{E}[dW(\omega) dW^\star(\omega')] = H(\omega) H^\star(\omega') \mathrm{E}[dZ(\omega) dZ^\star(\omega')]$$

which, taking into account Equ.(8.31), implies

$$\Phi_{yy}(\omega) = |H(\omega)|^2 \Phi_{xx}(\omega)$$

If the process is non-stationary, the representation is still applicable, but the orthogonality property (8.31) cannot be established any longer, which means that the process $Z(\omega)$ does not have orthogonal increments.

8.4.2 Evolutionary spectrum

In order to keep the local interpretation of the energy decomposition, Priestley has proposed the following non-stationary harmonic representation, which maintains the ***orthogonal*** nature of the process $Z(\omega)$:

$$X(t) = \int_{-\infty}^{\infty} a(\omega,t)e^{j\omega t}dZ(\omega) \tag{8.32}$$

Here $Z(\omega)$ is a process with independent increments and $a(\omega,t)$ represents a family of slowly-varying amplitude-modulating functions whose physical meaning is close to that of the envelope of a narrow-band process. $a(\omega,t)$ has a harmonic representation

$$a(\omega,t) = \int_{-\infty}^{\infty} e^{j\nu t}dA_\omega(\nu) \tag{8.33}$$

where $|dA_\omega(\nu)|$ presents a maximum at $\nu = 0$ (here, ω is a simple parameter). For a stationary process, Equ.(8.32) reduces itself to Equ.(8.30) with $a(\omega,t) = 1$. In general, there is an infinity of families $a(\omega,t)$ leading to the same representation (8.32); the best one is that leading to lower cut-off frequency ν_c of $|dA_\omega(\nu)|$. $T_c = 2\pi/\nu_c$ is a measure of the duration over which $a(\omega,t)$ can be regarded as approximately constant.

Equation (8.32) expresses the non-stationary process $X(t)$ as the limit sum of exponentials with slowly varying uncorrelated amplitudes $a(\omega,t)\,dZ(\omega)$. The autocorrelation function reads

$$\phi_{xx}(t_1,t_2) = \mathrm{E}[X(t_1)X^\star(t_2)] \tag{8.34}$$

$$= \int\int_{-\infty}^{\infty} a(\omega,t_1)a^\star(\omega',t_2)\, e^{j(\omega t_1-\omega' t_2)}\,\mathrm{E}[dZ(\omega)\,dZ^\star(\omega')]$$

and, taking into account Equ.(8.31),

$$\phi_{xx}(t_1,t_2) = \int_{-\infty}^{\infty} a(\omega,t_1)a^\star(\omega,t_2)e^{j\omega(t_1-t_2)}\Phi(\omega)\,d\omega \tag{8.35}$$

This relationship reduces itself to Equ.(3.50) for a stationary process. The variance is obtained by substituting $t_1 = t_2 = t$

$$\mathrm{E}[X^2(t)] = \int_{-\infty}^{\infty} |a(\omega,t)|^2\Phi_{xx}(\omega)\,d\omega \tag{8.36}$$

From this equation, the ***evolutionary spectrum*** is defined as the frequency decomposition of the variance at t:

$$S_x(\omega,t) = |a(\omega,t)|^2\Phi_{xx}(\omega) \tag{8.37}$$

It follows that

$$E[X^2(t)] = \int_{-\infty}^{\infty} S_x(\omega, t)\, d\omega \tag{8.38}$$

By definition, $S_x(\omega, t)$ is a non-negative, even function of ω. The above definition was proposed by Priestley. In fact, it is formally simpler to merge $a(\omega, t)$ and $\Phi_{xx}(\omega)$ by defining

$$\alpha(\omega, t) = a(\omega, t)\Phi_{xx}(\omega)^{1/2} \tag{8.39}$$

The non-stationary harmonic representation becomes

$$X(t) = \int_{-\infty}^{\infty} \alpha(\omega, t) e^{j\omega t} dW(\omega) \tag{8.40}$$

where the process with orthogonal increments $W(\omega)$ is such that

$$E[dW(\omega)\, dW(\omega')] = \delta(\omega - \omega')\, d\omega\, d\omega' \tag{8.41}$$

From Equ.(8.37) and (8.39),

$$S_x(\omega, t) = |\alpha(\omega, t)|^2 \tag{8.42}$$

We shall call $\alpha(\omega, t)$ the ***evolutionary amplitude*** of the scalar non-stationary process $X(t)$.

8.4.3 Vector process

The harmonic representation of non-stationary oscillatory scalar processes (8.40) can be extended to vector processes as follows:

$$X(t) = \int_{-\infty}^{\infty} \alpha(\omega, t)\, e^{j\omega t} dW(\omega) \tag{8.43}$$

where $W(\omega)$ is also a vector process, not necessarily of the same dimension as $X(t)$, with statistically independent components and orthogonal increments, in the sense

$$E[dW(\omega) dW^{\star}(\omega')] = I\delta(\omega - \omega')\, d\omega\, d\omega' \tag{8.44}$$

Here I is the identity matrix and $\star$ stands for the conjugate transposed. We shall refer to $\alpha(\omega, t)$ as the ***evolutionary amplitude matrix*** of the process. With the foregoing definition, the ***evolutionary spectral matrix*** is defined as

$$S_x(\omega, t) = \alpha(\omega, t)\alpha^{\star}(\omega, t) \tag{8.45}$$

It is the non-stationary generalization of the power spectral density matrix; it is Hermitian and positive semi-definite. It follows that given $S_x(\omega, t)$, one can always find some matrix $\alpha(\omega, t)$ such that Equ.(8.45) is satisfied. It is in general

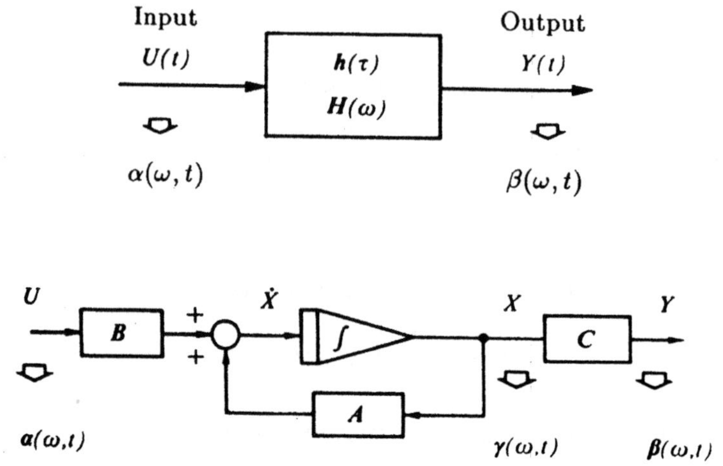

Figure 8.6: Evolutionary amplitude matrix input-output relationship for a linear time-invariant system. (a) Impulse response/transfer matrix representation. (b) State-space representation.

rectangular; the number of rows is equal to the dimension of X and the number of columns is equal to the maximum rank of $S_x(\omega, t)$. Note that there is no loss of generality in assuming that the components of dW are independent; a change of coordinates can always be performed in such a way that Equ.(8.44) is satisfied. The only restriction on the structure of $\alpha(\omega, t)$ is that it varies slowly with t.

Differentiating Equ.(8.43), one finds easily that the evolutionary amplitude matrix $\delta(\omega, t)$ of $\dot{X}(t)$

$$\dot{X}(t) = \int_{-\infty}^{\infty} \delta(\omega, t) e^{j\omega t} dW(\omega)$$

is related to $\alpha(\omega, t)$ by

$$\delta(\omega, t) = \dot{\alpha}(\omega, t) + j\omega\alpha(\omega, t) \tag{8.46}$$

8.4.4 Input-output relationship

If $h(t)$ is the impulse response matrix of a multivariate linear time-invariant system (Fig.8.6.a), the input-output relationship consists of the convolution integral

$$Y(t) = \int_{-\infty}^{\infty} h(t-\tau) U(\tau)\, d\tau \tag{8.47}$$

Substituting the spectral representation of $U(\tau)$

$$U(\tau) = \int_{-\infty}^{\infty} \alpha(\omega, \tau) e^{j\omega\tau} dW(\omega) \tag{8.48}$$

one gets

$$Y(t) = \int_{-\infty}^{\infty} \beta(\omega, t) e^{j\omega t} dW(\omega) \tag{8.49}$$

with

$$\beta(\omega, t) = \int_{-\infty}^{\infty} h(\tau) e^{-j\omega\tau} \alpha(\omega, t - \tau)\, d\tau \tag{8.50}$$

In the particular case where $\alpha(\omega, t)$ varies slowly, as compared to the memory of the system [i.e. the effective duration of $h(t)$], this equation can be approximated by

$$\beta(\omega, t) = H(\omega)\alpha(\omega, t) \tag{8.51}$$

(quasi-stationary approximation). For a second order vibrating system governed by

$$M\ddot{Y} + C\dot{Y} + KY = U \tag{8.52}$$

it is readily established, substituting Equ.(8.48) and (8.49), that the evolutionary amplitude matrix of the response satisfies the matrix differential equation

$$M\ddot{\beta} + (C + 2j\omega M)\dot{\beta} + (K + j\omega C - \omega^2 M)\beta = \alpha \tag{8.53}$$

8.4.5 State variable form

The simplest way to write the input-output relationship for linear systems is in state variable form (Fig.8.6.b). If the system is described by

$$\dot{X} = AX + BU, \qquad Y = CX \tag{8.54}$$

and if the evolutionary amplitude matrices of U, X and Y are respectively α, γ and β, substituting the spectral decompositions in (8.54), one finds that the evolutionary amplitude matrix of the state vector is governed by the matrix differential equation

$$\dot{\gamma}(\omega, t) = (A - j\omega I)\gamma(\omega, t) + B\alpha(\omega, t) \tag{8.55}$$

$$\beta(\omega, t) = C\gamma(\omega, t) \tag{8.56}$$

This equation applies for time-varying linear systems and for arbitrary $\alpha(\omega, t)$. If the excitation is a white noise, α does not depend of ω and the state vector is Markovian, as we shall see in chapter 9. For a time-invariant system, Equ.(8.55) can be solved efficiently with matrix exponentials, by noting that the eigenvalues of

$$D(\omega) = A - j\omega I \tag{8.57}$$

can be calculated directly from those of A. In fact, if λ_i and P are the eigenvalues and the eigenvectors of A, so that $P^{-1}AP = \text{diag}(\lambda_i)$, then

$$P^{-1}D(\omega)P = \text{diag}(\lambda_i - j\omega) \tag{8.58}$$

This means that the eigenvectors of $D(\omega)$ do not depend on ω and the eigenvalues are simply those of A, translated by $-j\omega$. It can be verified by substitution that the general solution of (8.55) is

$$\gamma(\omega,t) = e^{D(\omega)t}\gamma(\omega,0) + \int_0^t e^{D(\omega)(t-\tau)}B\,\alpha(\omega,\tau)\,d\tau \tag{8.59}$$

where $\gamma(\omega,0)$ is the initial condition. The matrix exponential is defined as

$$e^{Dt} = \sum_{k=0}^{\infty} D^k \frac{t^k}{k!} \tag{8.60}$$

Since P diagonalizes D, it also diagonalizes e^{Dt}:

$$e^{D(\omega)t} = e^{-j\omega t}P\text{diag}(e^{\lambda_i t})P^{-1} = e^{-j\omega t}e^{At} \tag{8.61}$$

From Equ.(8.59), the following recursive formula is readily obtained:

$$\gamma(\omega,t+\Delta t) = e^{D\Delta t}[\gamma(\omega,t) + \int_0^{\Delta t} e^{-D\tau}B\,\alpha(\omega,t+\tau)\,d\tau] \tag{8.62}$$

This equation can be used to devise approximate integration schemes, by making assumptions on the variation of the evolutionary amplitude of the excitation between successive time steps. For example, if one assumes that $\alpha(\omega,\tau)$ is constant in $(t, t+\Delta t]$.

$$\gamma(\omega,t+\Delta t) = e^{D\,\Delta t}\gamma(\omega,t) + D^{-1}(e^{D\,\Delta t} - I)B\,\alpha(\omega,t) \tag{8.63}$$

8.4.6 Remarks

The physical spectrum can be expressed as a weighted average of the evolutionary spectrum; the weighting function depends on the window used. The algebra to show this is rather lenghty and is left as an exercise (Problem P.8.2).

In the differential equation (8.55), ω acts as a parameter: The integration is made independently for each frequency. This excludes any energy transfer between frequencies.

8.5 Applications

8.5.1 Structural response to a sweep sine

This problem has already been considered in section 8.3.4, where we computed the physical spectra of the excitation and the response of a 2 d.o.f. system. Since

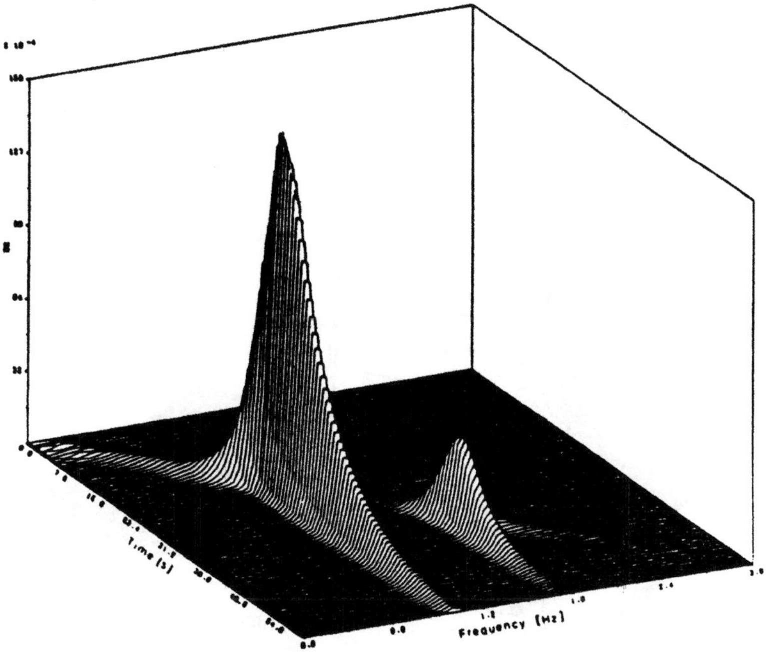

Figure 8.7: Evolutionary spectrum of the displacement y_1, in response to the excitation spectrum of Fig.8.5.a.

the physical spectrum is a weighted average of the evolutionary spectrum, it is interesting to investigate the analytical prediction of the evolutionary spectrum of the response when the spectrum of Fig.8.5.a is used as evolutionary spectrum of the input. Figure 8.7 shows the predicted response of y_1, obtained by numerical integration of Equ.(8.55). Comparing with Fig.8.5, one sees that the computed evolutionary spectrum is very similar to the physical spectrum computed from the time-history of the response. This confirms the physical meaning of the evolutionary spectrum.

8.5.2 Transient response of an oscillator

Figure 8.8 shows the analytical prediction of the evolutionary spectrum of the transient response of a single d.o.f. oscillator starting from rest, to a white noise excitation of limited duration (10 s). The damping ratio is $\xi = 0.05$. This example is useful to test numerical techniques, because the analytical solution is available (Problem 8.3).

8.5.3 Earthquake records

The non-stationary behaviour of earthquake records is known to affect both the amplitude and the frequency content of the time-history; there are more high

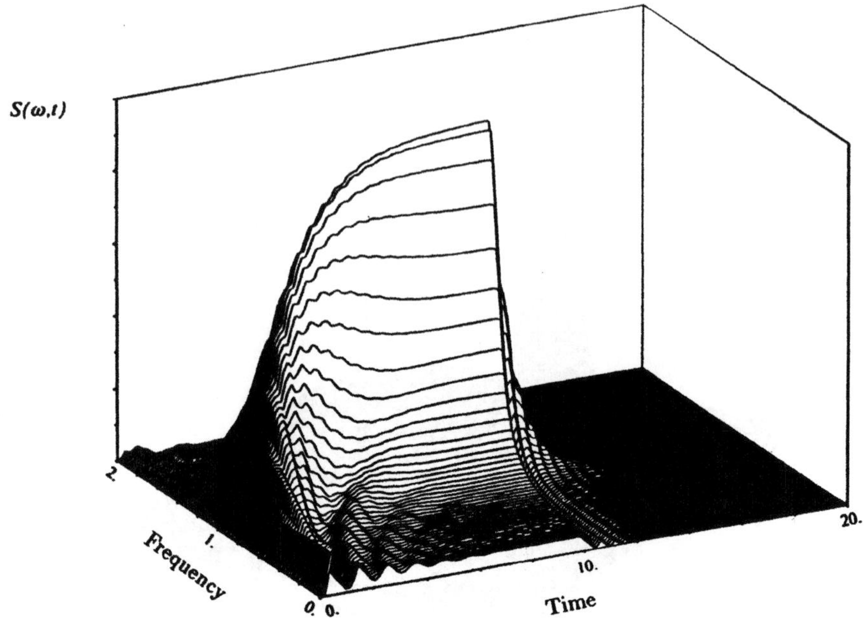

Figure 8.8: Evolutionary spectrum of the transient response of a single d.o.f. oscillator to a finite duration white noise excitation.

frequency components at the beginning than at the end of the record. These features can be represented by the following evolutionary spectrum

$$S_x(\omega,t) = \psi(\omega)\, t^b e^{-a(\omega)t} \tag{8.64}$$

where b and $a(\omega)$ take care of the transient behaviour in the time and frequency domains, and $\psi(\omega)$ is a stationary *PSD* providing an adequate energy distribution over ω. Examples of simulated records corresponding to various values of b and $a(\omega)$ are shown in Fig.8.9.

8.6 Summary

Various spectral descriptions of non-stationary oscillatory processes have been reviewed. They aim at supplying a mapping of the energy of the signal in the (frequency-time) plane. Priestley's ***evolutionary spectrum*** appears as the most appropriate ***analytical*** tool; it enjoys a simple input-output relationship for linear systems. Mark's ***physical spectrum*** is convenient for ***estimation***; it is a local weighted average of the evolutionary spectrum of which it can be used as an estimator (the weighting function is related to the window function used in the definition of the physical spectrum). The resolution in the time and frequency domains are not independent; they are related by the ***uncertainty principle***.

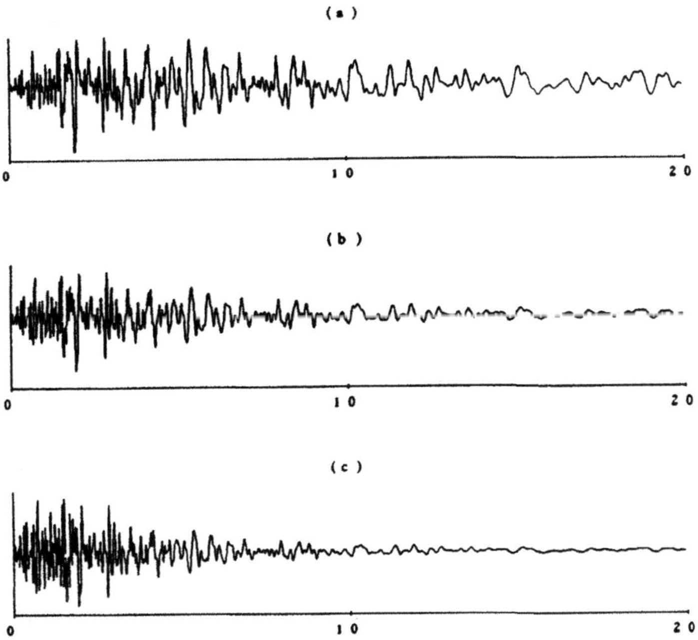

Figure 8.9: Simulated earthquake records for various values of b and $a(\omega)$.

8.7 References

J.S.BENDAT & A.G.PIERSOL, *Random Data: Analysis and Measurement Procedures*, Wiley-Interscience, 1966 & 1971.

J.K.HAMMOND, On the response of single and multidegree of freedom systems to non-stationary random excitations, *Journal of Sound and Vibration 7(3), pp.393-416, 1968.*

R.M.LOYNES, On the concept of the spectrum for non-stationary processes, *J. Roy. Stat. Soc., Series B,30(1), pp.1-30, 1968.*

W.D.MARK, Spectral analysis of the convolution and filtering of non-stationary stochastic processes, *Journal of Sound and Vibration 11(1), pp.19-63, 1970.*

A.PAPOULIS, *The Fourier Integral and its Applications*, McGraw-Hill, 1962.

M.B.PRIESTLEY, Evolutionary spectra and non-stationary processes, *J. Roy. Stat. Soc., Series B,27(2), pp.204-237, 1965.*

M.B.PRIESTLEY, Power spectral analysis of random processes, *Journal of Sound and Vibration 6(1), pp.86-97, 1967.*

S.SHIHAB & A.PREUMONT, Non-stationary random vibrations of linear multi-degree-of freedom systems, *Journal of Sound and Vibration 132(3), pp.457-471, 1989.*

8.8 Problems

P.8.1 *Assume that the evolutionary amplitude matrix $\alpha(\omega,t)$ varies linearly between t and $t+\Delta t$. Show that Equ.(8.62) leads to the numerical integration scheme*

$$\gamma(\omega,t+\Delta t) = e^{D\,\Delta t}\gamma(\omega,t) + D^{-1}[e^{D\,\Delta t}B\alpha(\omega,t) - B\alpha(\omega,t+\Delta t)]$$

$$+(1/\Delta t)D^{-2}(I - e^{D\,\Delta t})B[\alpha(\omega,t) - \alpha(\omega,t+\Delta t)]$$

P.8.2 *Show that the physical spectrum can be expressed as a weighted average of the evolutionary spectrum*

$$S_x(\omega_0,t;w) = \int_{-\infty}^{\infty} S_x(\omega,t)|\Gamma(\omega,t)|^2 d\omega$$

where $\Gamma(\omega,t)$ is the generalized transfer function

$$\Gamma(\omega,t) = \int_{-\infty}^{\infty} h(\tau)\, e^{-j\omega\tau}\frac{\alpha(\omega,t-\tau)}{\alpha(\omega,t)}d\tau$$

and $h(\tau)$ is the impulse response of the narrow-band filter

$$h(\tau) = \frac{1}{\sqrt{2\pi}}w(\tau)e^{j\omega_0\tau}$$

based on the window $w(\tau)$ used in the definition of the physical spectrum.

P.8.3 *Consider a single d.o.f. oscillator starting from rest at $t=0$ and excited with a constant amplitude in time*

$$\alpha_1(\omega,t) = \phi(\omega) \qquad (t \geq 0)$$

Show that the evolutionary amplitude of the response is given by

$$\beta_1(\omega,t) = \phi(\omega)\frac{H(\omega)}{\omega_d}\{\omega_d - e^{-(\xi\omega_n+j\omega)t}[\omega_n \sin(\omega_d t+\theta) + j\omega \sin(\omega_d t)]\}$$

with $\omega_d = \omega_n\sqrt{1-\xi^2}$, $\tan\theta = \sqrt{1-\xi^2}/\xi$ and $H(\omega)$ is the frequency response function of the oscillator.

P.8.4 *Same as above with*

$$\alpha_2(\omega,t) = \phi(\omega)\, e^{-a(\omega)t} \qquad (t \geq 0)$$

Show that the response is

$$\beta_2(\omega,t) = \phi(\omega)\frac{H(\omega+ja)}{\omega_d}\{\omega_d e^{-at} - e^{-(\xi\omega_n+j\omega)t}[\omega_n \sin(\omega_d t+\theta) + (j\omega - a)\sin(\omega_d t)]\}$$

P.8.5 *Since the relationship between $\alpha(\omega,t)$ and $\beta(\omega,t)$ is linear, use the result of the previous problem to evaluate the evolutionary amplitude for the following excitations (both can be used to model earthquake records)*

$$\alpha_3(\omega,t) = \phi(\omega)[e^{-a(\omega)t} - e^{-b(\omega)t}] \qquad b(\omega) > a(\omega) \geq 0$$

$$\alpha_4(\omega,t) = \phi(\omega)\, t\, e^{-a(\omega)t}$$

[Hint: Since $\alpha_4(\omega,t) = -\partial\alpha_2(\omega,t)/\partial a$, $\beta_4(\omega,t) = -\partial\beta_2(\omega,t)/\partial a$.]

Chapter 9

Markov Process

9.1 Conditional probability

By definition, the conditional probability density function is such that

$$p_n(x_n, t_n | x_{n-1}, t_{n-1}; ...; x_1, t_1)\, dx_n$$

represents the probability that the value of the random process $X(t)$ at t_n belongs to the interval $(x_n, x_n + dx_n]$, knowing that at the previous times $t_1 < t_2 < ... < t_{n-1}$ its values were respectively x_1, x_2, ..., x_{n-1}. By definition,

$$p_2(x_1, t_1; x_2, t_2) = p_1(x_1, t_1) p_2(x_2, t_2 | x_1, t_1)$$

$$p_3(x_1, t_1; x_2, t_2; x_3, t_3) = p_2(x_1, t_1; x_2, t_2) p_3(x_3, t_3 | x_2, t_2; x_1, t_1) \tag{9.1}$$

etc. The conditional probability density satisfies the following conditions

$$p_n(x_n, t_n | x_{n-1}, t_{n-1}; ...; x_1, t_1) \geq 0 \tag{9.2}$$

$$\int_{-\infty}^{\infty} p_n(x_n, t_n | x_{n-1}, t_{n-1}; ...; x_1, t_1)\, dx_n = 1 \tag{9.3}$$

$$\int_{-\infty}^{\infty} \cdots \int_{-\infty}^{\infty} p_n(x_n, t_n | x_{n-1}, t_{n-1}; ...; x_1, t_1)$$

$$\cdot p_{n-1}(x_1, t_1; ...; x_{n-1}, t_{n-1})\, dx_1 ... dx_{n-1} = p_1(x_n, t_n) \tag{9.4}$$

The conditional density of the second order, $p_2(x_2, t_2 | x_1, t_1)$ plays a key role in the theory of Markov processes; it is called ***transition probability density***; it is often denoted $q(x_2, t_2 | x_1, t_1)$ and satisfies

$$\lim_{t_2 \to t_1} q(x_2, t_2 | x_1, t_1) = \delta(x_2 - x_1) \tag{9.5}$$

which simply expresses the fact that when $t_2 \to t_1$, $X(t_2) = x_1$ with probability 1. Conversely, the statistical dependence between the values of a random process vanishes when the time separation goes to infinity:

$$\lim_{t_2 - t_1 \to \infty} q(x_2, t_2 | x_1, t_1) = p_1(x_2, t_2) \tag{9.6}$$

This equation states that the condition on the value at t_1 ceases to affect the value at t_2 when the time separation becomes large.

9.2 Classification of random processes

As we discussed in chapter 3, the complete specification of a random process requires the probability density functions of all orders, $n = 1, 2, 3, \ldots$. In fact, the probability density function of order n contains all the information contained in the probability density functions of lower orders. $p_m(\,)$ can always be recovered from $p_n(\,)$ $(n > m)$ by partial integration on the variables which do not appear in p_m.

Since the joint probability density functions of increasing orders offer an increasingly complete description of a random process, one possible classification of random processes is according to the order n necessary to characterize the process completely. The simplest class is that entirely defined by the probability density of order $n = 1$; such processes are called *purely random* or *without memory*. The transition probability density of a purely random process is identical to the first order density:

$$q(x_2, t_2 | x_1, t_1) = p_1(x_2, t_2) \tag{9.7}$$

It follows that the joint probability density functions can be factorized into that of the first order

$$p_2(x_1, t_1; x_2, t_2) = p_1(x_1, t_1) . p_1(x_2, t_2) \tag{9.8}$$

and so on for any order. The first order density function describes the process completely. It is readily observed that the values for different times of a purely random process are uncorrelated. Such a process can only be an idealization, because when the time interval decreases, the values of all physical processes become correlated, as expressed by Equ.(9.5). A white noise is an example of a process without memory.

Next in the classification based on the joint density functions are the processes with *one step memory* or *Markov processes*. A Markov process is such that of the values of the process at $n - 1$ previous times $t_1 < \ldots < t_{n-1}$, only the latest, i.e. the most recent one at t_{n-1} influences the future values of the process at $t_n > t_{n-1}$:

$$p_n(x_n, t_n | x_{n-1}, t_{n-1}; \ldots; x_1, t_1) = q(x_n, t_n | x_{n-1}, t_{n-1}) \tag{9.9}$$

This ensures the factorization of the joint probability density functions:

$$p_n(x_1,t_1;...;x_n,t_n) = q(x_n,t_n|x_{n-1},t_{n-1})...q(x_2,t_2|x_1,t_1).p_1(x_1,t_1) \quad (9.10)$$

Since both the first order density and the transition probability density functions can be derived from the second order joint probability density function, a Markov process can be regarded as a process *entirely specified by the second order joint probability density function.*

One could continue the classification, considering processes specified by the third order joint density, etc..., but this is not useful in practice, because most non-Markovian processes involved in practical applications can be considered as a *component of an appropriate vector Markov process.* In particular, we shall see that the *state vector* of a system excited by a purely random process is Markovian. This is true whether the system is linear or not.

9.3 Smoluchowski equation

Equation (9.9) can be regarded as the definition of a Markov process. The factorization of the higher order joint density functions into a product of the transition probability densities is a direct consequence of the definition. In order to represent a Markov process, the transition probability density cannot be an arbitrary function of its arguments. In addition to the classical condition recalled earlier in this chapter, it must satisfy a compatibility condition known as the *Smoluchowski equation* (also called the *Chapman-Kolmogorov equation*) which reads

$$q(x_2,t_2|x_1,t_1) = \int q(x,t|x_1,t_1)q(x_2,t_2|x,t)\,dx \qquad t_1 \le t \le t_2 \quad (9.11)$$

This equation is obtained by partitioning the transition from x_1 to x_2 into x_1 to x and x to x_2 and taking into account the definition (9.9). For a process with stationary increments, the transition probability depends only on the time difference, $q(x_2,t_2|x_1,t_1) = q(x_2,t_2-t_1|x_1)$ and the Smoluchowski equation reads

$$q(x_2,t_2|x_1) = \int q(x,t|x_1)q(x_2,t_2-t|x)\,dx \qquad 0 \le t \le t_2 \quad (9.12)$$

9.4 Process with independent increments

Clearly, the central issue in Markov processes is the statistical independence between disjoint time intervals: the influence of former times on later values of the process is restricted to the *latest* value available.

Let Y_i be mutually independent random variables, and x_0 be an initial state known with probability 1. Consider the sequence

$$X_m = x_0 + \sum_{i=1}^{m} Y_i \tag{9.13}$$

By construction, the differences $X_2 - X_1 = Y_2, \ldots, X_n - X_{n-1} = Y_n$ are statistically independent. This is also true for any arbitrary non overlapping intervals. Such a process is called ***a process with independent increments.*** The Poisson process is an example of a counting process with independent increments. By construction, a process with independent increments is Markovian; the joint probability densities can be factorized as in Equ.(9.10).

9.4.1 Random Walk

Consider the repeated tossing of a fair coin and assume that when the outcome is ***head***, one player wins $1 while when the outcome is ***tail***, he loses it. Clearly, if one associates a discrete random variable W with the outcome of the experiment, which is such that $W(head) = +1$ and $W(tail) = -1$, the total gain after k tossings of the player betting on ***head*** is given by

$$X_k = X_{k-1} + W_k \qquad (x_0 = 0) \tag{9.14}$$

The random variable W is such that

$$\mathrm{E}[W] = \sum w_i p_i = 0 \qquad \mathrm{E}[W^2] = \sum w_i^2 p_i = 1$$

The sequence of the values W_k for different tossings is purely random.

The gain X_k is, by construction, a process with independent increments and therefore Markovian; It is nonstationary and, from Equ.(2.61),

$$\mathrm{E}[X_k] = 0 \qquad \mathrm{E}[X_k^2] = k \tag{9.15}$$

A few samples are represented in Fig.9.1. X_k is distributed according to the binomial distribution; the probability that one of the players wins $\$\nu$ after k tossings is

$$P[X_n = \nu] = \frac{k!}{(\frac{k+\nu}{2})!(\frac{k-\nu}{2})!} \cdot (\frac{1}{2})^k \tag{9.16}$$

if $|\nu| < k$ and $k + \nu$ even ($P[X_n = \nu] = 0$ otherwise). The proof of this is based on the same argument as in section 4.2.2; it is left as an exercise (Problem P.9.1). The distribution becomes Gaussian at the limit, as the number of tossings $k \to \infty$.

9.4.2 Wiener process

The ***Wiener process*** is the continuous generalization of the random walk:

$$\dot{X} = F(t) \qquad (x_0 = 0) \tag{9.17}$$

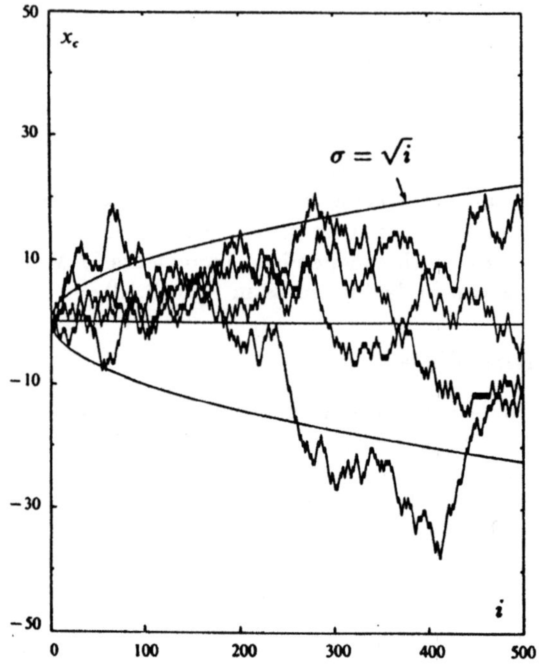

Figure 9.1: Samples of Random Walk.

where $F(t)$ is a Gaussian white noise of zero mean and intensity D

$$E[F(t_1)F(t_2)] = D\delta(t_1 - t_2) \tag{9.18}$$

The increments

$$Y_1 = \int_{t_1}^{t_2} F(t)\,dt \qquad Y_2 = \int_{t_3}^{t_4} F(t)\,dt$$

corresponding to disjoint intervals are orthogonal:

$$E[Y_1 Y_2] = \int_{t_1}^{t_2}\int_{t_3}^{t_4} E[F(t)F(t')]dtdt' = D\int_{t_1}^{t_2} dt \int_{t_3}^{t_4} \delta(t - t')dt' = 0$$

because the intervals are disjoint. Since $F(t)$ is Gaussian, so is Y and, as discussed in section 4.3.1, the increments are indeed independent. The joint distribution of $X(t)$ can be factorized, and the process is Markovian. Using $t_3 = t_1$ and $t_4 = t_2$, we may easily establish from the previous equation that

$$E[Y^2] = E\{[X(t_2) - X(t_1)]^2\} = D(t_2 - t_1) \tag{9.19}$$

A process in which the distribution of the increment depends only on the time difference is called ***a process with stationary increment.***

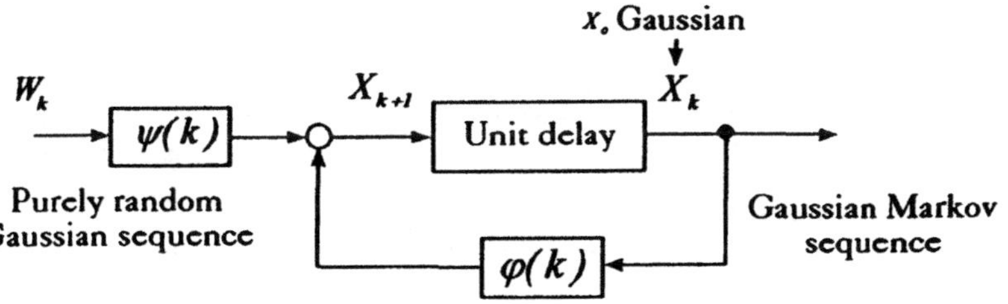

Figure 9.2: Construction of a Markov sequence from a purely random sequence.

9.5 Markov process and state variables

The Markovian property of the Random Walk defined by Equ.(9.14) can be readily extended to the class of processes defined by the following recursive formula (Fig.9.2):

$$X_k = \phi(k)X_{k-1} + \psi(k)W_k \tag{9.20}$$

where $\phi(k)$ and $\psi(k)$ are sequences of known numbers. If W_k is purely random, X_k is Markovian. Since a linear transformation preserves the Gaussian property, if W_k and the initial state are Gaussian, so is X_k. A further generalization consists of considering the same equation in vector form:

$$X_k = \Phi(k)X_{k-1} + \Psi(k)W_k \tag{9.21}$$

where $\Phi(k)$ and $\Psi(k)$ are matrices of known coefficients. If W_k is a purely random vector process, X_k is a vector Markov process. People familiar with the use of state variables know that the foregoing representation is quite general and includes the response of any discrete system governed by a finite difference equation excited by a purely random process (Problem P.9.2).

Similarly, in the continuous case, any vector process described by a system of first order linear differential equations

$$\dot{X} = A(t)X + B(t)W \tag{9.22}$$

excited by a purely random process W is Markovian.

Rather than establishing this formally, let us try to relate the physical meaning of the state variables with that of a Markov process: If one knows the state x_0 of the system at some time t_0, the state of the system at a later time $t > t_0$ depends only on x_0 and the excitation during $[t_0, t]$. This means that the state at t does not depend on the state before t_0. On the other hand, the purely random excitation during $[t_0, t]$ is, of course, statistically independent from that occuring before t_0. Thus, the knowledge of the current state at t_0 brings

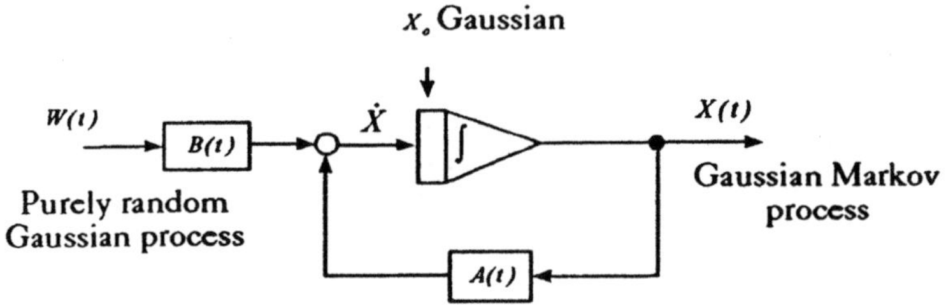

Figure 9.3: State vector representation of a linear system excited by a purely random process.

a *complete decoupling between the future* ($t > t_0$) *and the past* ($t < t_0$). This is exactly the Markov property.

What we have said here applies also to nonlinear systems under purely random excitation. Since a differential equation of order n (linear or not) can always be rewritten as a system of n first order differential equations, the corresponding state vector is Markovian and the response of the system is the projection (one component) of the vector Markov process. *Any system of differential equations of finite order excited by a purely random process can always, by defining an appropriate state vector, be converted into a vector Markov process.*

Any physical excitation has a finite correlation time; *the white noise (purely random) approximation is appropriate if the correlation time of the excitation,* τ_{cor}*, is small compared to the time constant of the system to which it is applied.* The correlation time of the excitation can be defined as

$$\tau_{cor} = \frac{\int_{-\infty}^{\infty} |\tau| R_{ww}(\tau)\, d\tau}{\int_{-\infty}^{\infty} R_{ww}(\tau)\, d\tau} \tag{9.23}$$

Often, the white noise approximation is not directly applicable, because the foregoing condition is not satisfied. The excitation is said to be *colored.* In that case, the excitation process can be modelled in such a way that the actual input to the system is the output of a filter excited by a white noise (Fig.9.4). If one couples the system with the filter, the augmented sytem is excited by a white noise and its state vector is Markovian.

Figure 9.4: Augmented system for a colored excitation.

9.6 Gaussian Markov process

9.6.1 Covariance matrix

Once again, consider the linear state space equation (9.22) excited by a Gaussian white noise vector process such that

$$E[W(t)] = v(t)$$

$$\mathrm{Cov}[W(t)] = E\{[W(t) - v(t)][W(\tau) - v(\tau)]^T\} = \nu(t)\delta(t - \tau) \tag{9.24}$$

The initial state is assumed Gaussian with mean $E[X(t_0)] = \mu_0$ and covariance matrix $E\{[X(0) - \mu_0][X(0) - \mu_0]^T\} = \sigma_0$. The state vector $X(t)$ is Markovian. Since the system is linear, the response is also Gaussian; it is entirely characterized by the mean and autocorrelation matrix. Taking the expectation of Equ.(9.22), we see that the mean satisfies the same differential equation as the system:

$$\dot{\mu}_x = A(t)\mu_x + B(t)v(t) \tag{9.25}$$

Next, consider the covariance matrix at t, $\sigma(t) = \kappa(t,t)$. After some lengthy algebra, it can be shown that it satisfies the matrix differential equation

$$\dot{\sigma} = A(t)\sigma + \sigma A(t)^T + B(t)\nu(t)B^T(t) \tag{9.26}$$

with the initial condition $\sigma(0) = \sigma_0$. The development of this equation can be found in the classical control literature (e.g. Bryson & Ho, 1975). If the system is stable and time-invariant, and the excitation is stationary, the covariance matrix tends to a steady state value which is governed by the *Lyapunov equation*:

$$A\sigma + \sigma A^T + B\nu B^T = 0 \tag{9.27}$$

It can also be shown that the covariance matrix for different times is given by

$$\kappa(t + \tau, t) = \Phi(t + \tau, t)\sigma(t) \qquad \tau \geq 0 \tag{9.28}$$

where $\Phi(t, t_0)$ is the transition matrix of the system, governing its free response at t from initial conditions at t_0 according to

$$x(t) = \Phi(t, t_0)x_0$$

Substituting this equation into the differential equation of the system, one sees that the transition matrix satisfies

$$\frac{d}{dt}\Phi(t, t_0) = A(t)\Phi(t, t_0) \qquad \Phi(t_0, t_0) = I \tag{9.29}$$

For a time-invariant system, the transition matrix depends only on the difference of its arguments

$$\Phi(t + \tau, t) = \Phi(\tau) = \exp(A\tau) \qquad \tau \geq 0 \tag{9.30}$$

Substituting this into Equ.(9.28), one gets

$$\kappa(\tau) = \exp(A\tau)\,\sigma \qquad \tau \geq 0 \tag{9.31}$$

and, for negative times, one gets similarly

$$\kappa(\tau) = \sigma\, \exp(-A^T\tau) \qquad \tau < 0 \tag{9.32}$$

Equations (9.31) and (9.32) are the general form of the covariance matrix of a stationary Gaussian vector Markov process. For a scalar process of zero mean,

$$R_{xx}(\tau) = \sigma_x^2 e^{-\beta|\tau|} \tag{9.33}$$

where β is a positive constant; this is the process with exponential correlation that we have already met in section 3.8.3. The autocorrelation function and the corresponding *PSD* are illustrated in Fig.3.4. From the previous discussion, one easily observes that this process is in fact the output of the filter

$$\dot{X} = -\beta X + W(t) \tag{9.34}$$

excited by a white noise of zero mean and autocorrelation $R_{ww}(\tau) = 2\beta\sigma_x^2\delta(\tau)$.

If we consider the transient response of the foregoing system from a known initial state $X(0) = x_0$, we can easily check that the solutions of Equ.(9.25) and (9.26) are respectively

$$\mu_x(t) = x_0 e^{-\beta t} = x_0 \varrho$$

and

$$\sigma^2(t) = \sigma_x^2[1 - e^{-2\beta t}] = \sigma_x^2(1 - \varrho^2) \tag{9.35}$$

with the notation $\varrho = \exp(-\beta t)$. If $W(t)$ is Gaussian, so is $X(t)$ and its first order probability density function reads

$$p_x(x, t) = \frac{1}{\sigma_x\sqrt{2\pi(1 - \varrho^2)}} \exp[-\frac{(x - x_0\varrho)^2}{2\sigma_x^2(1 - \varrho^2)}] \tag{9.36}$$

Since $p_x(x,t)$ is the probability density function with the initial condition $X(0) = x_0$, it is indeed the conditional probability at t under the condition x_0 at 0. This is, by definition, the ***transition probability*** $q_x(x,t|x_0,0)$. Knowing that $X(t)$ is Markovian, we can use it to construct the joint probability density functions of any order according to Equ.(9.10).

Note that a form similar to (9.36) applies for more general excitations, with the appropriate expressions for the conditional mean and covariance. In these cases, however, the factorization of higher order density functions does not apply, because the process is no longer Markovian.

9.6.2 Wide sense Markov process

If one defines the ***normalized correlation matrix*** as

$$\varrho(\tau) = \kappa(\tau)\sigma^{-1} \tag{9.37}$$

one observes from (9.31) that it can be factorized in the following way

$$\varrho(t_3 - t_1) = \varrho(t_3 - t_2)\varrho(t_2 - t_1) \qquad t_3 \geq t_2 \geq t_1 \tag{9.38}$$

This is a *necessary and sufficient condition for a Gaussian stationary process to be Markovian (Doob's theorem).* A weakly stationary process which satisfies Equ.(9.38) is called *Markovian in the wide sense.* If it is Gaussian, it is also Markovian in the strict sense. Note, however, that being Markovian in the strict sense does not imply that the process is Markovian in the wide sense if it is not Gaussian.

9.6.3 Power spectral density matrix

The general form of the *PSD* matrix of the state vector of the Gaussian Markov process can be obtained by Fourier transforming Equ.(9.31). However, it is more convenient to use the differential equation to determine the transfer matrix between the excitation and the state vector and apply the general equation (6.80). Upon Fourier transforming Equ.(9.22), one easily gets

$$(j\omega I - A)X = BW \quad \text{or} \quad X = (j\omega I - A)^{-1}BW \tag{9.39}$$

The general form of the transfer matrix of a linear time-invariant system in state variable form is therefore

$$H(j\omega) = (j\omega I - A)^{-1}B \tag{9.40}$$

It is in general rectangular, with a number of rows equal to the dimension of the state vector and a number of columns equal to the size of the excitation vector. If the excitation is a white noise of zero mean and intensity matrix ν, then

$$R_{ww}(\tau) = \nu\delta(\tau) \qquad \Phi_{ww}(\omega) = \frac{\nu}{2\pi}$$

Applying Equ.(6.80), one gets the general form of the *PSD* matrix

$$\Phi_{xx}(\omega) = \frac{1}{2\pi} H(j\omega) B \nu B^T H^\star(j\omega) \tag{9.41}$$

where $H^\star$ stands for the Hermitian (conjugate transpose) of H. This form of the *PSD* matrix implies that the *PSD* function of any quantity which depends linearly on the state vector is a ***rational function of*** ω^2 (the numerator and the denominator are polynomials in ω^2). Conversely, any Gaussian random process whose *PSD* is a rational function of ω^2 can be seen as the projection of a Gaussian Markov vector process. A consequence of the foregoing discussion is that, since an arbitrary *PSD* can be approximated by a rational function with an arbitrarily small error, a Gaussian but non Markovian process can always be approximated by the projection of a vector Markov process. The quality of the approximation improves with the size of the vector process.

To illustrate this, consider the process defined by

$$R_{ff}(\tau) = \sigma^2 e^{-\beta^2\tau^2/4} \qquad \Phi_{ff}(\omega) = \frac{\sigma^2}{\beta\sqrt{\pi}} e^{-\omega^2/\beta^2} \tag{9.42}$$

From the second of the above equations,

$$e^{\omega^2/\beta^2} \Phi_{ff}(\omega) = \frac{\sigma^2}{\beta\sqrt{\pi}} \tag{9.43}$$

Expanding the exponential with a limited number of terms, we replace the non Markovian process $F(t)$ by $F_n(t)$ such that

$$[1 + (\frac{\omega}{\beta})^2 + \ldots + \frac{1}{n!}(\frac{\omega}{\beta})^{2n}]\Phi_n(\omega) = \frac{\sigma^2}{\beta\sqrt{\pi}} \tag{9.44}$$

The *PSD* $\Phi_n(\omega)$ is a rational function of ω^2; therefore, $F_n(t)$ is the projection of a vector Markov process. In fact, the polynomial in the left hand side of Equ.(9.44) can be factorized into a product

$$P_n(j\omega).P_n(-j\omega)$$

where $P_n(\,)$ is a polynomial of degree n. As a result, the approximation of order n can be seen as the response of the system governed by the nth order differential equation

$$P_n(\frac{d}{dt})F_n = W \tag{9.45}$$

where $W(t)$ is a white noise of *PSD* $\Phi_{ww}(\omega) = \sigma^2/\beta\sqrt{\pi}$. For $n = 1$, $|P_1(j\omega)|^2 = 1 + \omega^2/\beta^2$, which corresponds to the differential equation

$$\frac{1}{\beta}\dot{F}_1 + F_1 = W(t)$$

$F_1(t)$ is the first order Markovian approximation of the process $F(t)$. Higher order approximations are left as exercises (Problem P.9.7).

9.7 Random walk and diffusion equation

Before developing the *Fokker-Planck equation* which governs the diffusion of the transition probabilty density of a Markov process, we consider a couple of examples of Random Walk. The difference equation governing the conservation of probability is written and transformed into a diffusion equation by taking the the limit.

Here, we consider a discrete Markov sequence taking the discrete values $n\Delta$ at discrete times $s\tau$. If one denotes by $P_2(n,s|m) = P_2(n\Delta, s\tau|m\Delta)$ the conditional probability that the process takes the value $x = n\Delta$ at the time $t = s\tau$ knowing that $x = m\Delta$ at $t = 0$, one finds the stationary Smoluchowski equation

$$P_2(n, s+1|m) = \sum_k P_2(k,s|m)Q(n|k) \tag{9.46}$$

where $Q(n|k) = P_2(n,1|k)$ is the transition probability over one time step. $Q(n|k)$ depends on the physical mechanism under consideration; it satisfies the identity

$$\sum_n Q(n|k) = Q(k|k) + \sum_{n\neq k} Q(n|k) = 1$$

If one splits Equ.(9.46) into its contributions relative to $k = n$ and $k \neq n$ and uses the above relationship, one gets

$$P_2(n,s+1|m) - P_2(n,s|m) = \sum_{k\neq n} P_2(k,s|m)Q(n|k) - P_2(n,s|m)\sum_{k\neq n} Q(k|n) \tag{9.47}$$

This relationship expresses the balance of probability at s. The change of the conditional probability $P_2(n,s|m)$ between s and $s+1$ is equal to the difference of two terms: the first represents the probabilty of arriving at n at time $s+1$ from all the states $k \neq n$ at time s; the second is the probabilty of leaving the state n at time s for any state $k \neq n$. The initial condition for this equation is $P_2(n,0|m) = \delta(n,m)$, where $\delta(n,m) = \delta_{nm}$ is the Kronecker delta index. The transition probability depends on the physical mechanism; two examples are analysed below.

9.7.1 Random walk of a free particle

This is the simplest case where at any discrete time $s\tau$, a particle can move one step Δ either to the left or to the right, with equal probability. The transition probability reads

$$Q(m|k) = \frac{1}{2}\delta(m, k-1) + \frac{1}{2}\delta(m, k+1) \tag{9.48}$$

This form states that leaving the state k, in one step, the process can only go either to the state $k-1$ or $k+1$, each one with a probability $1/2$. Introducing this into Equ.(9.47), one gets the following difference equation

$$P_2(n, s+1|m) = \frac{1}{2}P_2(n+1, s|m) + \frac{1}{2}P_2(n-1, s|m) \tag{9.49}$$

with the initial condition $P_2(n, 0|m) = \delta(n, m)$. The solution has already been given in section 9.4.1; $P_2(n, s|m)$ is equal to the probability that, after s tossings, the gain of one player is $n - m$. It follows the binomial distribution (9.16), with $k = s$ and $\nu = n - m$.

If we substract $P_2(n, s|m)$ from both sides of Equ.(9.49), we can rewrite it (in full)

$$\frac{P_2[n\Delta, (s+1)\tau|m\Delta] - P_2[n\Delta, s\tau|m\Delta]}{\tau} =$$

$$\frac{\Delta^2}{2\tau} \cdot \frac{P_2[(n+1)\Delta, s\tau|m\Delta] - 2P_2[n\Delta, s\tau|m\Delta] + P_2[(n-1)\Delta, s\tau|m\Delta]}{\Delta^2}$$

One recognizes the finite difference discretization of the partial differential equation involving a first order time derivative (forward difference) and a second order space derivative (central difference). With the notations $t = s\tau$, $x = n\Delta$, $x_0 = m\Delta$, if one takes the limit as $\tau \to 0$ and $\Delta \to 0$ in such a way that

$$D = \lim_{\Delta,\tau \to 0} \frac{\Delta^2}{2\tau}$$

be finite, one gets the one dimensional diffusion equation

$$\frac{\partial p_2}{\partial t} = D\frac{\partial^2 p_2}{\partial x^2} \tag{9.50}$$

The initial condition is $p_2(x, 0|x_0) = \delta(x - x_0)$.

9.7.2 Random walk of an elastically bound particle

Next, consider the case where the particle is not free to move, but is subject to an elastic restoring force proportional to the distance to the equilibrium position $m = 0$. More specifically, assume that when the particle is at k (Fig.9.5), the probability that it moves towards $k+1$ and $k-1$ are respectively

$$p = \frac{1}{2}(1 - \frac{k}{K}) \qquad q = \frac{1}{2}(1 + \frac{k}{K}) \qquad (-K \leq k \leq K)$$

Notice that the positions $-K\Delta$ and $K\Delta$ act as barriers which cannot be crossed by the process. In this case, the transition probability reads

$$Q(n|k) = \frac{1}{2}(1 - \frac{k}{K})\delta(n, k+1) + \frac{1}{2}(1 + \frac{k}{K})\delta(n, k-1) \tag{9.51}$$

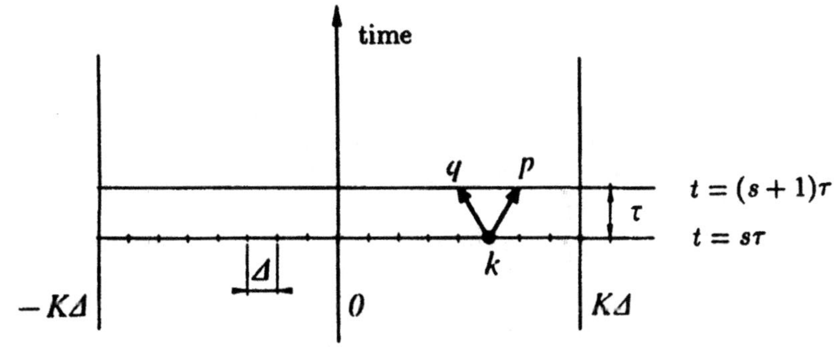

Figure 9.5: Random walk of an elastically bound particle.

Substituting into Equ.(9.47), one gets

$$P_2(n, s+1|m) = P_2(n+1, s|m)\frac{K+n+1}{2K} + P_2(n-1, s|m)\frac{K-n+1}{2K} \quad (9.52)$$

with the initial condition $P_2(n, 0|m) = \delta(n, m)$. The solution of this problem is considerably more complicated than that of the previous problem (see e.g. Kac, 1947); we shall not consider it in detail here. It can be shown that the average value goes to zero according to

$$\mathrm{E}[n(s+1)] = \sum_{n=-K}^{K} nP_2(n, s+1|m) = m(1-\frac{1}{K})^{s+1}$$

If, as for the free particle, one uses the notations $x = n\Delta$, $x_0 = m\Delta$ and $t = s\tau$ and if one takes the limit for $\tau \to 0$, $\Delta \to 0$ and $K \to \infty$ in such a way that

$$D = \lim_{\Delta,\tau\to 0}\frac{\Delta^2}{2\tau} \qquad \beta = \lim_{\tau\to 0}\frac{1}{K\tau}$$

be finite, one transforms the difference equation into the diffusion equation

$$\frac{\partial p_2}{\partial t} = D\frac{\partial^2 p_2}{\partial x^2} + \beta\frac{\partial}{\partial x}(xp_2) \quad (9.53)$$

with the same initial condition as for the free particle.

9.8 One-dimensional Fokker-Planck equation

In the previous section we considered examples of continuous Markov processes obtained as the limit of a random walk; the conditional probability density was found to be governed by a diffusion equation. That observation can be generalized to any Markov process: The transition probability density of a Markov

process is governed by a diffusion type partial differential equation known as the ***Fokker-Planck equation***, of which the examples that we met in the foregoing section are particular cases. In this section, it is shown that the Fokker-Planck equation is a direct consequence of the Smoluchowski equation. The proof is made for a one dimensional process, but it will be extended to multi-dimensional processes in section 9.9.
In this section, we consider a stationary one-dimensional Markov process $X(t)$ satisfying the Smoluchowski equation

$$q(x,t+\Delta t|x_0) = \int_{-\infty}^{\infty} q(z,t|x_0)q(x,\Delta t|z)dz \qquad 0 \le t \le t+\Delta t \tag{9.54}$$

Before deriving the Fokker-Planck equation, consider the rate of change of the moments of the space coordinate:

$$A_n(z) = \lim_{\tau\to 0} \frac{\mathrm{E}[(X-z)^n]}{\tau} = \lim_{\tau\to 0} \frac{1}{\tau} \int_{-\infty}^{\infty} (x-z)^n q(x,\tau|z)dx \tag{9.55}$$

They are sometimes called ***derivate moments***; they exist if $\mathrm{E}[(X - z)^n] = A_n(z)\tau + O(\tau^2)$. This form was obtained for the second moment of the Wiener process, Equ.(9.19). According to section 2.8, it implies that

$$\kappa_n[X-z] = A_n(z)\tau + O(\tau^2) \tag{9.56}$$

The increment $X - z$ is related to the derivative process $\xi(t) = \dot{X}(t)$ by

$$X(\tau) - z = \int_t^{t+\tau} \xi(t')\,dt'$$

The corresponding cumulants are related by

$$\kappa_n[X-z] = \int_t^{t+\tau} \dots \int_t^{t+\tau} \kappa_n[\xi(t_1),\dots,\xi(t_n)]dt_1\dots dt_n \tag{9.57}$$

Equation (9.56) implies that the cumulants of $\xi(t)$ have delta function singularities of the form

$$\kappa_n[\xi(t_1),\dots,\xi(t_n)] = A_n(z)\delta(t_1-t_2)\dots\delta(t_1-t_n) + B_n \tag{9.58}$$

where B_n denotes other possible functions with singularities of lower order than the first term, so that their contribution to $\kappa_n[X - z]$ is of the order $0(\tau^2)$ and more. Such processes are said to be ***delta correlated***; they play an important role in Markov processes, because the derivative of a Markov process is always a delta correlated process. The derivate moments, $A_n(z)$, are identical to the ***intensity coefficients*** of the derivative $\xi(t)$.

As discussed in chapter 4, a zero mean Gaussian white noise is such that

$$\kappa_1 = 0, \qquad \kappa_2(t_1-t_2) = K\delta(t_1-t_2), \qquad \kappa_n = 0 \quad (n>2)$$

In other words, of all the intensity coefficients, only the second one is different from zero. In what follows, we *assume* that

$$A_1(z) \text{ and } A_2(z) \text{ are finite}$$

$$A_n(z) = 0 \ (n > 2) \tag{9.59}$$

This puts some restriction on the level of discontinuity of $\xi(t)$; the assumption is satistied by a Gaussian white noise.

9.8.1 Derivation of the Fokker-Planck equation

Consider the integral

$$I = \int_{-\infty}^{\infty} R(x) \frac{\partial q(x,t|x_0)}{\partial t} dx$$

where $R(x)$ is an arbitrary function which goes to zero fast enough at infinity so that the integral exists. Of course,

$$I = \lim_{\Delta t \to 0} \int_{-\infty}^{\infty} R(x) \frac{q(x,t+\Delta t|x_0) - q(x,t|x_0)}{\Delta t} dx$$

Substituting the Smoluchowski equation (9.54), we find

$$I = \lim_{\Delta t \to 0} \frac{1}{\Delta t} \{ \int_{-\infty}^{\infty} R(x) [\int_{-\infty}^{\infty} q(z,t|x_0) q(x,\Delta t|z) dz] dx - \int_{-\infty}^{\infty} R(x) q(x,t|x_0) dx \}$$

Interchange the order of integration and develop $R(x)$ in Taylor series in $(x-z)$; the double integral can be rewritten

$$\frac{1}{\Delta t} \int_{-\infty}^{\infty} q(z,t|x_0) [\sum_{n=0}^{\infty} \frac{R^{(n)}(z)}{n!} \int_{-\infty}^{\infty} (x-z)^n q(x,\Delta t|z) dx] dz$$

where $R^{(n)}(z)$ stands for the nth derivative of $R(z)$. We recognize the derivate moments that we introduced in the previous section. From assumption (9.59), the terms of order higher than 2 in the sum vanish. Substituting into the integral, one easily gets

$$I = \int_{-\infty}^{\infty} q(z,t|x_0) [R^{(1)}(z) A_1(z) + \frac{1}{2} R^{(2)}(z) A_2(z)] dz$$

The derivative of $R(z)$ can be eliminated by partial integration. Writing x for z and substracting from the original expression of I, one gets

$$\int_{-\infty}^{\infty} R(x) \{ \frac{\partial q(x,t|x_0)}{\partial t} + \frac{\partial}{\partial x} [A_1(x) q(x,t|x_0)] - \frac{1}{2} \frac{\partial^2}{\partial x^2} [A_2(x) q(x,t|x_0)] \} dx = 0$$

Since this must hold for any $R(x)$, the expression $\{\}$ must be zero

$$\frac{\partial q(x,t|x_0)}{\partial t}+\frac{\partial}{\partial x}[A_1(x)q(x,t|x_0)]-\frac{1}{2}\frac{\partial^2}{\partial x^2}[A_2(x)q(x,t|x_0)]=0 \tag{9.60}$$

with the initial condition

$$\lim_{t\to t_0} q(x,t|x_0,t_0)=\delta(x-x_0) \tag{9.61}$$

This is the *Fokker-Planck equation*, of which Equ.(9.50) and (9.53) are particular cases. Introducing the *probability current*

$$G(x,t)=A_1(x)q(x,t|x_0)-\frac{1}{2}\frac{\partial}{\partial x}[A_2(x)q(x,t|x_0)] \tag{9.62}$$

we can write it

$$\dot{q}+\frac{\partial G}{\partial x}=0 \tag{9.63}$$

which expresses the conservation of probability at x, in the same way as the heat conduction equation expresses the conservation of the thermal energy. If the process is defined over the whole real axis, the boundary conditions are usually

$$G(-\infty,t)=G(\infty,t)=0 \quad \text{and} \quad q(-\infty,t|x_0)=q(\infty,t|x_0)=0 \tag{9.64}$$

The first of these equations states that the trajectories cannot appear or disappear at infinity. If the domain is bounded ($x_1 \le X(t) \le x_2$), the boundary conditions express the fact that the probability current vanishes at the limits

$$G(x_1,t)=G(x_2,t)=0 \tag{9.65}$$

The solution of the one dimensional Fokker-Planck equation is discussed in (Stratonovich, 1963).

9.8.2 Kolmogorov equation

The Fokker-Planck equation is often called *forward*, because the derivatives are taken with respect to t and x, the time and space variables at the forward time $t \ge t_0$. In fact, if the transition probability $q(x,t|x_0,t_0)$ is regarded as a function of x_0 and t_0, the space and time variables at the backward time, it can be shown that it satisfies the partial differential equation (e.g. Bharucha-Reid, 1960)

$$\frac{\partial q}{\partial t_0}+A_1(x_0)\frac{\partial q}{\partial x_0}+\frac{1}{2}A_2(x_0)\frac{\partial^2 q}{\partial x_0^2}=0 \tag{9.66}$$

which is the adjoint of Equ.(9.60). This is called the *Kolmogorov equation*; it must be solved backward in time from the initial condition x at t.

For a stationary random process, the transition probability $q(x,\tau|x_0)$ depends only on the time difference $\tau = t - t_0$ and

$$\frac{\partial q}{\partial t_0} = -\frac{\partial q}{\partial t} = -\frac{\partial q}{\partial \tau}$$

Therefore, the Kolmogorov equation can be rewritten

$$\frac{\partial q}{\partial \tau} = A_1(x_0)\frac{\partial q}{\partial x_0} + \frac{1}{2}A_2(x_0)\frac{\partial^2 q}{\partial x_0^2} \tag{9.67}$$

This equation is homogeneous in q and, as a result, any linear operation on q which does not involve the independent variables τ and x_0 satisfies the same equation. In particular, this is the case for the integral

$$P(\Omega,\tau|x_0) = \int_\Omega q(x,\tau|x_0)dx \tag{9.68}$$

which represents the probability that the value of the process belongs to the domain Ω at τ under the condition $X(0) = x_0$. If $\Omega \equiv (-\infty, x]$, $P(\Omega,\tau|x_0)$ is the transition probability distribution function. The initial condition is

$$P(\Omega,0|x_0) = 1 \text{ if } x_0 \in \Omega; \qquad P(\Omega,0|x_0) = 0 \text{ if not}$$

9.9 Multi-dimensional Fokker-Planck equation

If $X = (X_1, ..., X_n)^T$ is a n-dimensional vector Markov process, it can be shown in a similar manner (see e.g. Lin, 1967) that its transition probability density $q(x,t|x_0)$ is governed by the multi-dimensional Fokker-Planck equation

$$\frac{\partial q}{\partial t} + \sum_{i=1}^{n}\frac{\partial}{\partial x_i}[A_1^i(x)q] - \frac{1}{2}\sum_{k,l=1}^{n}\frac{\partial^2}{\partial x_k \partial x_l}[A_2^{kl}(x)q] = 0 \tag{9.69}$$

where the derivate moments are defined as

$$A_1^i(x) = \lim_{\tau\to 0}\frac{\mathrm{E}[(X_i - x_i)|x]}{\tau}$$

$$= \lim_{\tau\to 0}\frac{1}{\tau}\int_{-\infty}^{\infty}...\int (y_i - x_i)q(y,\tau|x)dy_1...dy_n \tag{9.70}$$

$$A_2^{kl}(x) = \lim_{\tau\to 0}\frac{\mathrm{E}[(X_k - x_k)(X_l - x_l)|x]}{\tau}$$

$$\lim_{\tau\to 0}\frac{1}{\tau}\int_{-\infty}^{\infty}...\int (y_k - x_k)(y_l - x_l)q(y,\tau|x)dy_1...dy_n \tag{9.71}$$

The initial condition is, as usual

$$q(x,0|x_0) = \delta(x - x_0) = \prod \delta(x_i - x_{0i})$$

The corresponding Kolmogorov (backward) equation reads

$$\frac{\partial q}{\partial t_0} + \sum_{i=1}^{n} A_1^i(x_0)\frac{\partial q}{\partial x_{0i}} + \frac{1}{2}\sum_{k,l=1}^{n} A_2^{kl}(x_0)\frac{\partial^2 q}{\partial x_{0k}\partial x_{0l}} = 0 \tag{9.72}$$

9.10 The Brownian motion of an oscillator

As an illustration of the previous section, consider the response of a single d.o.f. oscillator

$$\ddot{X} + 2\xi\omega_n\dot{X} + \omega_n^2 X = F(t) \tag{9.73}$$

to a Gaussian white noise excitation of zero mean and intensity $2D$: $R_{ff}(\tau) = 2D\delta(\tau)$. This problem is often referred to as the *Brownian motion* of a single d.o.f. oscillator, because it was observed that the random force generated by the impact of the fluid molecules on an immersed microscopic particle can be considered as a Gaussian white noise. As discussed earlier, the second order differential equation can be recast into state variable form as a set of two first order differential equations. With the notation $P = \dot{X}$, one gets

$$\begin{pmatrix} \dot{X} \\ \dot{P} \end{pmatrix} = \begin{pmatrix} 0 & 1 \\ -\omega_n^2 & -2\xi\omega_n \end{pmatrix}\begin{pmatrix} X \\ P \end{pmatrix} + \begin{pmatrix} 0 \\ F(t) \end{pmatrix} \tag{9.74}$$

This equation shows that the state vector $(X, P)^T$ is a vector Markov process. From Equ.(9.70) and (9.71), the derivate moments are

$$A_1^1(x,p) = p$$

$$A_1^2(x,p) = -\omega_n^2 x - 2\xi\omega_n p$$

$$A_2^{11} = A_2^{12} = 0$$

$$A_2^{22} = 2D$$

and the Fokker-Planck equation reads

$$\frac{\partial q}{\partial t} + p\frac{\partial q}{\partial x} = \frac{\partial}{\partial p}[(2\xi\omega_n p + \omega_n^2 x)q] + D\frac{\partial^2 q}{\partial p^2} \tag{9.75}$$

This equation can be solved analytically (see e.g.Wang & Uhlenbeck, 1945). However, since the system is linear and the excitation is Gaussian, we know that the response is also Gaussian; it is entirely determined by its mean and its

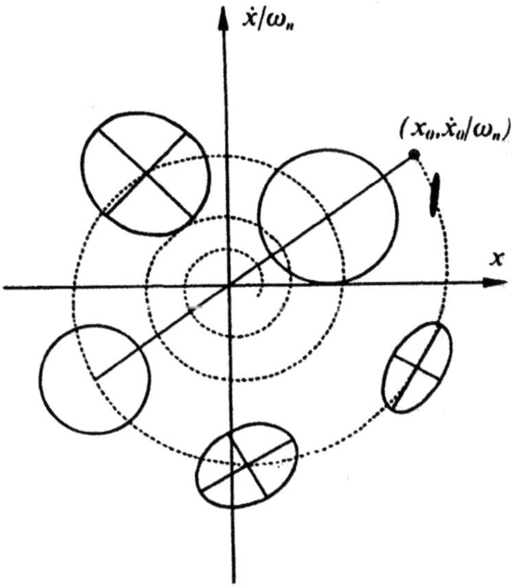

Figure 9.6: Single d.o.f. oscillator. Evolution of the transition probability in the phase plane (from Wang & Uhlenbeck).

covariance matrix. Since the excitation has a zero mean, the conditional mean follows the same trajectory as the free response of the system:

$$\mu = \Phi(t)\mu_0 \tag{9.76}$$

from the initial conditions $\mu_0 = (x_0, p_0)^T$. The transition matrix of a single d.o.f. oscillator is given by

$$\Phi(t) = e^{-\xi\omega_n t}\begin{pmatrix} \cos\omega_d t + \frac{\xi}{\sqrt{1-\xi^2}}\sin\omega_d t & \frac{1}{\omega_d}\sin\omega_d t \\ -\frac{\omega_n^2}{\omega_d}\sin\omega_d t & \cos\omega_d t - \frac{\xi}{\sqrt{1-\xi^2}}\sin\omega_d t \end{pmatrix} \tag{9.77}$$

In the phase plane $(x, \dot{x}/\omega_n)$, the trajectory consists of a spiral rotating clockwise and converging to zero with a decay rate depending on the damping of the system (Fig.9.6). The covariance matrix σ can be obtained by solving Equ.(9.26) from the initial condition $\sigma_0 = 0$. It can be shown that

$$\sigma_x^2 = \frac{D}{2\xi\omega_n^3}\{1 - \frac{e^{-2\xi\omega_n t}}{\omega_d^2}[\omega_d^2 + 2(\xi\omega_n \sin\omega_d t)^2 + \xi\omega_n\omega_d \sin 2\omega_d t]\}$$

$$\sigma_p^2 = \frac{D}{2\xi\omega_n}\{1 - \frac{e^{-2\xi\omega_n t}}{\omega_d^2}[\omega_d^2 + 2(\xi\omega_n \sin\omega_d t)^2 - \xi\omega_n\omega_d \sin 2\omega_d t]\}$$

$$\sigma_x\sigma_p\varrho_{xp} = \frac{D}{\omega_d^2}e^{-2\xi\omega_n t}\sin^2\omega_d t$$

Note that as t increases,

$$\sigma_x^2 \rightarrow \frac{D}{2\xi\omega_n^3} \qquad \sigma_p^2 \rightarrow \frac{D}{2\xi\omega_n} \qquad \varrho_{xp} \rightarrow 0$$

Thus, the distribution tends towards a steady state and X and $\dot{X}$ eventually become independent. For small t,

$$\sigma_x^2 \approx \frac{2}{3}Dt^3 \qquad \sigma_p^2 \approx 2Dt \qquad \sigma_x\sigma_p\varrho_{xp} \approx Dt^2$$

This indicates that, in the phase plane, the initial two-dimensional Dirac delta function $\delta(x-x_0)\delta(p-p_0)$ will first become a narrow ellipse elongated along the p axis. Next it will turn and broaden until $t = \pi/\omega_d$ where it becomes a circle (Fig.9.6). After that, the same pattern will repeat itself with a larger and larger amplitude and a period π/ω_d while the center of the distribution goes to the origin.

9.11 Replacement of an actual process by a Markov process

9.11.1 One-dimensional process

We have seen that the Markov property is related to that of having independent increments. This implies that the derivative has no memory (purely random process). Roughly speaking, a Markov process is such that its first derivative is a white noise. We know that a white noise itself can only be an idealization of a broad band process with a finite bandwidth. Now consider the process

$$\dot{X} = F(t) \tag{9.78}$$

where $F(t)$ is a zero mean stationary process with a finite correlation time τ_{cor} [see Equ.(9.23)]. Since $F(t)$ is not purely random, $X(t)$ is not exactly Markovian. However, it can be shown (Stratonovich, 1963, p.83 and followings) that for time intervals with length much greater than the correlation time ($\Delta t >> \tau_{cor}$), the increments can be treated as independent. Therefore, the long term behaviour of the process $X(t)$ is that of a Markov process. The joint probability density functions can be partitioned as in Equ.(9.10), where the transition probability density is the solution of a Fokker-Planck equation. That equation can be obtained by approximating the real process $F(t)$ by a delta correlated process (white noise) $F_0(t)$ with the same intensity coefficients as the actual process $F(t)$:

$$\kappa_n[F_0(t_1), ..., F_0(t_n)] = K_n\delta(t_2-t_1)...\delta(t_n-t_1)$$

with

$$K_n = \int_{-\infty}^{\infty} ... \int \kappa_n[F(\tau_1), ..., F(\tau_n)]d\tau_2...d\tau_n \tag{9.79}$$

For the more general equation

$$\dot{X} = f(X) + g(X)F(t) \tag{9.80}$$

it can be shown (see Stratonovich, 1963, p.96) that, for time intervals much greater than the correlation time of $F(t)$, $X(t)$ can be considered as Markovian, with a transition probability density governed by the Fokker-Planck equation (9.60) with the derivate moments

$$A_1(x) = f(x) + \frac{1}{2}g(x)\frac{\partial g(x)}{\partial x}\kappa \tag{9.81}$$

$$A_2(x) = g^2(x)\kappa \tag{9.82}$$

where κ is the second intensity coefficient of the excitation $F(t)$. The second term appearing in the first derivate moment accounts for the correlation between $X(t)$ and $F(t)$, which is responsible for

$$\lim_{\tau \to 0} \mathrm{E}[\frac{1}{\tau}\int_0^\tau g(X)F(t)dt] \neq g(x) \lim_{\tau \to 0} \mathrm{E}[\frac{1}{\tau}\int_0^\tau F(t)dt]$$

9.11.2 Stochastically equivalent systems

A given system equation leads to a unique Fokker-Planck equation; however, different system equations may lead to the same Fokker-Planck equation. Thus, the inverse problem of finding the system equation leading to a given Fokker-Planck equation does not have a unique solution. It becomes unique if we restrict ourselves to equations of the form (9.80) excited by a Gaussian white noise of zero mean and unit intensity ($\kappa = 1$). In fact, from Equ.(9.81) and (9.82), one readily obtains

$$g(x) = \sqrt{A_2(x)}$$

$$f(x) = A_1(x) - \frac{1}{4}\frac{\partial A_2(x)}{\partial x}$$

Therefore, the *arbitrary* Fokker-Planck equation (9.60) corresponds to the system equation

$$\dot{X} = A_1(X) - \frac{1}{4}\frac{\partial A_2(X)}{\partial X} + \sqrt{A_2(X)}F_0(t) \tag{9.83}$$

where $F_0(t)$ is a zero mean Gaussian white noise of unit intensity. Two systems leading to the same Fokker-Planck equation are said to be *stochastically equivalent*.

9.11.3 Multi-dimensional process

The foregoing discussion can be generalized to vector processes (Stratonovich, 1963). If

$$\dot{X} = f(X) + g(X)F(t) \tag{9.84}$$

where $f(X)$ is a vector function, $g(X)$ a matrix function and $F(t)$ is a vector of zero mean independent white noise processes such that

$$\kappa_2[F_i(t), F_j(t+\tau)] = \delta_{ij}\delta(\tau)$$

the derivate moments of the multi-dimensional Fokker-Planck equation are

$$A_1^i(x) = f_i(x) + \frac{1}{2}\sum_{m,j} \frac{\partial g_{ij}(x)}{\partial x_m} g_{mj}(x) \tag{9.85}$$

$$A_2^{kl}(x) = \sum_i g_{ki}(x) g_{li}(x) = (gg^T)_{kl} \tag{9.86}$$

As for the one dimensional process, if $F(t)$ is not white, but its correlation time is small compared to the time constants of the system, it can be approximated by a Gaussian white noise with the same intensity matrix.

9.12 References

J.D.ATKINSON, Eigenfunction expansions for randomly excited non-linear systems, *Journal of Sound and Vibration 30(2), pp.153-172,* 1973.

A.T.BHARUCHA-REID, *Elements of Theory of Markov Processes and their Applications,* McGraw-Hill, 1960.

A.E.BRYSON & Y.C.HO, *Applied Optimal Control (Optimization, Estimation and Control),* J.Wiley, 1975.

T.K.CAUGHEY, Nonlinear theory of random vibrations, *Advances in Applied Mechanics 11, pp.209-253,* 1971.

M.KAC, Random walk and the theory of Brownian motion, *American Mathematical Monthly 54, No 7, pp.369-391,* 1947. Reprinted in *Selected Papers on Noise and Stochastic Processes,* N.WAX ed., Dover, 1954.

Y.K.LIN, *Probabilistic Theory of Structural Dynamics,* McGraw-Hill, 1967.

A.PAPOULIS, *Probability, Random Variables and Stochastic Processes,* McGraw-Hill, 1965.

R.L.STRATONOVICH, *Topics in the Theory of Random Noise,* 1, Gordon & Breach, N-Y, 1963.

M.C.WANG & G.E.UHLENBECK, On the theory of Brownian motion II, *Review of Modern Physics, Vol.17, No 2 and 3, April-July, pp.323-342,* 1945. Reprinted in *Selected Papers on Noise and Stochastic Processes,* N.WAX ed., Dover, 1954.

9.13 Problems

P.9.1 *Show that the probability distribution of the random walk follows the binomial distribution (9.16). [Hint: Follow the same lines as in section 4.2.2].*
P.9.2 *Write the following difference equation in state variable form*

$$X_k = \phi_1(k)X_{k-1} + \phi_2(k)X_{k-2} + \psi(k)W_k$$

P.9.3 *Consider the stationary response of a single d.o.f. oscillator excited by a white noise of intensity $2D$ [$R_{ww}(\tau) = 2D\delta(\tau)$]. Using the Lyapunov equation, show that*
(a) the variance of the response is given by $\sigma_x^2 = D/2\xi\omega_n^3$;
(b) the variance of the velocity is $\sigma_{\dot{x}}^2 = \omega_n^2\sigma_x^2$;
(c) $X(t)$ and $\dot{X}(t)$ are uncorrelated random variables.
P.9.4 *Show that the Wiener process defined as $\dot{X} = \xi(t)$, where $\xi(t)$ is a zero mean Gaussian white noise of intensity $2D$ [$R_{\xi\xi}(\tau) = 2D\delta(\tau)$] is governed by the Fokker-Planck equation*

$$\frac{\partial q}{\partial t} = D\frac{\partial^2 q}{\partial x^2}$$

P.9.5 *If the process $X(t)$ is governed by the first order differential equation $\dot{X} + \beta X = \xi(t)$, where $\xi(t)$ is a zero mean Gaussian white noise of intensity $2D$, show that its transition probability density is governed by the Fokker-Planck equation*

$$\frac{\partial q}{\partial t} = \beta\frac{\partial}{\partial x}(xq) + D\frac{\partial^2 q}{\partial x^2}$$

P.9.6 *Consider the system $\dot{X} = f(X) + F(t)$ where $F(t)$ is a first order Markov process of autocorrelation function $R_{ff}(\tau) = \exp(-\beta|\tau|)$. Assume that the time constant of the system, $\tau_0 \sim (\partial f/\partial x)^{-1}$ is much greater than the correlation time of the excitation.*
(a) Using a Gaussian white noise approximation $F_0(t)$ of $F(t)$, write the corresponding Fokker-Planck equation.
(b) Show that the system can be described by the second order equation

$$\ddot{X} + [\beta - f'(X)]\dot{X} - \beta f(X) = W$$

where $W(t)$ is a Gaussian white noise of intensity 2β.
(c) Rewrite the foregoing equation in state variable form and write the corresponding Fokker-Planck equation.
P.9.7 *Find the second order Markov approximation of the Gaussian process defined by Equ.(9.42).*

Chapter 10

Threshold Crossings, Maxima, Envelope and Peak Factor

10.1 Introduction

In the preceding chapters, we have learned how to predict the statistics of the structural response (displacements, stresses, etc...) from the statistics of the random excitation. Most of the time, if the structure is linear, the response statistics are available in the form of *PSD* functions. From them, it is straightforward to evaluate the *RMS* response, but this is rarely enough to assess the reliabilty of the system, which depends on the failure mode of the structure.

In some situations, the designer will mainly be concerned with avoiding vibrations of excessive amplitude, which could either lead to major problems in the operation of the system (e.g. vibration amplitude of a rotor exceeding the gap in the casing), or exceed regulatory limits (e.g. yield stress for an *Operating Basis Earthquake* in a nuclear power plant). In both cases, the designer will want to evaluate the probability distribution of the *largest* value of the response, which is related to the *RMS* value by the *peak factor*. This mode of failure by limit exceedance will be considered in this chapter.

In other situations, especially when the stress level is high and the structure is exposed to random excitation for a large number of cycles, the failure may result from *fatigue* damage. Random fatigue will be considered in the next chapter, based on linear damage theory.

As a prerequisite for the study of both failure modes, this chapter will start with two related problems, the statistics of threshold crossings and the number of maxima with amplitude exceeding some threshold. The concept of *envelope*

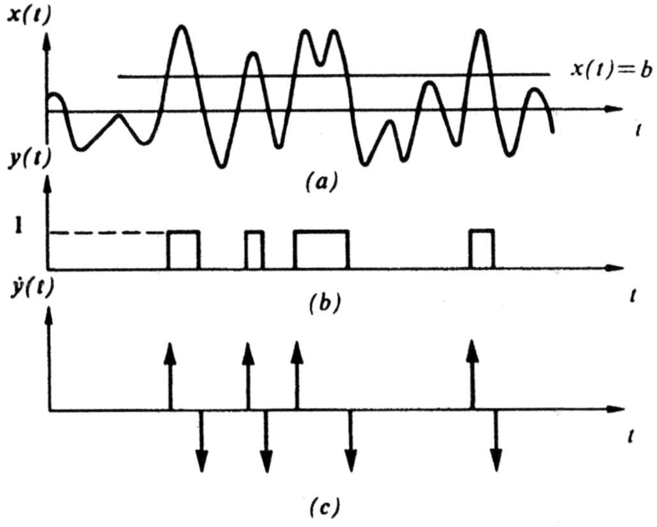

Figure 10.1: Construction of a counting process for the crossings of a level b

will also be discussed in detail. Throughout this chapter, we will assume that the process is Gaussian with zero mean.

10.2 Threshold crossings

10.2.1 Up-crossings of a level b

Consider the zero mean Gaussian process $X(t)$, a sample of which is represented in Fig.10.1. We wish to evaluate the average number of crossings of some level b during the time period $[t_1, t_2]$. To do that, we construct a counting process $\tilde{N}(b, t_1, t_2)$ in the following way (Middleton, 1960): First, we define the process

$$Y(t) = \mathbf{1}[X(t) - b] \tag{10.1}$$

where $\mathbf{1}[\ ]$ is Heaviside's step function. A sample of $Y(t)$ is represented in Fig.10.1.b. $y(t)$ is such that its value is 1 wherever $x(t) > b$ and 0 elsewhere. Next, since we know that differentiating unit step functions supplies Dirac delta functions with unit intensity, a set of alternating positive and negative unity delta functions is generated by differentiating $Y(t)$:

$$\dot{Y}(t) = \dot{X}(t)\ \delta[X(t) - b] \tag{10.2}$$

The corresponding sample is represented in Fig.10.1.c. We see that every up-crossing generates a positive unit impulse while every down-crossing generates a negative one. Integrating the absolute value $|\dot{Y}|$ provides exactly the total number of crossings for the period of integration. Thus, the counting process can be expressed by

$$\tilde{N}(b,t_1,t_2) = \int_{t_1}^{t_2} |\dot{X}(t)|\ \delta[X(t) - b]\, dt \tag{10.3}$$

From this equation, we can define the *rate of threshold crossings*

$$N(b,t) = |\dot{X}(t)|\ \delta[X(t) - b] \tag{10.4}$$

Since $N(b,t)$ depends on $\dot{X}$ and X, its expected value requires the knowledge of the joint probability density $p(x,\dot{x},t)$. From Equ.(2.53),

$$\mathrm{E}[N(b,t)] = \int_{-\infty}^{\infty} \int |\dot{x}|\delta(x-b)p(x,\dot{x},t)\, dx\, d\dot{x}$$

or

$$\mathrm{E}[N(b,t)] = \nu_b = \int_{-\infty}^{\infty} |\dot{x}|\, p(b,\dot{x},t)\, d\dot{x} \tag{10.5}$$

The expected rate of threshold crossings with positive slope (*up-crossings*), $N_+(b,t)$ is obtained by restricting the foregoing integral to the positive values of the velocity

$$\mathrm{E}[N_+(b,t)] = \nu_b^+ = \int_0^{\infty} \dot{x}\, p(b,\dot{x},t)\, d\dot{x} \tag{10.6}$$

If the process $X(t)$ is stationary, Gaussian with zero mean, the joint probability density is given by Equ.(5.59) and one gets

$$\nu_b^+ = \frac{1}{2}\nu_b = \frac{1}{2\pi}\frac{\sigma_{\dot{x}}}{\sigma_x}\exp(-\frac{b^2}{2\sigma_x^2}) \tag{10.7}$$

10.2.2 Central frequency

An important particular case is that where $b = 0$. In this case, Equ.(10.7) can be simplified to

$$\nu_0^+ = \frac{1}{2\pi}\frac{\sigma_{\dot{x}}}{\sigma_x} = \frac{1}{2\pi}(\frac{m_2}{m_0})^{1/2} \tag{10.8}$$

where the spectral moments m_0 and m_2 are defined by Equ.(5.39) and (5.40). This formula is known as *Rice's formula* and ν_0^+ is called the *central frequency*. In fact, if one considers a sample of a narrow-band process as represented in Fig.10.2, its general shape is that of a sine function with slowly varying amplitude and frequency. By analogy with the sine, the part of the sample between two successive zero up-crossings (i.e. with positive velocity) can be regarded as an *equivalent cycle*. Rice's formula applies irrespective of the bandwidth of the process; for a wide-band process, the central frequency must be interpreted in the sense of average rate of zero up-crossings. For a narrow-band process, it also indicates the frequency where most of the power is concentrated in the process.

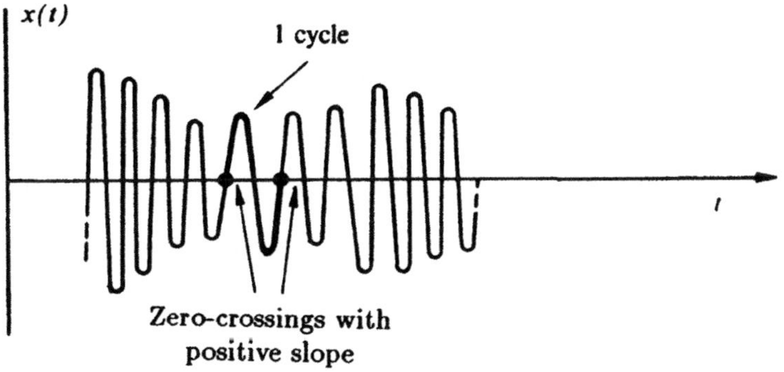

Figure 10.2: Sample of a narrow-band process. Definition of a cycle.

10.3 Maxima

In a manner similar to that used for the threshold crossings, a counting process for the maxima can be constructed in the following way (Fig.10.3): The process $Y(t) = \mathbf{1}[\dot{X}(t)]$ is such that its value is 1 wherever the slope of $X(t)$ is positive and 0 if it is negative. As in the previous section, the derivative

$$\dot{Y}(t) = \ddot{X}\,\delta[\dot{X}] \tag{10.9}$$

generates a set of alternating unity delta functions at the *extrema*, where $\dot{x} = 0$.

Now, if we are interested in the extrema *above the threshold b*, they can be isolated by multiplying $\dot{Y}(t)$ by $\mathbf{1}[X - b]$. Therefore, the number of extrema above b is given by

$$\tilde{E}(b, t_1, t_2) = \int_{t_1}^{t_2} |\ddot{X}(t)|\;\delta[\dot{X}(t)]\;\mathbf{1}[X(t) - b]\,dt \tag{10.10}$$

If one is interested in the *maxima*, the integral must be restricted to the negative values of $\ddot{X}$; this can be achieved by multiplying the expression inside the integral by $\mathbf{1}[-\ddot{X}]$. Finally, we can defined the ***rate of maxima above the threshold b*** as

$$M(b, t) = -\ddot{X}(t)\;\mathbf{1}[-\ddot{X}]\;\delta[\dot{X}(t)]\;\mathbf{1}[X(t) - b] \tag{10.11}$$

This expression involves X, $\dot{X}$ and $\ddot{X}$. Therefore, to calculate the expected value, we need to know the joint distribution $p(x, \dot{x}, \ddot{x}, t)$. We easily get

$$\mathrm{E}[M(b,t)] = -\int_{-\infty}^{\infty}\int_{-\infty}^{\infty}\int_{-\infty}^{\infty} p(x, \dot{x}, \ddot{x}, t)\;\ddot{x}\;\mathbf{1}[-\ddot{x}]\;\delta(\dot{x})\;\mathbf{1}[x - b]\;dx\,d\dot{x}\,d\ddot{x}$$

and finally, accounting for the delta and Heaviside functions,

$$\mathrm{E}[M(b,t)] = -\int_{b}^{\infty} dx \int_{-\infty}^{0} \ddot{x}\;p(x, 0, \ddot{x}, t)\;d\ddot{x} \tag{10.12}$$

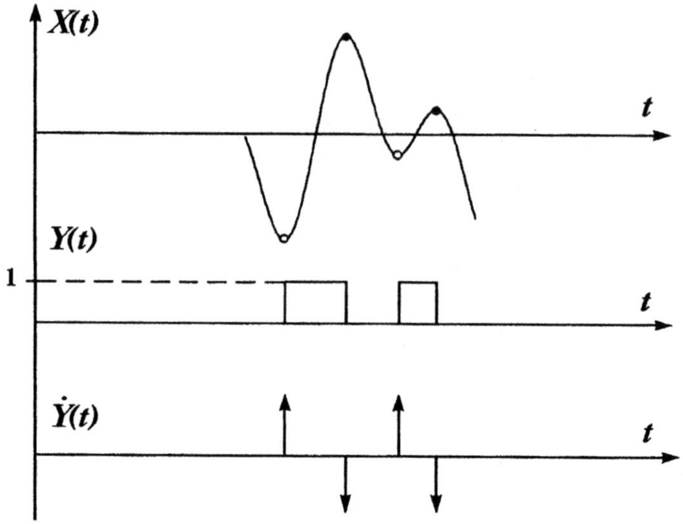

Figure 10.3: Construction of a counting process for the extrema.

The total number of maxima, regardless of their magnitude, is obtained from the previous expression by setting $b = -\infty$.

$$E[M_T(t)] = -\int_{-\infty}^{\infty} dx \int_{-\infty}^{0} \ddot{x}\; p(x, 0, \ddot{x}, t)\; d\ddot{x} \tag{10.13}$$

The ratio

$$\frac{E[M(b,t)]}{E[M_T(t)]}$$

represents the fraction of maxima above b at t, that is the probability that a maximum occuring at t be larger than b. Therefore, the probability distribution function of the maxima is given by

$$F(b,t) = 1 - \frac{E[M(b,t)]}{E[M_T(t)]} \tag{10.14}$$

and the corresponding [conditional] probability density is

$$q(b,t) = \frac{\partial}{\partial b} F(b,t) = -\frac{1}{E[M_T(t)]} \frac{\partial}{\partial b} E[M(b,t)]$$

$$q(b,t) = -\frac{1}{E[M_T(t)]} \int_{-\infty}^{0} \ddot{x}\; p(b, 0, \ddot{x}, t)\; d\ddot{x} \tag{10.15}$$

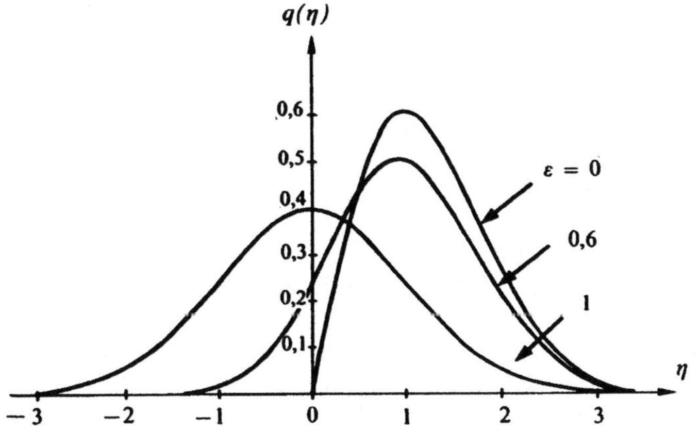

Figure 10.4: Probability density function of the maxima of a zero mean stationary Gaussian process, for various values of ε.

If the process $X(t)$ is stationary, Gaussian with zero mean, the joint probability density of X, $\dot{X}$ and $\ddot{X}$ has the standard form

$$p(z) = \frac{1}{(2\pi)^{3/2}|S|^{1/2}} \exp(-\frac{1}{2} z^T S^{-1} z) \tag{10.16}$$

where $z = (x, \dot{x}, \ddot{x})^T$ and S is the covariance matrix

$$S = \mathrm{E}[ZZ^T] = \begin{pmatrix} m_0 & 0 & -m_2 \\ 0 & m_2 & 0 \\ -m_2 & 0 & m_4 \end{pmatrix} \tag{10.17}$$

In this formula, m_0, m_2 and m_4 are the spectral moments defined according to Equ.(5.39) to (5.41). Introducing this into Equ.(10.13), one gets

$$\nu_1 = \mathrm{E}[M_T(t)] = \frac{1}{2\pi} \frac{\sigma_{\ddot{x}}}{\sigma_{\dot{x}}} = \frac{1}{2\pi} (\frac{m_4}{m_2})^{1/2} \tag{10.18}$$

This result is also due to Rice; it could have been derived directly from Equ.(10.8), because the maxima correspond to zero crossings with negative slope of the derivative $\dot{X}$. Combining Equ.(10.15) to (10.18), we can establish the following result for the probability density function of the maxima (Cartwright & Longuet-Higgins, 1956)

$$q(\eta) = \frac{1}{(2\pi)^{1/2}} [\varepsilon e^{-\eta^2/2\varepsilon^2} + (1-\varepsilon^2)^{1/2} \eta e^{-\eta^2/2} \int_{-\infty}^{\eta(1-\varepsilon^2)^{1/2}/\varepsilon} e^{-x^2/2} dx] \tag{10.19}$$

where η stands for the normalized amplitude,

$$\eta = \frac{b}{\sigma_x} = \frac{b}{m_0^{1/2}}$$

and ε is a parameter depending on the bandwidth of the process

$$0 \leq \varepsilon^2 = 1 - \frac{m_2^2}{m_0 m_4} \leq 1 \tag{10.20}$$

In fact, combining Equ.(10.8) and (10.18), one gets

$$\frac{\nu_0^+}{\nu_1} = \frac{m_2}{(m_0 m_4)^{1/2}} = (1-\varepsilon^2)^{1/2} \tag{10.21}$$

As already discussed in section 5.6.3, this ratio is always smaller than 1; it is close to 1 for a narrow band process because nearly every cycle contains a single maximum. As the bandwidth of the process increases, this ratio decreases because some cycles tend to contain several maxima. For wide band processes, the maxima can even have negative amplitude as illustrated in Fig.5.10. The probability density function $q(\eta)$ is illustrated in Fig.10.4 for various values of ε. When $\varepsilon \to 1$ (wide-band process), the distribution is Gaussian and for $\varepsilon = 0$ (narrow-band process), it is identical to the Rayleigh distribution

$$q(\eta) = \eta \exp(-\frac{\eta^2}{2}) \qquad (\eta \geq 0) \tag{10.22}$$

10.4 Envelope

10.4.1 Crandall & Mark's definition

Crandall & Mark's definition of the envelope was already discussed in section 5.7.1 when we analysed the random response of a single d.o.f. oscillator. The envelope $A(t)$ was defined as the radius of the image point of the process in the phase plane:

$$A^2(t) = X^2(t) + (\frac{\dot{X}(t)}{\omega_0})^2 \tag{10.23}$$

We have seen that $A(t)$ follows the same Rayleigh distribution (10.22) as the maxima of a narrow-band process.

10.4.2 Rice's definition

The foregoing definition was based on the assumption of a narrow-band process. Such a process has its *PSD* concentrated about some representative (carrier) value ω_m with a bandwidth $\Delta\omega$ such that $\Delta\omega \ll \omega_m$. Since a narrow-band process appears as a sine wave with slowly varying amplitude and phase, it can be written in the form

$$X(t) = A(t) \cos[\omega_m t + \theta(t)] \tag{10.24}$$

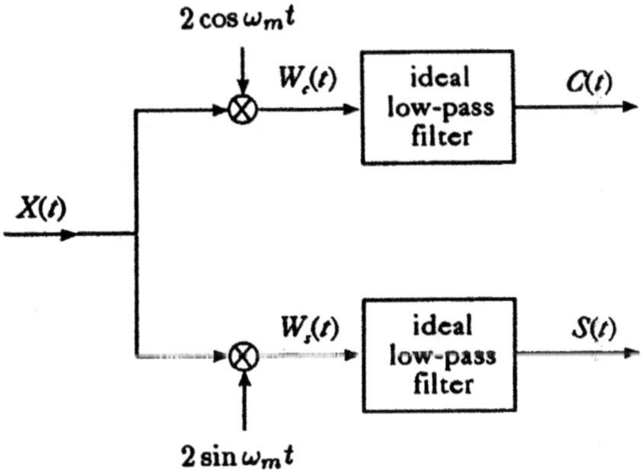

Figure 10.5: Construction of the sine and cosine components of a narrow band process.

where $A(t)$ and $\theta(t)$ are random processes with spectral content concentrated about $\omega = 0$. $A(t)$ is the envelope of the process according to Rice. Expanding this equation, we can write alternatively

$$X(t) = C(t)\cos\omega_m t - S(t)\sin\omega_m t \tag{10.25}$$

where

$$C(t) = A(t)\cos\theta(t) \qquad \text{and} \qquad S(t) = A(t)\sin\theta(t) \tag{10.26}$$

are called respectively the *cosine component* and the *sine component* of $X(t)$. They too are slowly varying processes and the envelope is related to them according to

$$A(t) = [C(t)^2 + S(t)^2]^{1/2} \tag{10.27}$$

The sine and cosine components can be constructed from $X(t)$ as indicated in Fig.10.5 (W.B.Davenport, 1970). Multiplying $X(t)$ by $2\cos\omega_m t$, one gets

$$W_c(t) = 2X(t)\cos\omega_m t = 2C(t)\cos^2\omega_m t - 2S(t)\sin\omega_m t\cos\omega_m t$$

$$= C(t) + [C(t)\cos 2\omega_m t - S(t)\sin 2\omega_m t]$$

Since $C(t)$ and $S(t)$ are slowly varying functions, $W_c(t)$ has a frequency component centered about $\omega = 0$ and components centered about twice the carrier frequency ω_m. The latter can be eliminated by low-pass filtering to isolate the cosine component $C(t)$. A similar procedure based on multiplying by a sine at the carrier frequency is used to isolate the sine component $S(t)$. By construction,

$$C(t) = 2\int_{-\infty}^{\infty} h(u)\,X(t-u)\,\cos\omega_m(t-u)\,du \tag{10.28}$$

$$S(t) = 2\int_{-\infty}^{\infty} h(u)\,X(t-u)\,\sin\omega_m(t-u)\,du$$

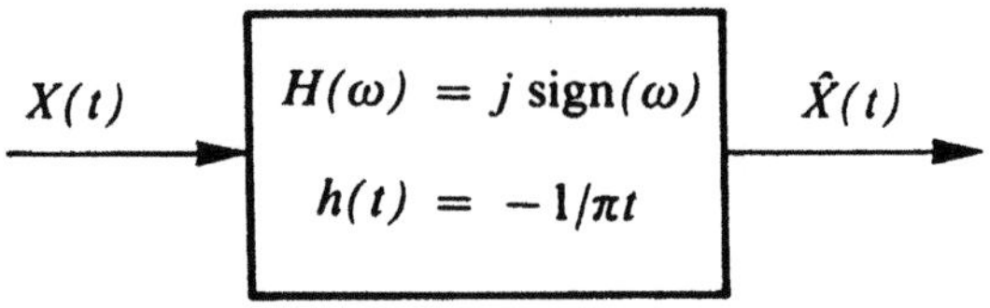

Figure 10.6: The Hilbert transform.

where $h(u)$ is the impulse response of the low-pass filter.

Since $C(t)$ and $S(t)$ are obtained from a linear transformation on $X(t)$, they are jointly Gaussian. Furthermore, it can be shown that they are orthogonal and that they have the same variance as $X(t)$ (Problem P.10.1). As a result, their joint probability density reads

$$p_{cs}(c, s) = \frac{1}{2\pi\sigma^2} \exp\left(-\frac{c^2 + s^2}{2\sigma^2}\right)$$

From section 2.6.4, we therefore conclude that the envelope $A(t)$ follows also the Rayleigh distribution (10.22). Note that it is independent of the carrier frequency ω_m.

10.4.3 The Hilbert transform

Before discussing an alternative definition attributed to Cramer & Leadbetter, we introduce the ***Hilbert transform***, which transforms a process $X(t)$ into a ***quadrature*** process $\hat{X}(t)$. The Hilbert transform is defined as the result of the linear transformation with the following impulse response and frequency response function (Fig.10.6)

$$h(t) = \frac{-1}{\pi t} \quad \Leftrightarrow \quad H(\omega) = j\,\mathrm{sign}(\omega) \tag{10.29}$$

Thus, $H(\omega)$ produces a phase shift of $+90^0$ for positive frequencies and -90^0 for negative frequencies. From the convolution theorem,

$$\hat{X}(t) = \frac{1}{\pi} \int_{-\infty}^{\infty} \frac{X(\tau)}{\tau - t} d\tau \tag{10.30}$$

The integral is evaluated as a Cauchy principal value. The Hilbert transform of a real function is also a real function. All sine components transform into cosine components and vice versa:

$$x(t) = \sin\omega_0 t \quad \rightarrow \quad \hat{x}(t) = \cos\omega_0 t \tag{10.31}$$

$$x(t) = \cos\omega_0 t \quad \rightarrow \quad \hat{x}(t) = -\sin\omega_0 t$$

A consequence of this property is that the Hilbert transforms of even functions are odd and those of odd functions are even. Two successive transformations of a signal restitute the original signal, with a negative sign. It can be shown that at the same time, $X(t)$ and $\hat{X}(t)$ are orthogonal and have the same variance (Problem P.10.2). Moreover, since the transformation is linear, $\hat{X}(t)$ is Gaussian with $X(t)$.

10.4.4 Cramer & Leadbetter's definition

From $X(t)$ and its quadrature process $\hat{X}(t)$, we construct the complex random process

$$X_+(t) = X(t) - j\hat{X}(t) \tag{10.32}$$

Note that if $x(t) = \cos\omega t$, $\hat{x}(t) = -\sin\omega t$ and $x_+(t) = \exp(j\omega t)$. Its image point in the complex plane rotates on the unit circle with a constant angular velocity ω. From this observation, an alternative definition of the envelope is the amplitude of $X_+(t)$, that is

$$A(t) = |X_+(t)| = [X^2(t) + \hat{X}^2(t)]^{1/2} \tag{10.33}$$

It is known as Cramer & Leadbetter's definition of the envelope. It is not restricted to narrow-band processes, because each harmonic component in $X(t)$ has its quadrature component in $\hat{X}(t)$.

10.4.5 Discussion

In Rice's definition of the envelope and phase processes, Equ.(10.24) can be supplemented by

$$\hat{X}(t) = -A(t)\sin[\omega_m t + \theta(t)] \tag{10.34}$$

The envelope is the amplitude of the slowly varying complex valued process

$$V(t) = C(t) + jS(t) = A(t)e^{j\theta(t)}$$

$V(t)$ is related to $X_+(t)$ according to (Problem P.10.3)

$$V(t) = X_+(t)e^{-j\omega_m t}$$

It follows that

$$|V(t)| = |X_+(t)|$$

for ***any*** value of ω_m, which makes Rice's and Cramer & Leadbetter's definitions ***equivalent***.

Returning to Crandall & Mark's definition, the reader will observe that the derivative $\dot{X}/\omega_m$ generates a signal in quadrature with X in the vicinity of ω_m. For a narrow-band process, the result is equivalent to that obtained with the Hilbert transform; the corresponding envelope is therefore equivalent. When the

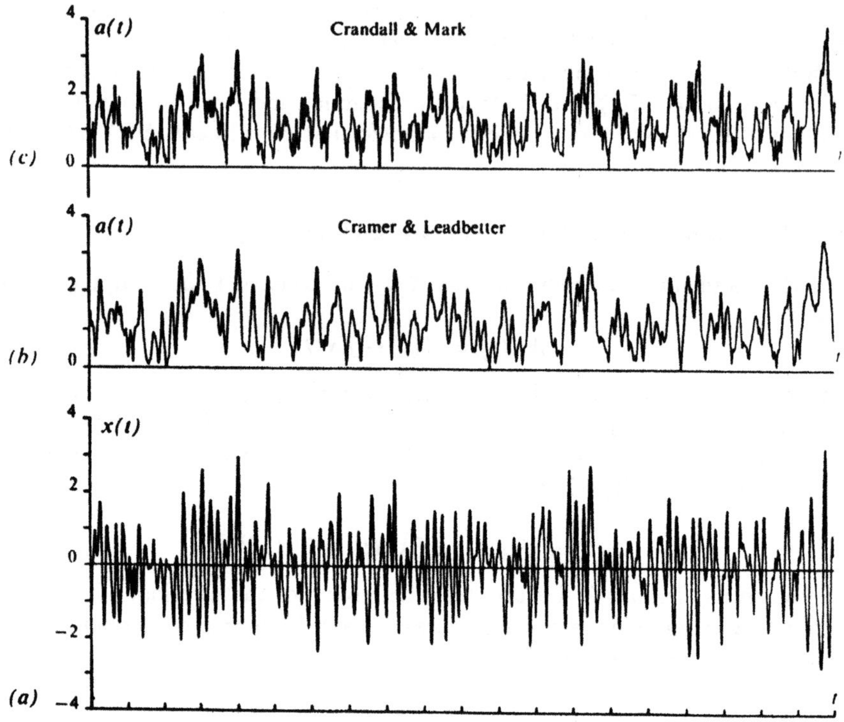

Figure 10.7: Comparison of various envelope definitions for a wide band process. (a) Sample of the process; (b) $[X^2(t) + \hat{X}^2(t)]^{1/2}$; (c) $[X^2(t) + \dot{X}^2/\omega_0^2]^{1/2}$

bandwidth of the process increases, the derivative tends to act in a significantly different way from the Hilbert transform, as illustrated in Fig.10.7. Crandall & Mark's envelope is less appropriate for wide-band processes because it tends to contain higher frequency components than that based on the Hilbert transform. On the other hand, being defined from *local* values of the process and its derivative, Crandall & Mark's definition is more appropriate for *non-stationary* narrow-band processes.

Finally, let us mention the *energy envelope* which is useful for non-linear oscillators (Crandall, 1963). It is defined by

$$V(A) = m\frac{\dot{X}^2}{2} + V(X) \tag{10.35}$$

where $V(x)$ is the potential energy stored in the elastic restoring device for the displacement x. The envelope $A(t)$ is defined as the displacement resulting from the conversion of the total energy of the system into potential energy. For the linear oscillator, this definition is equivalent to Equ.(10.23).

10.4.6 Second order joint distribution of the envelope

The second order density function of the envelope at different times t and $t+\tau$, depends on the definition which is used. The procedure for deriving it is similar in each case and we shall illustrate it with Rice's definition. The derivation for Cramer & Leadbetter's definition can be found in (Sveshnikov, 1966, Ch.5). We start from Equ.(10.26) which indicates a one-to-one relationship between the random vectors

$$[A(t), \theta(t), A(t+\tau), \theta(t+\tau)]^T \Leftrightarrow [C(t), S(t), C(t+\tau), S(t+\tau)]^T$$

The fourth order joint density of the envelope can therefore be derived from that of C and S according to Equ.(2.50). The determinant of the Jacobian of the transformation is $a_1 a_2$ and one gets

$$p_a(a_1, \theta_1, t; a_2, \theta_2, t+\tau) = a_1 a_2 p_{cs}(a_1 \cos\theta_1, a_1 \sin\theta_1, t; a_2 \cos\theta_2, a_2 \sin\theta_2, t+\tau) \tag{10.36}$$

If the process is Gaussian, the fourth order distribution of C and S is the standard Gaussian distribution with the following covariance matrix

$$S = \mathrm{E}[ZZ^T] = \begin{pmatrix} m_0 & 0 & \mu_{13} & \mu_{14} \\ 0 & m_0 & -\mu_{14} & \mu_{13} \\ \mu_{13} & -\mu_{14} & m_0 & 0 \\ \mu_{14} & \mu_{13} & 0 & m_0 \end{pmatrix} \tag{10.37}$$

where

$$m_0 = \mathrm{E}[C^2(t)] = \mathrm{E}[S^2(t)] = \mathrm{E}[X^2(t)] \tag{10.38}$$

$$\mu_{13} = \mathrm{E}[C(t)C(t+\tau)] = \mathrm{E}[S(t)S(t+\tau)] = 2\int_0^\infty \Phi_{xx}(\omega)\, \cos(\omega-\omega_m)\tau\, d\omega \tag{10.39}$$

$$\mu_{14} = \mathrm{E}[C(t)S(t+\tau)] = -\mathrm{E}[C(t+\tau)S(t)] = 2\int_0^\infty \Phi_{xx}(\omega)\, \sin(\omega-\omega_m)\tau\, d\omega \tag{10.40}$$

The demonstration of this result is left as an exercise (Problem P.10.1). Upon introducing this into Equ.(10.36) and eliminating the random variables θ_1 and θ_2 by partial integration over the complete range $[0, 2\pi]$, one gets the second order density of the envelope:

$$p_a(a_1, t; a_2, t+\tau) = \frac{a_1 a_2}{B} I_0[\frac{a_1 a_2}{B}(\mu_{13}^2 + \mu_{14}^2)^{1/2}] \exp[-\frac{m_0}{2B}(a_1^2 + a_2^2)] \tag{10.41}$$

with the notation

$$B = m_o^2 - \mu_{13}^2 - \mu_{14}^2 \tag{10.42}$$

and where $I_0[\,]$ is the modified Bessel function of order zero (e.g. see Abramowitz & Stegun, 1972, p.376).

One notices that the carrier frequency appears explicitely in the moments μ_{13} and μ_{14}; however, one can show easily that $\mu_{13}^2 + \mu_{14}^2$ is *independent* of ω_m and that, consequently, the joint density function of the envelope is independent of ω_m too. This is not surprising, since we have seen earlier that Rice's and Cramer & Leadbetter's definitions are equivalent.

10.4.7 Threshold crossings

The crossing rate of the threshold b by the process $X(t)$ has been studied earlier in this chapter; the expected rate of up-crossings is given by Equ.(10.6). Similarly, the expected rate of up-crossings of the level b by the envelope process $A(t)$ is given by

$$n_b^+ = \int_0^\infty \dot{a}\, p_{a\dot{a}}(b, \dot{a}, t)\, d\dot{a} \tag{10.43}$$

where $p_{a\dot{a}}(a, \dot{a}, t)$ is the joint probability density of the envelope process and its derivative. It is independent of t if the process is stationary.

$p_{a\dot{a}}(a, \dot{a})$ can be derived from $p_a(a_1, t; a_2, t+\tau)$ by noting that, for small τ, there is a one-to-one transformation

$$[A(t), A(t+\tau)] \Leftrightarrow [A(t), \dot{A}(t)]$$

because $a(t+\tau) \simeq a(t) + \tau\dot{a}(t)$. The determinant of the Jacobian of the transformation is τ. After some lenghty algebra (see e.g. Sveshnikov, 1966, p.266), it can be shown that the joint density can be factorized into the product of first order densities

$$p_{a\dot{a}}(a, \dot{a}) = \{\frac{a}{\sigma^2} \exp(-\frac{a^2}{2\sigma^2})\}.\{\frac{1}{\sqrt{2\pi}\sigma\sqrt{\omega_0^2 - \omega_1^2}} \exp[-\frac{\dot{a}^2}{2\sigma^2(\omega_0^2 - \omega_1^2)}]\} \tag{10.44}$$

where $\sigma = \sigma_x$, $\omega_0 = (m_2/m_0)^{1/2}$ is the central frequency and ω_1 is defined as

$$\omega_1 = \frac{m_1}{m_0} \tag{10.45}$$

This result shows that

- at the same time, $A(t)$ and $\dot{A}(t)$ are *independent* random variables.
- $A(t)$ follows a *Rayleigh distribution* (as we already know).
- $\dot{A}(t)$ follows a *Gaussian distribution* of zero mean and standard deviation

$$\sigma_{\dot{a}} = \sigma\sqrt{\omega_0^2 - \omega_1^2} = \sigma\,\omega_0\,\delta \tag{10.46}$$

where δ is a bandwidth parameter defined as

$$\delta = \sqrt{1-\omega_1^2/\omega_0^2} = (1 - \frac{m_1^2}{m_0 m_2})^{1/2} \tag{10.47}$$

Schwarz's inequality implies that $0 \le m_1^2/m_0 m_2 \le 1$ (see Problem P.10.5), hence $0 \le \delta \le 1$. δ is small for narrow-band processes and close to 1 for wide-band processes. Since $\sigma_{\dot{x}} = \omega_0 \sigma_x$, it follows from Equ.(10.46) that

$$\delta = \sigma_{\dot{a}}/\sigma_{\dot{x}} \tag{10.48}$$

Thus, δ *is the ratio between the RMS value of the slope of the envelope and that of the process.*

Another bandwidth parameter ε was defined by Equ.(10.20); both are close to zero if the process is narrow-band and close to 1 if it is wide-band. However, δ may be more convenient than ε, because its definition involves the spectral moments of orders up to 2 while that of ε involves also m_4 which does not exist for many processes used in practice (e.g. the response of a single d.o.f. oscillator to a white noise).

An alternative interpretation of δ is that of *relative width of the spectrum*. Indeed, the spread of the *PSD* about ω_1 can be written

$$\frac{2}{m_0}\int_0^\infty (\omega-\omega_1)^2 \Phi_{xx}(\omega)\, d\omega = \omega_0^2 (1 - \frac{\omega_1^2}{\omega_0^2}) = \omega_0^2 \delta^2 \tag{10.49}$$

The physical meaning of the frequency ω_1 is the following: In Cramer & Leadbetter's representation, $X(t) = A(t)\cos\phi(t)$, it can be shown that the phase derivative $Z(t) = \dot{\phi}(t)$ [which can be regarded as the instantaneous frequency] is distributed according to

$$p_Z(z) = \frac{\omega_0^2 - \omega_1^2}{2[(z-\omega_1)^2 + (\omega_0^2 - \omega_1^2)]^{3/2}} \tag{10.50}$$

(see e.g. Sveshnikov, 1966, p.267). This distribution is symmetric with respect to ω_1 where it is also maximum. Hence,

$$E[\dot{\phi}(t)] = \omega_1$$

Let us now return to the threshold crossing rate. Upon substituting Equ.(10.44) into (10.43), one gets

$$n_b^+ = \frac{1}{\sqrt{2\pi}}\,\delta\,\frac{b}{\sigma}\,\omega_0\, e^{-b^2/2\sigma^2} = \sqrt{2\pi}\,\delta\,\frac{b}{\sigma}\,\nu_b^+ \tag{10.51}$$

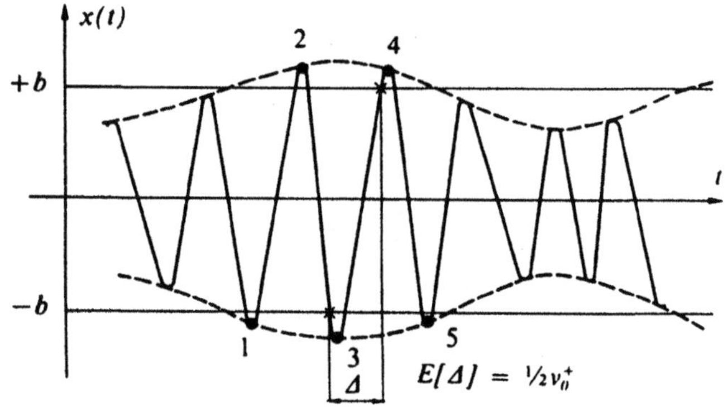

Figure 10.8: Threshold crossings of a narrow-band process.

10.4.8 Clump size

A typical sample of a zero mean narrow-band process is represented in Fig.10.8. Due to the slow variation of the envelope process, the threshold crossings of the process appear in clumps. The ***average clump size***, $< CS >$ is defined as the average number of crossings of the threshold $\pm b$ by the process $X(t)$ corresponding to a single crossing of the envelope $A(t)$ (Lyon, 1961). From Equ.(10.51), we get

$$< CS >= \frac{2\nu_b^+}{n_b^+} = \frac{\sqrt{2}}{\sqrt{\pi}\delta\eta} \tag{10.52}$$

with the usual notation $\eta = b/\sigma$. This interpretation appears to be correct for low values of the threshold, when $\nu_b^+ \gg n_b^+$. However, because the value of the envelope always exceeds that of the process, some of the envelope crossings may occur without any crossing of the process. This becomes significant when η is large, and it is reflected in Equ.(10.52) by $< CS >$ becoming smaller than 1.

To understand that, we refer to Fig.10.9 where various types of crossings are defined on the basis of the corresponding safety domain in the phase plane (Crandall, Chandiramani & Cook, 1966). A ***type** B* crossing corresponds to a one-sided barrier with a safety domain defined as $x < b$. *Type D* refers to a two-sided barrier, $|x| < b$ and ***type** E* corresponds to envelope crossings, $a < b$.

Comparing the safety domains for type D and type E crossings, one easily sees that some of the type E crossings may not be followed by type D crossings. Equ.(10.52) must be corrected to account for them. The fraction of envelope crossings which are not followed by type D crossings can be evaluated in the following way (Vanmarcke, 1975):

We construct a two state discrete process whose value is 1 when the envelope is above b and 0 when $A(t) < b$. Let T_0 and T_1 be the time intervals spent in state 0 and 1, respectively. Since $T_0 + T_1$ represents the time between two envelope

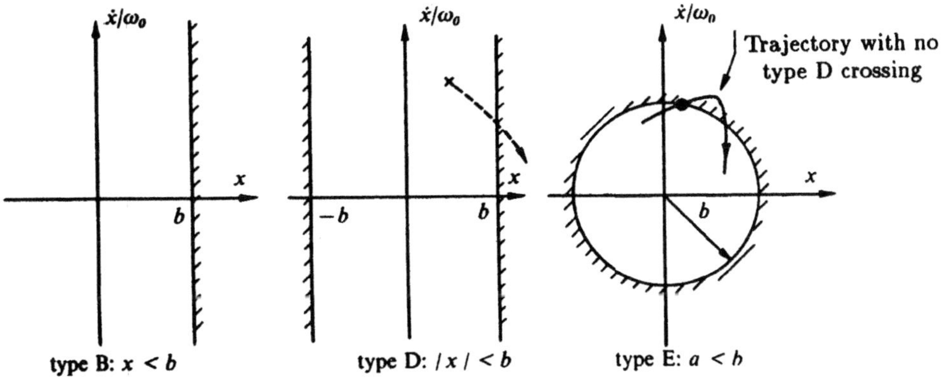

Figure 10.9: Definition of the various types of crossings and corresponding safety domains.

crossings (Fig.10.8),

$$\mathrm{E}[T_0 + T_1] = \frac{1}{n_b^+} \tag{10.53}$$

The fraction of time spent by the envelope above b can be evaluated from its probability density function according to

$$\frac{\mathrm{E}[T_1]}{\mathrm{E}[T_0 + T_1]} = \int_b^\infty p_a(a)\, da \tag{10.54}$$

Substituting the Rayleigh distribution (10.22), we find

$$\frac{\mathrm{E}[T_1]}{\mathrm{E}[T_0 + T_1]} = e^{-b^2/2\sigma^2} = \frac{\nu_b^+}{\nu_0^+} \tag{10.55}$$

It follows that

$$\mathrm{E}[T_1] = \frac{1}{n_b^+}\frac{\nu_b^+}{\nu_0^+} \qquad \mathrm{E}[T_0] = \frac{1}{n_b^+}\left(1 - \frac{\nu_b^+}{\nu_0^+}\right) \tag{10.56}$$

The fraction r of the type E crossings (of the envelope) which are directly followed by type D crossings is obtained as follows: If the time interval T_1 is larger than a half-cycle ($T_1 > 1/2\nu_0^+$), a type D crossing must take place, because the time Δ between two successive type D crossings is about the duration of a half-cycle. If $T_1 < 1/2\nu_0^+$, the probability that a type D crossing occurs is $2\nu_0^+ T_1$. Hence,

$$1 - r = \int_0^{1/2\nu_0^+} (1 - 2\nu_0^+ t) p_{T_1}(t)\, dt \tag{10.57}$$

In this expression, $1 - r$ is the fraction of type E crossings without type D crossings; $1 - 2\nu_0^+ t$ is the probability that a type D crossing does not occur during the time interval $T_1 = t$ in state 1. $p_{T_1}(t)$ is the [unknown] probability density function of T_1. It is computationally convenient to assume exponential distributions for T_0 and T_1:

$$P(T_0 < t) = 1 - e^{-\alpha t} \qquad P(T_1 < t) = 1 - e^{-\beta t} \tag{10.58}$$

The exponents are easily determined from Equ.(10.56):

$$\alpha = n_b^+ \left(\frac{\nu_0^+}{\nu_0^+ - \nu_b^+}\right) \qquad \beta = n_b^+ \frac{\nu_0^+}{\nu_b^+} \tag{10.59}$$

Introducing $p_{T_1}(t) = \beta e^{-\beta t}$ into Equ.(10.57), we find

$$r = \frac{2\nu_b^+}{n_b^+}\left[1 - \exp\left(-\frac{n_b^+}{2\nu_b^+}\right)\right] \tag{10.60}$$

An improved estimate of the average clump size can be obtained by correcting the denominator of Equ.(10.52) to include only those type E crossings associated with type D ones:

$$< CS > = \frac{2\nu_b^+}{r\, n_b^+} = \frac{1}{1 - \exp\left(-\frac{n_b^+}{2\nu_b^+}\right)} \tag{10.61}$$

As expected, Equ.(10.61) converges to (10.52) for small values of the threshold, and to 1 as b becomes very large. An alternative approach for determining the clump size can be found in (Racicot & Moses, 1971).

To illustrate the foregoing discussion, consider the stationary response of a lightly damped single d.o.f. oscillator to a white noise; the bandwidth parameter δ reads (Problem P.10.6)

$$\delta \approx \left(\frac{4\xi}{\pi}\right)^{1/2}$$

It follows that $n_b^+/2\nu_b^+ = \eta\sqrt{2\xi}$. For $\xi = 0.02$ and $\eta = 3$, formulae (10.52) and (10.61) give $< CS >= 1.67$ and $< CS >= 2.22$, respectively. For $\xi = 0.04$ and $\eta = 4$, we get $< CS >= 0.88$ ($< 1!$) and $< CS >= 1.48$.

Next consider the band-limited white noise $[\omega_n, \Delta\omega]$ (Problem P.10.7). If $\Delta\omega \ll \omega_n$, the bandwidth parameter reads

$$\delta \approx \frac{\Delta\omega}{2\sqrt{3}\,\omega_n}$$

If one chooses the same bandwidth as in the previous example, $\Delta\omega = 2\xi\omega_n$, one finds $n_b^+/2\nu_b^+ = \eta\xi\sqrt{\pi/6}$. For $\xi = 0.02$ and $\eta = 3$, Equ.(10.52) and (10.61) give respectively $< CS >= 23.03$ and $< CS >= 23.54$. These values are much larger than those obtained in the previous example; this shows that *the clump size depends critically of the shape of the PSD in the vicinity of the central frequency* ω_n. This observation is illustrated in Fig.10.10 where samples corresponding to various spectral shapes are shown, together with their envelope.

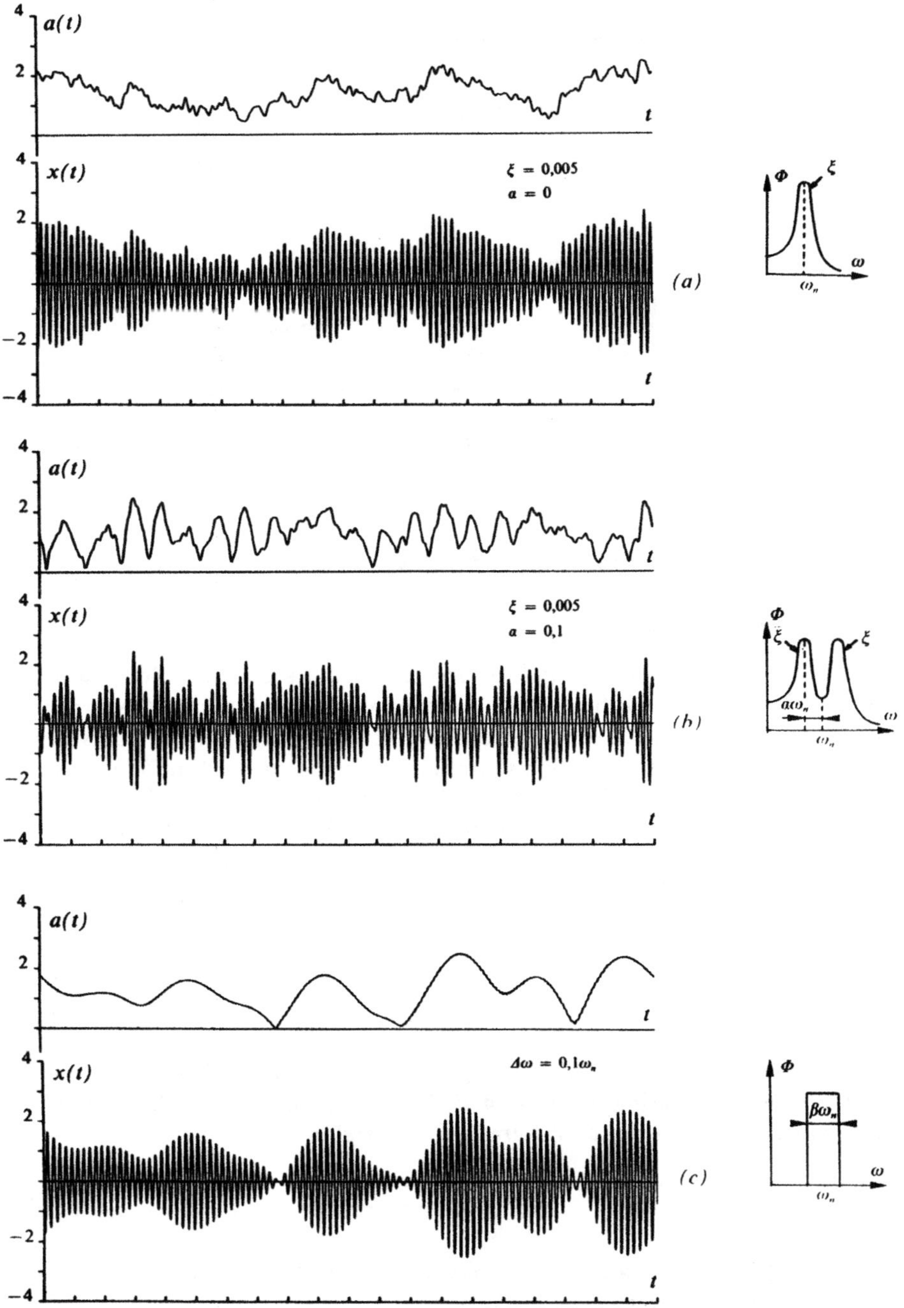

Figure 10.10: Sample of stationary Gaussian processes and corresponding envelope for various spectral shapes ($\sigma = 1$). (a) Linear oscillator. (b) Bimodal spectrum. (c) Band-limited white noise.

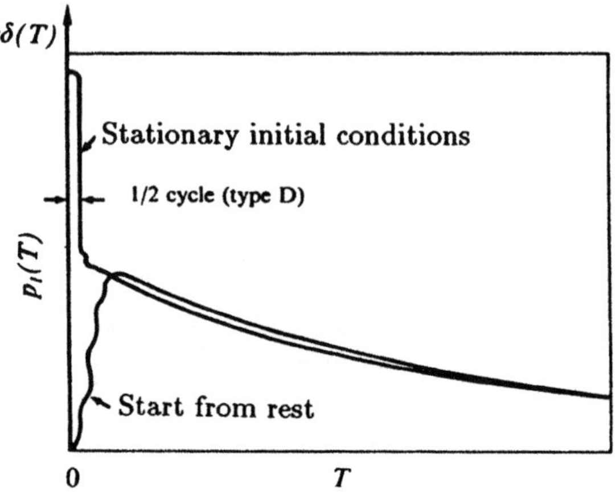

Figure 10.11: First-crossing density of a single d.o.f. oscillator. The earlier part of the distribution depends on the initial conditions.

10.5 First-crossing problem

10.5.1 Introduction

Consider a structure whose failure occurs when its response (displacement, stress, ...) exceeds some limit. The designer wishes to evaluate the probability that the limit value be exceeded over a given operating period. Equivalently, he can evaluate the probability distribution of the operating time T after which the limit is exceeded for the first time (***First-crossing problem***). This problem is extremely difficult and not fully solved, even for fairly simple situations.

Let $X(t)$ be the response of a lightly damped single d.o.f. oscillator to a Gaussian white noise. For each type of crossings of Fig.10.9, one can define the *reliability* $W(T)$ as the probability that the process remains within the safe domain. For type D crossings, it reads

$$W(T) = \text{Prob}\{|X(t)| < b, \quad 0 \le t < T\} \tag{10.62}$$

$W(T)$ represents the fraction of samples which have not left the safe domain after T. It is a decreasing function of T. The probability density function of the first-crossing time (in short ***first-crossing density***) is

$$p_1(T) = -\frac{\mathrm{d}W(T)}{\mathrm{d}T} \tag{10.63}$$

because $p_1(T)\,dT$ represents the probability that the first-crossing occurs in $[T, T + dT]$. It is illustrated in Fig.10.11.

After some time, $p_1(T)$ appears as a decaying exponential. The earlier portion of the curve depends very much on the initial conditions: If the system starts from rest, $p_1(T)$ starts from 0 and increases gradually until the system response reaches its steady state. If the oscillator starts from stationary initial conditions, $p_1(T)$ exhibits larger amplitudes near the origin for about half a cycle (type D crossings). They result from two contributions: *(i)* A Dirac delta function at the origin corresponds to the probability that the initial state be outside the safe domain. *(ii)* The larger amplitudes during the first half-cycle correspond to the probability that the initial conditions lead to immediate crossing of the process. Such a trajectory is shown in the centre of Fig.10.9. This part of the first crossing density would last a complete cycle for type B crossings.

For large times, the first crossing density can be written

$$p_1(T) = A\alpha\, e^{-\alpha T} \tag{10.64}$$

A accounts for the initial conditions: $A > 1$ for zero initial conditions, $A < 1$ for steady state initial conditions and $A \to 1$ for high threshold levels, because the probability of crossing within the first half-cycle becomes small. α is called *limiting decay rate.* If $A = 1$,

$$p_1(T) = \alpha\, e^{-\alpha T} \tag{10.65}$$

The mean time until first-crossing is

$$\mathrm{E}[T] = \frac{1}{\alpha} \tag{10.66}$$

and the standard deviation is also $\sigma_T = 1/\alpha$. In what follows, we examine some approximate solutions for $p_1(T)$ resulting from various assumptions on the crossing process.

10.5.2 Independent crossings

The simplest approximation is obtained by assuming that the up-crossings of the threshold b can be considered as *independent* events. Accordingly, the number of crossings within $[0, T)$ constitutes a Poisson process of arrival rate $\lambda = \nu_b^+$ for type B crossings or $\lambda = 2\nu_b^+$ for type D crossings. We shall focuse on type D crossings in what follows. The probability that n crossings occur within the observation period T is given by Equ.(4.23):

$$P\{n \text{ crossings in } [0, T)\} = e^{-2\nu_b^+ T} \frac{(2\nu_b^+ T)^n}{n!} \tag{10.67}$$

The reliability $W(T)$ corresponds to $n = 0$:

$$W(T) = e^{-2\nu_b^+ T} \tag{10.68}$$

and, from Equ.(10.63),

$$p_1(T) = 2\nu_b^+\, e^{-2\nu_b^+ T} \tag{10.69}$$

Combining with Equ.(10.7), one finds the limiting decay rate

$$\alpha = 2\nu_b^+ = 2\nu_0^+ \, e^{-b^2/2\sigma^2} \tag{10.70}$$

The assumption of independent crossings may be criticized, especially for narrow-band processes, because it ignores the tendency of the crossings to occur in clumps. It is conservative, because it tends to underestimate the duration between successive crossings. It can be shown that the assumption of independent crossings is asymptotically correct when $b \to \infty$ (Cramer & Leadbetter, 1967).

10.5.3 Independent envelope crossings

The principal weakness of the foregoing model is the assumption that the crossings occur according to a Poisson process. Since the crossings of a narrow-band process tend to occur in clumps and an envelope crossing occurs before each clump, another approximation can be obtained by assuming that *the envelope crossings are independent events.* Doing that, we have substituted type E crossings for type D crossings. Following the same procedure as in the previous section, the number of envelope crossings in $[0, T)$ constitutes a Poisson process with arrival rate $\lambda = n_b^+$. Taking into account Equ.(10.51), one finds

$$\alpha = n_b^+ = \sqrt{2\pi}\, \delta\, \eta\, \nu_b^+ \tag{10.71}$$

Note that this estimator of α depends on the bandwidth of the process. It is better than Equ.(10.70) for narrow-band processes and low values of the threshold. However, for large b, it may be over-conservative, because type E crossings occur more often than type D crossings. This weakness can be removed as discussed below.

10.5.4 Approach based on the clump size

The weakness of the foregoing model comes from the fact that, for large values of b, some envelope crossings may not be followed by crossings of the process. As in section 10.4.8, this drawback can be removed by replacing the arrival rate of the envelope crossings, n_b^+, by the arrival rate of the clumps, $2\nu_b^+/ < CS >$. From Equ.(10.61) and (10.52), the limiting decay rate reads

$$\alpha = \frac{2\nu_b^+}{< CS >} = 2\nu_b^+ [1 - \exp(-\sqrt{\frac{\pi}{2}}\delta\eta)]. \tag{10.72}$$

Note that this estimator tends to $2\nu_b^+$ as $\eta \to \infty$.

10.5.5 Vanmarcke's model

For stationary initial conditions, the term A in Equ.(10.64) represents the probability that the initial value of the envelope be smaller than b or, equivalently, that the initial state of the discrete process defined in section 10.4.8 be 0. From Equ.(10.55),

$$A = \frac{\mathrm{E}[T_0]}{\mathrm{E}[T_0 + T_1]} = 1 - \frac{\nu_b^+}{\nu_0^+} = 1 - e^{-\eta^2/2} \qquad (10.73)$$

Combining this with the limiting decay rate of Equ.(10.72), one gets the following form of the reliability:

$$W(T) = Ae^{-\alpha T} = (1 - e^{-\eta^2/2}) \exp\{-2\nu_0^+ T . \frac{[1 - \exp(-\sqrt{\frac{\pi}{2}}\delta\eta)]}{\exp(\eta^2/2)}\} \qquad (10.74)$$

This expression corresponds to type E crossings; the following modified form has been established for type D crossings (Vanmarcke, 1975)

$$W(T) = Ae^{-\alpha T} = (1 - e^{-\eta^2/2}) \exp\{-2\nu_0^+ T . \frac{[1 - \exp(-\sqrt{\frac{\pi}{2}}\delta\eta)]}{\exp(\eta^2/2) - 1}\} \qquad (10.75)$$

This is an *explicit* form of the reliability as a function of the reduced level $\eta = b/\sigma$, the number of half-cycles $2\nu_0^+ T$ and the bandwidth parameter δ. Simulations have shown that the accuracy can be improved with the modified bandwidth parameter (Vanmarcke, 1975)

$$\delta_e = \delta^{1.2} \qquad (10.76)$$

10.5.6 Extreme point process

The discrete time process $Y(i)$ constituted by the absolute extrema of the process $X(t)$ is called the *extreme point process* (Fig.10.12). If $X(t)$ is narrow-band, the time interval Δ between successive extrema is nearly constant

$$\Delta = \frac{1}{2\nu_0^+} \qquad (10.77)$$

The first-passage problem of the extreme point process $Y(i)$ is analysed as follows. If $h(n)$ stands for the probability that the nth extremum be the first one beyond the threshold b:

$$h(n) = P[Y(n) \geq b | \bigcap_{i=1}^{n-1} Y(i) < b] \qquad (10.78)$$

the reliability can be written as

$$W(T) = \prod_{n=1}^{N} [1 - h(n)] \qquad (10.79)$$

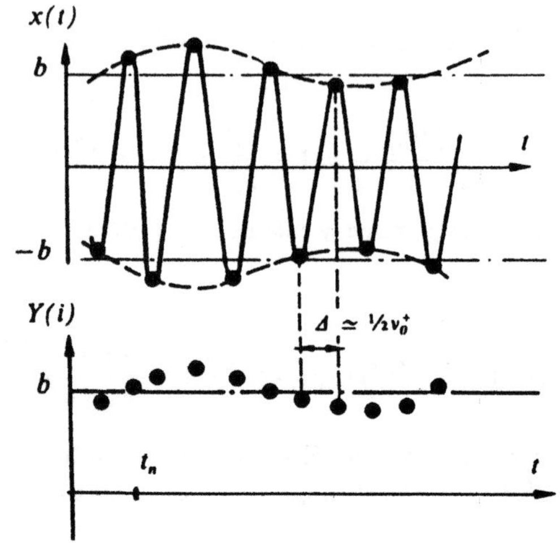

Figure 10.12: Definition of the extreme point process.

where $N = 2\nu_0^+ T$ is the number of half-cycles during T. Various models can be built by making assumptions on the way the extrema occur.

The simplest assumption is that of *independent extrema*. In that case, the condition disappears from Equ.(10.78). For a narrow-band process, the extrema follow a Rayleigh distribution and $h(n)$ reads

$$h(n) = q_0 = P[Y(n) \geq b] = \int_\eta^\infty x e^{-x^2/2} dx = e^{-\eta^2/2} \tag{10.80}$$

After substitution into Equ.(10.79), one gets

$$W(T) = (1 - e^{-\eta^2/2})^N = (1 - \frac{\nu_b^+}{\nu_0^+})^{2\nu_0^+ T} \tag{10.81}$$

This result is very close to Equ.(10.68) for $\eta > 2.5$ (Problem P.10.8). This is not surprising, because the assumption of independent extrema is quite close to that of independent crossings.

The next step consists of assuming that the extreme point process is *Markovian*. In that case, the conditional probability $h(n)$ depends only on the latest of the previous times:

$$h(n) = P[Y(n) \geq b | Y(n-1) < b] = \frac{P[Y(n) \geq b \cap Y(n-1) < b]}{P[Y(n-1) < b]} \tag{10.82}$$

Its evaluation requires the joint distribution of the maxima. For a narrow-band process, it can be approximated by the joint distribution of the envelope for instant of times separated by a half-cycle, $\Delta = 1/2\nu_0^+$. If we denote

$$\bar{q} = P[Y(n) \geq b \cap Y(n-1) < b] = \int_b^\infty da_2 \int_0^b p_a(a_1; a_2, \Delta)\, da_1 \tag{10.83}$$

$$h(n) = \frac{\bar{q}}{1 - q_0} \qquad (n > 1) \tag{10.84}$$

and

$$W(T) = (1 - q_0)[1 - \frac{\bar{q}}{1 - q_0}]^{N-1} \tag{10.85}$$

For narrow-band processes, the Markov assumption leads to a substantial improvement with respect to that of independent extrema. Although formally simple, Equ.(10.85) requires the numerical evaluation of $\bar{q}$ from Equ.(10.83).

10.6 First-passage problem and Fokker-Planck equation

10.6.1 Multidimensional Markov process

All the approximate models of the reliability developed in the previous section use in a more or less adequate way the statistics of the process and its envelope. The quality of the approximation depends on the adequacy of the assumption involved. A more rigorous formulation can be based on a vector Markov process as discussed in section 9.9. Indeed, the reliability satisfies a Kolmogorov equation. However, no analytical solution has been found, even for a single d.o.f. oscillator excited by a white noise. Numerical solutions do exist.

For a lightly damped single d.o.f. oscillator, the narrow-band property can be exploited to reduce the problem to a one-dimensional Markov process. This can be done by the method of ***stochastic averaging*** (Stratonovich, 1967).

10.6.2 Fokker-Planck equation of the envelope

Consider the lightly damped linear oscillator

$$\ddot{X} + 2\xi\omega_n \dot{X} + \omega_n^2 X = F(t) \tag{10.86}$$

excited by a zero mean stationary wide-band process $F(t)$ ($\omega_c \gg \omega_n$). If we define the amplitude $A(t)$ and phase $\theta(t)$ by

$$X = A\cos(\omega_n t + \theta) \qquad \dot{X} = -- A\omega_n \sin(\omega_n t + \theta)$$

Equ.(10.86) can be replaced by the two first order equations

$$\dot{A} = -2\xi\omega_n A \sin^2(\omega_n t + \theta) - \frac{F\sin(\omega_n t + \theta)}{\omega_n} \tag{10.87}$$

$$\dot{\theta} = -2\xi\omega_n \sin(\omega_n t + \theta)\cos(\omega_n t + \theta) - \frac{F\cos(\omega_n t + \theta)}{A\omega_n}$$

The right hand side of these equations contains oscillatory terms at ω_n and at the double frequency $2\omega_n$. Since A and θ are *slowly varying* functions of t, the latter can be eliminated by averaging their contribution over one period of the system. This leads to

$$\dot{A} = -\xi\omega_n A - \frac{F\sin(\omega_n t + \theta)}{\omega_n} \tag{10.88}$$

$$\dot{\theta} = -\frac{F\cos(\omega_n t + \theta)}{A\omega_n}$$

If the bandwidth of the excitation is much larger than the natural frequency of the oscillator, it can be shown (Stratonovich, 1967; Ariaratnam & Pi, 1973) that this system is approximately stochastically equivalent (in the sense that they lead to the same Fokker-Planck equation) to the system

$$\dot{A} = -\xi\omega_n A + \frac{\pi\Phi_f(\omega_n)}{2A\omega_n^2} + \frac{\sqrt{\pi\Phi_f(\omega_n)}}{\omega_n}F_1 \tag{10.89}$$

$$\dot{\theta} = -\frac{\sqrt{\pi\Phi_f(\omega_n)}}{A\omega_n}F_2$$

where F_1 and F_2 are independent Gaussian white noises of unit intensity and $\Phi_f(\omega_n)$ is the *PSD* of the excitation for ω_n. We observe that the amplitude equation is *decoupled* from that of the phase, which means that the envelope process is approximately Markovian. The corresponding one-dimensional Fokker-Planck equation reads

$$\frac{\partial q}{\partial t} = \frac{\partial}{\partial a}[(\xi\omega_n a - \frac{\pi\Phi_f(\omega_n)}{2a\omega_n^2})q] + \frac{\pi\Phi_f(\omega_n)}{2\omega_n^2}\frac{\partial^2 q}{\partial a^2} \tag{10.90}$$

Its analytical solution can be found in (Stratonovich, 1963, p.73). Thus, provided that the correlation time of the excitation is smaller than the period of the oscillator, the arbitrary wide-band *PSD* $\Phi_f(\omega)$ is replaced by a white noise $\Phi_0 = \Phi_f(\omega_n)$.

10.6.3 Kolmogorov equation of the reliability

In section 9.8.3, we have seen that the probability that the value of a one-dimensional Markov process belongs to some domain Ω at τ given the initial condition x_0 satisfies the Kolmogorov equation (9.67). Accordingly, the reliability associated with type E crossings of the threshold b

$$W(\tau|a_0) = Prob\{A(t) < b, \quad 0 < t \le \tau | A(0) = a_0\}$$

satisfies the Kolmogorov equation

$$\frac{\partial W}{\partial \tau} = -\xi\omega_n(a_0 - \frac{\sigma^2}{a_0})\frac{\partial W}{\partial a_0} + \xi\omega_n\sigma^2\frac{\partial^2 W}{\partial a_0^2} \tag{10.91}$$

where σ^2 denotes the variance of the steady state response. If one uses the reduced time $\tau' = \tau\xi\omega_n$, it can be further simplified into

$$\frac{\partial W}{\partial \tau'} = -(a_0 - \frac{\sigma^2}{a_0})\frac{\partial W}{\partial a_0} + \sigma^2\frac{\partial^2 W}{\partial a_0^2} \tag{10.92}$$

with the initial and boundary conditions

$$W(0|a_0) = 1 \qquad 0 \le a_0 < b$$

$$W(\tau|b) = 0 \qquad \tau > 0$$

The limit b acts as an absorbing barrier, while the origin acts as a reflecting one, for $a_0 \ge 0$. From the above equation, it is possible to deduce ordinary differential equations for the moments of the first-passage time (Ariaratnam & Pi, 1973). The solution of the Kolmogorov equation (10.92) can be found in (Lennox & Fraser, 1974; Solomos & Spanos, 1982).

10.7 Peak factor

10.7.1 Extreme value probability

Consider a zero mean stationary random process that we observe during a period T. We seek the probability distribution function $p_e(\eta, T)$ of the absolute extreme value during T. Its reduced value with respect to the standard deviation, $\eta = b/\sigma$ is called *peak factor* (Fig.10.13). The probability that the peak factor is smaller than η during the observation period is

$$P_e(T,\eta) = \int_0^\eta p_e(T,x)\,dx \qquad (\eta \ge 0) \tag{10.93}$$

It is, by construction, identical to the reliability that we have defined as

$$W(T,\eta) = Prob[|X(t)| < \eta\sigma, \quad 0 \le t < T] \tag{10.94}$$

(type D crossings). As a result, the probability distribution function of the peak factor can be deduced from the reliability according to

$$p_e(T,\eta) = \frac{\partial W(T,\eta)}{\partial \eta} \tag{10.95}$$

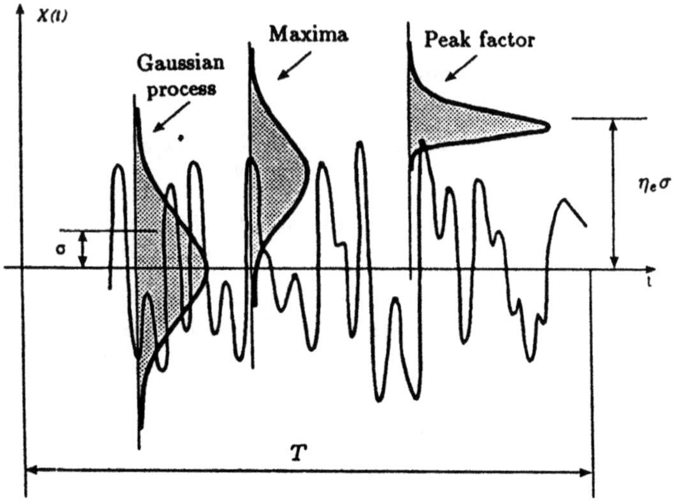

Figure 10.13: Definition of the peak factor.

All the reliability models of the previous sections are directly applicable. In order to illustrate the main features of the distribution, consider the model based on independent crossings (section 10.5.2):

$$W(T,\eta) = \exp[-2\nu_0^+ T e^{-\eta^2/2}] = \exp[-N e^{-\eta^2/2}] \tag{10.96}$$

where $N = 2\nu_0^+ T$ is the number of equivalent half-cycles during T. The corresponding probability density function is shown in Fig.10.14 for various N. As N increases, the distribution moves to the right and becomes increasingly peaked. This model depends only on two parameters: the reduced level η and the number of half-cycles N. It does not depend on the bandwidth of the process. As already noted, the Poisson model is conservative for narrow-band processes. This is illustrated in Fig.10.15, where it is compared to the models based on Equ.(10.75) and (10.85) which include a measure of the bandwidth.

10.7.2 Formulae for the peak factor

In Fig.10.14, we see that typical values of the peak factor range from 3 to 5, depending on the number of cycles. For engineering applications, it is important to have simple approximate formulae for the mean and standard deviation of the peak factor. Based on the Poisson model (10.96), the following formulae have been proposed (A.G.Davenport, 1964):

$$E[\eta_e] \approx (2\ln N)^{1/2} + \frac{\gamma}{(2\ln N)^{1/2}} \tag{10.97}$$

$$\sigma[\eta_e] \approx \frac{\pi}{\sqrt{6}} \frac{1}{(2\ln N)^{1/2}} \tag{10.98}$$

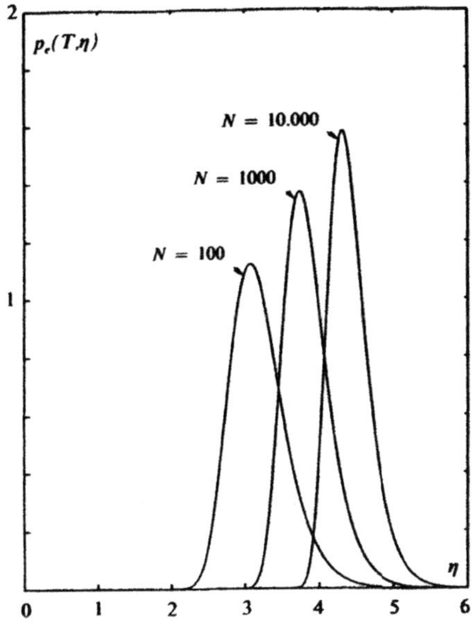

Figure 10.14: Poisson model. Probability density function of the peak factor for various values of N.

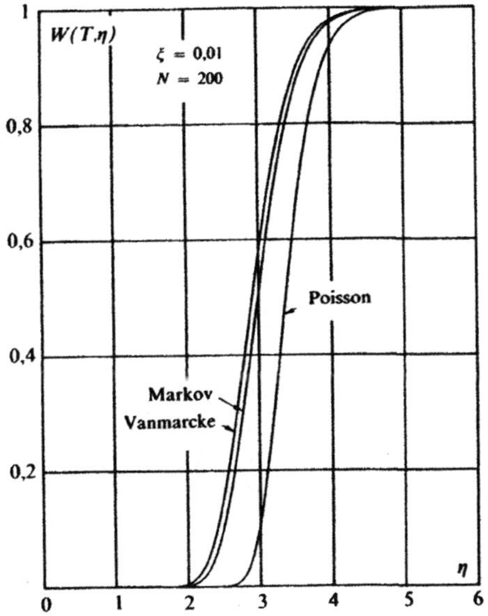

Figure 10.15: Probability distribution function of the peak factor of the response of a lightly damped linear oscillator ($\xi = 0.01$) to a white noise. Comparison of various models for $N = 200$.

where $\gamma = 0.5772$. These formulae involve the single parameter $N = 2\nu_0^+ T$. To account for the fact that the mean tends to decrease for narrow-band processes, formula (10.97) can be slightly modified according to

$$E[\eta_e] \approx (2\ln \kappa N)^{1/2} + \frac{\gamma}{(2\ln N)^{1/2}} \qquad (10.99)$$

where $\kappa < 1$ accounts for the bandwidth of the system. Based on extensive simulations (Preumont, 1985), κ can be chosen according to

$$\kappa = 1.5(1 - e^{-1.8\delta}) \qquad \delta < 0.5 \qquad (10.100)$$

$$\kappa = 0.94 \qquad \delta > 0.5$$

where δ is the bandwidth parameter defined by Equ.(10.47). The effect of the bandwidth on the standard deviation of the peak factor is fairly small.

10.8 References

S.T.ARIARATNAM & H.N.PI, On the first-passage time for envelope crossing for a linear oscillator, *Int. Journal of Control, Vol.18, No 1, pp.89-96,* 1973.

D.E.CARTWRIGHT & M.S.LONGUET-HIGGINS, The statistical distribution of the maxima of a random function, *Proc. Roy. Soc. Ser. A, 237, pp.212-232,* 1956.

H.CRAMER & M.R.LEADBETTER, *Stationary and Related Stochastic Processes,* Wiley, 1967.

S.H.CRANDALL, Zero crossings, peaks, and other statistical measures of random responses, *The Journal of the Acoustical Society of America, Vol.35, No 11, pp.1693-1699, November* 1963.

S.H.CRANDALL, K.L.CHANDIRAMANI & R.G.COOK, Some first passage problems in random vibration, *ASME Journal of Applied Mechanics, Vol.33, pp.532-538, September* 1966.

S.H.CRANDALL, First crossing probabilities of the linear oscillator, *Journal of Sound and Vibration 12(3), pp.285-299,* 1970.

A.G.DAVENPORT, Note on the distribution of the largest value of a random function with application to gust loading, *Proc. Inst. Civ. Eng., Vol 28, pp.187-196,* 1964.

W.B.DAVENPORT, *Probability and Random Processes,* McGraw-Hill, 1970.

S.KRENK, Nonstationary narrow-band response and first-passage probability, *ASME Journal of Applied Mechanics, Vol.46, pp.919-924, December* 1979.

W.C.LENNOX & D.A.FRASER, On the first-passage distribution for the envelope of a nonstationary narrow-band stochastic process. *ASME Journal of Applied Mechanics, Vol.41, pp.793-797, September* 1979.

Y.K.LIN, *Probabilistic Theory of Structural Dynamics,* McGraw-Hill, 1967.

Y.K.LIN, First-excursion failure of randomly excited structures, *AIAA Journal, Vol.8, No 4, pp.720-725,* 1970.

R.H.LYON, On the vibration statistics of a randomly excited hard-spring oscillator II, *The Journal of the Acoustical Society of America, Vol.33, No 10, pp.1395-1403, October* 1961.

D.MIDDLETON, *An Introduction to Statistical Communication Theory*, McGraw-Hill, 1960.

A.PREUMONT, On the peak factor of stationary Gaussian processes, *Journal of Sound and Vibration 100(1), pp.15-34,* 1985.

R.L.RACICOT & F.MOSES, A first-passage approximation in random vibration, *ASME Journal of Applied Mechanics, pp.143-147, March* 1971.

S.O.RICE, Mathematical analysis of random noise, *Bell System Technical Journal, 23, pp.282-332,* 1944; *24, pp.46-156,* 1945. Reprinted in *Selected Papers on Noise and Stochastic Processes*, N.WAX ed., Dover, 1954.

J.B.ROBERTS, First passage time for the envelope of a randomly excited linear oscillator, *Journal of Sound and Vibration, 46 (1), pp.1-14,* 1976.

G.P.SOLOMOS & P-T.D.SPANOS, Solution of the Backward-Kolmogorov equation for nonstationary oscillation problem. *ASME Journal of Applied Mechanics, Vol.49, pp.923-925, December* 1982.

R.L.STRATONOVICH, *Topics in the Theory of Random Noise*, Vol.2, Gordon & Breach, N-Y, 1967.

A.A.SVESHNIKOV, *Applied Methods of the Theory of Random Functions*, Pergamon Press, 1966.

E.H.VANMARCKE, Properties of spectral moments with applications to random vibration, *ASCE Journal of Engineering Mechanics Division, EM2, pp.425-446, April* 1972.

E.H.VANMARCKE, On the distribution of the first-passage time for normal stationary random processes, *ASME Journal of Applied Mechanics, Vol.42, pp.215-220, March* 1975.

J.N.YANG & M.SHINOZUKA, On the first excursion probability in stationary narrow-band random vibration, *ASME Journal of Applied Mechanics, Vol.38, pp.1017-1022, December* 1971.

J.N.YANG & M.SHINOZUKA, On the first excursion probability in stationary narrow-band random vibration, II, *ASME Journal of Applied Mechanics, Vol.39, pp.733-738, September* 1972.

10.9 Problems

P.10.1 *Show that the sine and cosine components of a narrow band process satisfy:*

(a) $C(t)$ and $S(t)$ are orthogonal with the same variance as $X(t)$.

(b)

$$E[C(t)C(t+\tau)] = E[S(t)S(t+\tau)] = 2\int_0^\infty \Phi_{xx}(\omega)\,\cos(\omega-\omega_m)\tau\,d\omega$$

(c)

$$E[C(t)S(t+\tau)] = 2\int_0^\infty \Phi_{xx}(\omega)\,\sin(\omega-\omega_m)\tau\,d\omega$$

[Hint: Start from Equ.(10.28) and use the fact that the filter is low-pass.]

P.10.2 *If $\hat{X}(t)$ is the Hilbert transform of $X(t)$, show that:*

(a) $X(t)$ and $\hat{X}(t)$ are orthogonal and have the same variance.

(b)

$$E[X(t)\hat{X}(t+\tau)] = 2\int_0^\infty \Phi_{xx}(\omega)\,\sin\omega\tau\,d\omega$$

(c)

$$E[\hat{X}(t)\hat{X}(t+\tau)] = R_{xx}(\tau)$$

[Hint: Start from Equ.(10.30).]

P.10.3 *For a narrow-band process of carrier frequency ω_m with the representation (10.24) and (10.34), show that*

$$V(t) = A(t)\exp[\theta(t)] = X_+(t)\exp(-j\omega_m t) = [X(t) - j\hat{X}(t)]\exp(-j\omega_m t)$$

P.10.4 *For a narrow-band process, show that the sine and cosine components $C(t)$ and $S(t)$ are related to the process $X(t)$ and its Hilbert transform $\hat{X}(t)$ by*

$$C(t) = X(t)\cos\omega_m t - \hat{X}(t)\sin\omega_m t$$

$$S(t) = -X(t)\sin\omega_m t - \hat{X}(t)\cos\omega_m t$$

$$X(t) = C(t)\cos\omega_m t - S(t)\sin\omega_m t$$

$$\hat{X}(t) = -C(t)\sin\omega_m t - S(t)\cos\omega_m t$$

P.10.5 *Show that $m_1^2 \leq m_0 m_2$.*

[Hint: Start from the results of problem P.10.2 and use Schwarz's inequality.]

P.10.6 *Consider the response of a lightly damped single d.o.f. oscillator to a white noise. Show that*

$$m_1 = \frac{\omega_n\sigma^2}{\sqrt{1-\xi^2}}\left[1 - \frac{1}{\pi}\arctan\left(\frac{2\xi\sqrt{1-\xi^2}}{1-2\xi^2}\right)\right]$$

and

$$\delta \approx (\frac{4\xi}{\pi})^{1/2}$$

[Hint: Start from Equ.(5.45).]

P.10.7 *Consider the band-limited white noise process*

$$\Phi_{xx}(\omega) = \Phi_0 \qquad \omega_n - \frac{\Delta\omega}{2} \leq |\omega| \leq \omega_n + \frac{\Delta\omega}{2}$$

[$\Phi_{xx}(\omega) = 0$ outside]. Show that

$$m_0 = 2\Phi_0\Delta\omega, \qquad m_1 = 2\Phi_0\omega_n\Delta\omega, \qquad m_2 = 2\Phi_0\Delta\omega(\omega_n^2 + \frac{\Delta\omega^2}{12})$$

$$\delta \approx \frac{\Delta\omega}{2\sqrt{3}\,\omega_n} \qquad (\Delta\omega \ll \omega_n)$$

P.10.8 *Compare the limiting decay rates α under the assumptions of independent crossings and independent extrema (draw a plot of $\alpha/2\nu_b^+$ as a function of η).*

P.10.9 *Consider the response of a lightly damped oscillator to a white noise. Draw a plot of $\alpha/2\nu_b^+$ based on the clump size [Equ.(10.71)] for various ξ. Compare with the Poisson assumption.*

P.10.10 *For a lightly damped oscillator observed during $N = 200$ cycles, compare formulae (10.97) and (10.99) of the mean peak factor for various ξ.*

Chapter 11

Random Fatigue

11.1 Introduction

Fatigue is probably the most frequent form of failure in mechanical structures. Most of the time, the failure occurs after a fairly large number of cycles ($N > 1000$) with a nominal stress never exceeding the yield stress of the material. The nominal structure remains linear and the stress field is Gaussian if the excitation is Gaussian. This is referred to as ***high-cycle fatigue***. On the contrary, low-cycle fatigue usually involves cyclic plastic strains and the Gaussian property is destroyed. Low-cycle fatigue will not be considered in this book.

With the increasing demand for high performance structures, the fatigue damage assessment has become more and more important, and it is desirable to integrate it within the finite element programs. This chapter proposes a numerical procedure allowing the evaluation of the high-cycle fatigue damage for multiaxial random stresses. A finite element implementation is proposed.

Usually, the information about the uniaxial fatigue behaviour of materials is available in the form of S-N curves, which provide the number of cycles N to failure under an alternating sine stress of constant amplitude S. N actually varies from one sample to another and is in fact a random variable. For a wide class of materials, the average of the distribution can be approximated by

$$NS^{\beta} = c \tag{11.1}$$

where the constants c and β depend on the material ($5 < \beta < 20$). This relationship implies that any stress level produces a damage (it does not account for the endurance limit). In what follows, we shall ignore the statistical scatter in the material behaviour and assume that Equ. (11.1) applies in the deterministic sense.

Fatigue life prediction for complex load histories can be treated by a cumulative damage analysis. The ***linear damage theory (Palmgren-Miner criterion)*** allows us to extrapolate constant amplitude tests by assuming that the damage associated with each stress cycle can be added linearly. Equation (11.1) is extended to stresses of varying amplitude as follows: If a sample is subjected to n_i cycles at the stress level S_i, it suffers a damage $d_i = n_i/N_i$, where $N_i = cS_i^{-\beta}$ is the number of cycles to failure corresponding to the constant amplitude S_i. The damages associated with various levels of loading add up linearly, producing a total damage

$$D = \sum_i \frac{n_i}{N_i} \tag{11.2}$$

The failure occurs when the total damage reaches $D = 1$. This criterion does not take into account the order of application of the various stress levels; it is known to be inaccurate, but it has the enormous advantage of being simple and relying on constant amplitude tests for which a lot of experimental data is available. At least, the criterion gives a good *relative* information and can be used for comparison purposes, as for example to check the influence of structural modifications. If the stress histories are available, the counting of the stress cycles can be done according to the *rainflow* method. The procedure requires the knowledge of the whole time signal before the count can start, but it becomes simple when the same stress history is repeatedly applied. In random vibration, the counting of the stress cycles must be derived from the *PSD* of the random stress.

11.2 Uniaxial loading with zero mean

Let $X(t)$ be a uniaxial Gaussian stress with zero mean and *PSD* $\Phi(\omega)$. We assume that the material behaves according to Equ.(11.1) and that the linear damage theory (11.2) applies.

In the classical theory of random fatigue, it is assumed that any *positive maximum* between b and $b + db$ contributes to the damage for one cycle, that is, according to the *S-N* curve, $b^\beta c^{-1}$. Since the fatigue damage is essentially related to tension stresses (and not to compression stresses), it is reasonable to assume that the negative maxima do not contribute to the damage. Accordingly, the expected damage per unit time is given by

$$E[D] = c^{-1} E[M_T] \int_0^\infty b^\beta q(b)\, db$$

where $E[M_T]$ is the expected [total] number of maxima per unit time and $q(b)$ is the probability density function of the maxima. Introduce the reduced stress $\eta = b/\sigma_x$,

$$E[D] = c^{-1} E[M_T]\, \sigma_x^\beta \int_0^\infty \eta^\beta q(\eta)\, d\eta \tag{11.3}$$

where $q(\eta)$ is given by Equ.(10.19). For a narrow-band process, the *central frequency* ν_0^+ can be substituted for $E[M_T]$ and $q(\eta)$ can be taken to be the *Rayleigh* distribution (10.22); Equ.(11.3) becomes

$$E[D] = c^{-1}\nu_0^+\sigma_x^\beta \int_0^\infty \eta^{\beta+1}e^{-\eta^2/2}\,d\eta$$

or

$$E[D] = c^{-1}\nu_0^+\,\sigma_x^\beta\,2^{\beta/2}\,\Gamma(1+\beta/2) \tag{11.4}$$

where $\sigma_x = m_0^{1/2}$ is the standard deviation of the stress and $\Gamma(.)$ is the *Gamma* function. This result was first derived by Miles (1954); it can be written alternatively as

$$E[D] = c^{-1}\,\frac{2^{\beta/2}}{2\pi}\,\Gamma(1+\beta/2)\,m_0^{\frac{\beta-1}{2}}\,m_2^{1/2} \tag{11.5}$$

where m_a is the spectral moment of order a, defined as before according to

$$m_a = 2\int_0^\infty \omega^a\Phi(\omega)\,d\omega \tag{11.6}$$

This result can also be used as an approximation for a wide-band process (Wirshing & Haugen, 1973), although simulations have shown that it is conservative, especially for bimodal spectra. Improved prediction models have been proposed (e.g.Wirshing & Light, 1980; Chaudhury & Dover, 1982; Kam & Dover, 1988), which correct the previous result by a factor λ depending on higher spectral moments and the exponent β of the *S-N* curve.

The *Single moment method* (Lutes & Larsen, 1990) has been formulated after extensive simulation and rainflow analysis; it assumes the following damage equation:

$$E[D] = c^{-1}\,\frac{2^{\beta/2}}{2\pi}\,\Gamma(1+\beta/2)\,(m_{2/\beta})^{\beta/2} \tag{11.7}$$

This equation uses only the spectral moment of order $2/\beta$; although it has no theoretical foundation, it is the only single moment method which can give the correct dependence on both ω and σ_x. Besides, it is equivalent to the Rayleigh approximation (11.5) for narrow-band processes (Problem P.11.1). It gives results in close agreement with rainflow simulations for various *PSD*, including bimodal spectra, for which the Rayleigh approximation leads to substantial errors.

11.3 Biaxial loading with zero mean

We now consider biaxial stress states. They are of great practical importance, because cracks often initiate at the surface, where the stress state is biaxial. The von Mises criterion correlates fairly well with a large amount of experimental data for biaxial stress states with constant principal directions (Sines & Ohgi, 1981) and it is generally regarded as conservative (Shigley & Mitchell, 1983). When the excitation is random, the principal directions change continuously with time. We *propose* to base the analysis on an equivalent von Mises stress constructed in the following way.

For a biaxial stress, the starting point for defining the von Mises stress S_c (it is a random process) is the quadratic relationship

$$S_c^2 = S_x^2 + S_y^2 - S_x S_y + 3S_{xy}^2 \tag{11.8}$$

where S_x, S_y and S_{xy} are the normal and tangential stresses, respectively. Defining the random stress vector as $S = (S_x, S_y, S_{xy})^T$, we can write Equ.(11.8) as

$$S_c^2 = S^T Q S = \text{Trace}\{Q[SS^T]\} \tag{11.9}$$

with

$$Q = \begin{pmatrix} 1 & -1/2 & 0 \\ -1/2 & 1 & 0 \\ 0 & 0 & 3 \end{pmatrix} \tag{11.10}$$

If we take the expectation, we find

$$\mathrm{E}[S_c^2] = \text{Trace}\{Q\,\mathrm{E}[SS^T]\} \tag{11.11}$$

where $\mathrm{E}[SS^T]$ is the covariance matrix of the stress vector, related to the *PSD* matrix of the stress vector by

$$\mathrm{E}[SS^T] = \int_{-\infty}^{\infty} \Phi_{ss}(\omega)\, d\omega \tag{11.12}$$

From these equations, we can *define* the *PSD* $\Phi_c(\omega)$ of the *equivalent von Mises stress* as a frequency decomposition of its mean square value:

$$\mathrm{E}[S_c^2] = \int_{-\infty}^{\infty} \Phi_c(\omega)\, d\omega = \int_{-\infty}^{\infty} \text{Trace}\{Q\Phi_{ss}(\omega)\} d\omega \tag{11.13}$$

where $\Phi_{ss}(\omega)$ is the *PSD* matrix of the stress vector. Equivalently,

$$\Phi_c(\omega) = \text{Trace}\{Q\Phi_{ss}(\omega)\} = \sum_{i,j} Q_{ij}\Phi_{s_i s_j}(\omega) \tag{11.14}$$

Note that, in the uniaxial case [where only $S_x \neq 0$, with a *PSD* $\Phi_x(\omega)$], we find $\Phi_c(\omega) = \Phi_x(\omega)$. Equation (11.14) defines the *equivalent uniaxial alternating stress (von Mises stress) as the scalar random process whose PSD is obtained from the PSD matrix of the stress components according to the von Mises quadratic combination rule.* Although the von Mises criterion is quadratic, since the definition of the equivalent stress has been based on the second moment only, it can be assumed Gaussian. This scalar process can therefore be used with the uniaxial prediction models of Equ.(11.5) or (11.7).

The methodology developed for biaxial stress states can readily be extended to triaxial stress states. It is formally the same, with different definitions of the stress vector and the Q matrix.

11.4 Finite element formulation

If α_{im} denotes the *modal stresses* within a specific finite element, the stress components in this element can be expanded according to

$$S_i = \sum_m \alpha_{im} y_m \tag{11.15}$$

where y_m is the vector of modal amplitudes. α_{im} varies from one finite element to another, while y_m is defined for the whole structure. It follows from equation (11.15) that

$$\Phi_{s_i s_j}(\omega) = \sum_{m,n} \alpha_{im} \alpha_{jn} \Phi_{mn}(\omega) \tag{11.16}$$

where $\Phi_{mn}(\omega)$ is the *PSD* matrix of the modal responses (it is Hermitian), which can be computed as discussed in chapter 6 [Equ.(6.82)]. Upon substituting Equ.(11.16) into Equ.(11.14) and exchanging the order of summation, one gets

$$\Phi_c(\omega) = \sum_{m,n} \Phi_{mn}(\omega) \sum_{i,j} Q_{ij} \alpha_{im} \alpha_{jn} = \sum_{m,n} \Phi_{mn}(\omega) A_{mn} \tag{11.17}$$

where

$$A_{mn} = \sum_{i,j} Q_{ij} \alpha_{im} \alpha_{jn} \tag{11.18}$$

The sums over m and n extend to all the modes and those over i and j extend to all the stress components. Note that A_{mn} does not depend on the frequency ω, but it varies from one element to another. A_{mn} is in fact the result of the application of the quadratic combination rule (11.9) to the modal stresses.

From $\Phi_c(\omega)$, it is easy to calculate the spectral moments according to equation (11.6) and apply any of the uniaxial prediction models element by element.

11.5 Fluctuating stresses

For uniaxial stress states, it has been observed that a constant compression stress does not affect the endurance limit, S_e, while a constant tension stress entails a reduction of S_e. A number of models describing the effect of a constant mean stress σ_m on S_e are available in the literature (e.g. see Shigley & Mitchell,

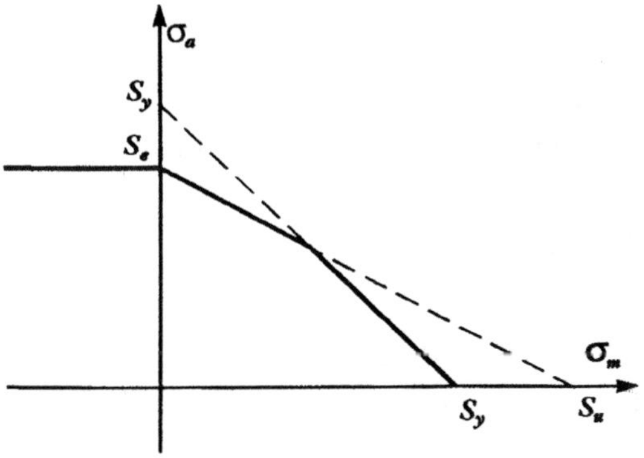

Figure 11.1: Goodman diagram.

1983). A simple rule is provided by the ***Goodman diagram*** (Figure 11.1). In this figure, S_y is the yield stress, S_u is the ultimate stress in tension and S_e is the endurance limit for purely alternating uniaxial stress. The diagram provides the modified endurance limit S'_e under a constant mean stress σ_m.

For multiaxial stress states, it appears that the admissible amplitude of the alternating von Mises stress depends on the ***first invariant*** of the static stress tensor, that is the ***sum of static principal stresses*** (e.g. Ellyin & Golos, 1988; Sines & Ohgi, 1981). This accounts for the fact that a constant torque does not affect the fatigue life in torsion.

Since the *S-N* curve does not explicitely refer to the endurance limit, the effect of the constant mean stress must be reflected by lowering it parallel to itself. This can be done by modifying the constant c appearing in Equ.(11.1) according to

$$c' = c\left(\frac{S'_e}{S_e}\right)^{\beta} \tag{11.19}$$

where S'_e is given by the Goodman diagram, with σ_m taken as the sum of principal stresses.

11.6 Recommended procedure

The suggested procedure for evaluating relative fatigue damage is the following:

- Perform a random vibration analysis of the structure in modal coordinates and get the *PSD* matrix of the modal responses $\Phi_{mn}(\omega)$.

- For each finite element,

  1. Compute A_{mn} according to Equ.(11.18),

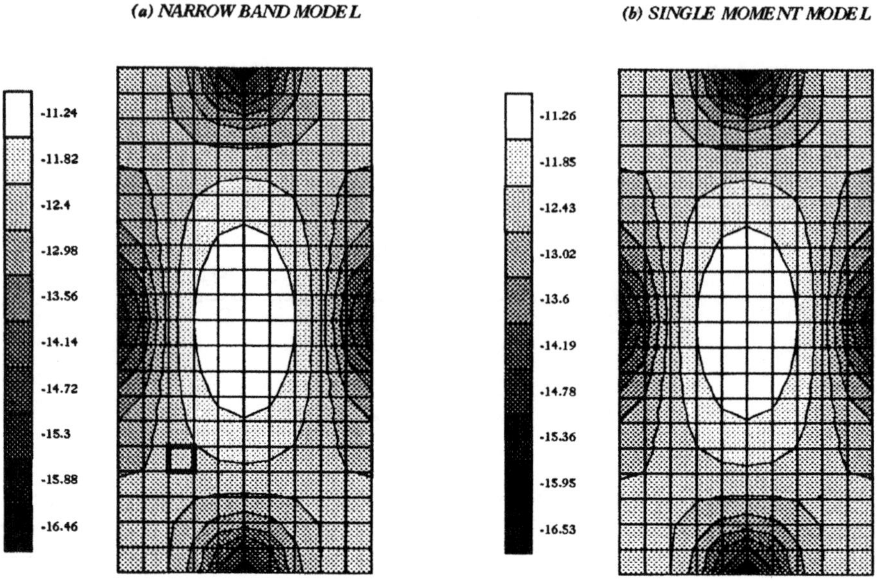

Figure 11.2: Map of relative damage per unit time of a rectangular plate. (a)Rayleigh (narrow-band) approximation. (b)Single moment method.

2. Compute $\Phi_c(\omega)$ from Equ.(11.17),
3. Compute the spectral moments m_a required for the selected uniaxial prediction model [e.g. Equ.(11.5) or (11.7)],
4. Use the *first invariant* of the static stress tensor to evaluate the new endurance limit from the Goodman diagram and calculate c' according to Equ.(11.19) ,
5. Compute the expected damage per unit time according to the selected uniaxial prediction model.

11.7 Example

The foregoing procedure is illustrated with a simply supported rectangular aluminium plate ($15.24cm \times 30.48cm$, $e = 0.8mm$) subjected to a band limited white noise random pressure field with perfect spatial coherence [$\Phi_{pp}(\omega) = 1\ Pa^2sec/rad$, $\omega_c = 6280\,rad/sec$]. The material constants are $c = 4.57\ 10^{55}$, $\beta = 6.09$. The first three modes ($\omega_1 = 663\,rad/sec$, $\omega_2 = 1061\,rad/sec$ and $\omega_3 = 1723\,rad/sec$) are within the bandwidth of the excitation, although the second mode is not excited (because it is anti-symmetrical, as we can see in Fig.6.6). A modal damping of $\xi_i = 0.02$ is assumed. 200 shell elements have

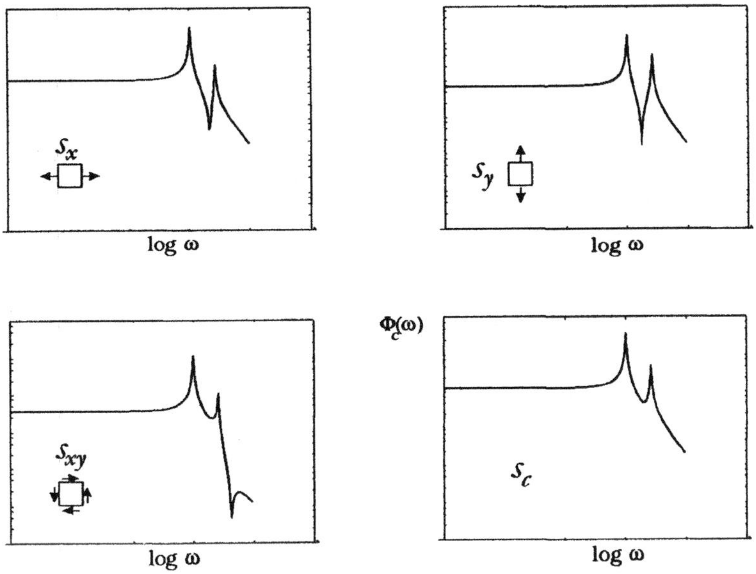

Figure 11.3: Power Spectral Density of the stress components and of the equivalent von Mises stress for the element framed in Fig.11.2.a

been used in the discretization. Figure 11.2.a shows the map of relative damage per unit time for the Rayleigh (narrow-band) approximation. Figure 11.2.b shows the prediction of the single moment method. The gray scale refers to the logarithm of the damage per unit time. The results of the two models are in very close agreement. Figure 11.3 shows the *PSD* of the component stresses and that of the equivalent von Mises stress for a typical element. Note that $\Phi_c(\omega)$ always envelopes the *PSD*'s of the stress components.

11.8 References

G.K.CHAUDHURY & W.D.DOVER, Fatigue analysis of offshore platforms subject to sea wave loadings. *J. Fatigue, 7, No 1, Mar., 13-19*, 1982.

S.H.CRANDALL & W.D.MARK, *Random Vibration in Mechanical Systems*, Academic Press, N.Y., 1963.

F.ELLYIN & K.GOLOS, Multiaxial fatigue damage criterion. *Trans. ASME, J. of Engineering Materials and Technology, Vol.110, January, 63-68*, 1988.

J.C.P.KAM & W.D.DOVER, Fast fatigue assessment procedure for offshore structures under random stress history. *Proc. Instn Civil Engrs, Part 2, 85, Dec., 689-700*, 1988.

C.E.LARSEN & L.D.LUTES, Predicting the fatigue life of offshore structures by

the single-moment spectral method. *Probabilistic Engineering Mechanics, Vol.6, No 2,* 1991.
Y.K.LIN, *Probabilistic Theory of Structural Dynamics*, McGraw-Hill, 1967.
L.D.LUTES & C.E.LARSEN, Improved spectral method for variable amplitude fatigue prediction. *J. Struct. Div., ASCE, 116 (4), 1149-1164*, 1990.
J.W.MILES, On structural fatigue under random loading, *J. of Aeronautical Sciences, 21, pp.753-762, 1954.*
J.E.SHIGLEY & L.D.MITCHELL, *Mechanical Engineering Design*, McGraw-Hill, 1983.
G.SINES & G.OHGI, Fatigue criteria under combined stresses or strains. *Trans. ASME, J. of Engineering Materials and Technology, Vol.103, April, 82-90,* 1981.
P.H.WIRSCHING & E.B.HAUGEN, Probabilistic design for random fatigue loads. *Journal of Engineering Mechanics Division, ASCE, Vol.99, EM6, December, 1165-1179*, 1973.
P.H.WIRSHING & M.C.LIGHT, Fatigue under wide band random stress. *J. Struct. Div., ASCE, 106 (7), 1593-1607*, 1980.

11.9 Problems

P.11.1 *Show that the single moment method is equivalent to the Rayleigh approximation for a narrow-band process.*

P.11.2 *A specimen must be subjected to a fatigue endurance test of duration T with a stationary random excitation of prescribed PSD $\Phi(\omega)$. In order to reduce the duration of the test, it is considered to scale up the excitation. Determine the scaling factor for the PSD, in order to produce the same damage in the reduced time T/a.*

Chapter 12

The Discrete Fourier Transform

12.1 Introduction

The Fourier transform has been used extensively throughout the previous chapters of this book. Its role is essentially related to the simplicity of the input-output relationship for *linear* systems, which is a consequence of the convolution theorem. The use of the continuous Fourier transform, however, is restricted to the cases where it is known analytically, most of the time from tables.

The *Discrete Fourier transform (DFT)* has been introduced to evaluate the Fourier transform of signals known numerically, at discrete times. Its use is now widespread, because of the *Fast Fourier Transform (FFT)* algorithm which drastically reduces the number of arithmetic operations. The algorithm organizes the calculation so that the number of arithmetic operations for computing the *FFT* is proportional to $N \log_2 N$, compared to N^2 for the traditional computation of the *DFT*. The Fast Fourier Transform algorithm was first developed in base 2 (Cooley & Tukey, 1965); many extensions have been proposed since that time and *FFT* subroutines are widely available in almost any language. Dedicated chips for real time applications are also on the market. The reader interested in *FFT* algorithms can refer to the specialized literature (e.g. *IEEE* Special Issue, 1967; Bergland, 1969; Brigham, 1974).

The quality of the approximation of the continuous Fourier transform by the *DFT* depends critically on the ***sampling rate***, the ***record length*** and the ***window*** used to reduce the leakage associated with the truncation. The correct understanding of these issues is the prime objective of the present chapter.

In what follows, in order to simplify the notations, we shall use the frequency f (in *Hz*) instead of the pulsation ω (in *rad/s*). Accordingly, the direct and inverse Fourier transform relationships become completely symmetrical in f

$h(t)$	$H(f)$
$\begin{cases} A & \lvert t\rvert < T_0 \\ A/2 & \lvert t\rvert = T_0 \\ 0 & \lvert t\rvert > T_0 \end{cases}$	$2AT_0 \frac{\sin(2\pi T_0 f)}{2\pi T_0 f}$
$2Af_0 \frac{\sin(2\pi f_0 t)}{2\pi f_0 t}$	$\begin{cases} A & \lvert f\rvert < f_0 \\ A/2 & \lvert f\rvert = f_0 \\ 0 & \lvert f\rvert > f_0 \end{cases}$
K	$K\delta(f)$
$k\delta(t)$	k
$A\cos(2\pi f_0 t)$	$\frac{A}{2}\delta(f - f_0) + \frac{A}{2}\delta(f + f_0)$
$A\sin(2\pi f_0 t)$	$-j\frac{A}{2}\delta(f - f_0) + j\frac{A}{2}\delta(f + f_0)$
$\sum_{n=-\infty}^{+\infty} \delta(t - nT)$	$\frac{1}{T} \sum_{n=-\infty}^{+\infty} \delta(f - \frac{n}{T})$
$(\frac{\alpha}{\pi})^{1/2} \exp(-\alpha t^2)$	$\exp\left(-\frac{\pi^2 f^2}{\alpha}\right)$

Table 12.1: Some Fourier transform pairs

and t:

$$H(f) = \int_{-\infty}^{\infty} h(t)e^{-j2\pi ft}dt \tag{12.1}$$

$$h(t) = \int_{-\infty}^{\infty} H(f)e^{j2\pi ft}df \tag{12.2}$$

Comprehensive and illustrated tables of Fourier transform pairs can be found in (Brigham, 1974; Bracewell, 1978). Some of the pairs which will be particularly useful in this chapter are summarized in Table 12.1. The reader will notice

- the duality between the time and frequency domains;
- a harmonic function is transformed into a Dirac delta function;
- a sequence of equi-distant impulses is transformed into a sequence of equi-distant impulses in the frequency domain; the frequency spacing is the reciprocal of that in the time domain.

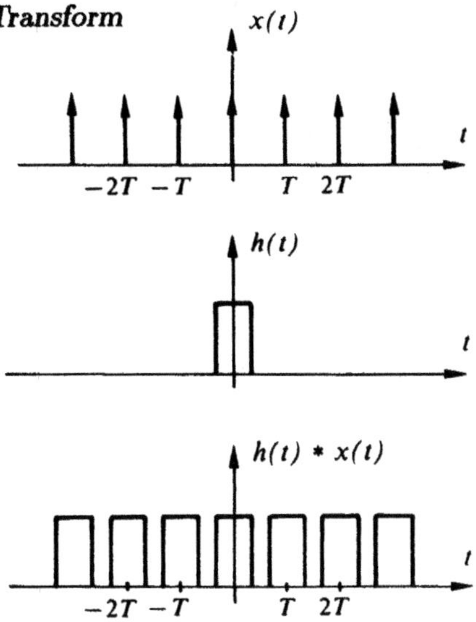

Figure 12.1: Periodic continuation is achieved by convolving the original signal by a sequence of impulses separated by the period of the signal.

All the results of chapter 1 can be transformed from ω to f in a straightforward manner; for example, Parseval's theorem (1.7) reads

$$E = \int_{-\infty}^{\infty} h^2(t)\,dt = \int_{-\infty}^{\infty} |H(f)|^2 df \qquad (12.3)$$

12.2 Consequences of the convolution theorem

12.2.1 Periodic continuation

The graphical interpretation of the convolution has been discussed in section 1.3.1. Using the procedure given there (folding, translation, ...) with one of the functions being a sequence of equi-distant impulses as in Fig.12.1, we can readily establish that the result of the convolution is the periodic continuation of the other signal. Consequently, ***a periodic signal can be regarded as the result of the convolution of one period of the signal by a sequence of unit impulses separated by the period*** T. Next, we consider the same problem in the frequency domain.

According to the convolution theorem, the convolution in one domain corresponds to a product in the dual domain. Besides, we know from Table 12.1 that the Fourier transform of a sequence of impulses of period T_0 is another sequence of impulses separated by $1/T_0$. Putting these two facts together, we conclude that the Fourier transform of a periodic signal is the product of the Fourier transform of one period by a sequence of impulses separated by $1/T_0$, as illustrated in Fig.12.2. This amounts to sampling $H(f)$ in the frequency domain, with a frequency resolution $1/T_0$.

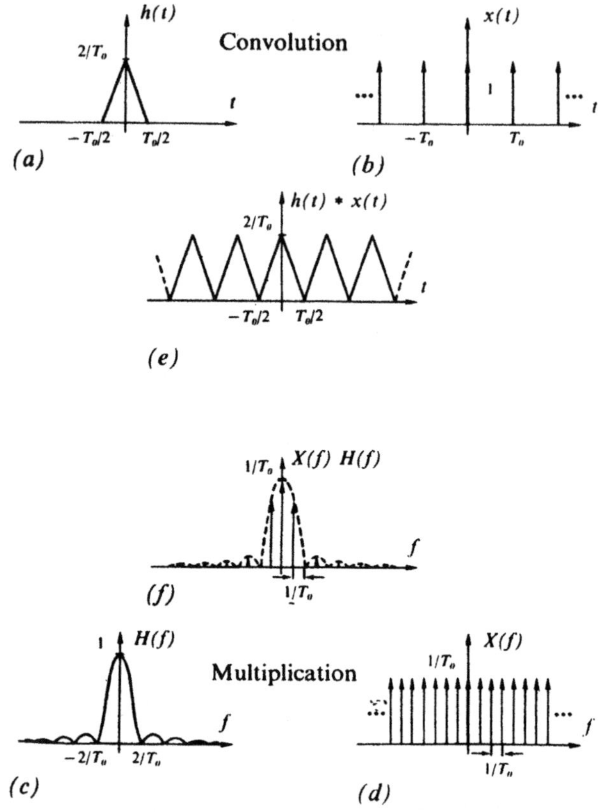

Figure 12.2: Periodic continuation of a triangular signal. In the time domain, signal *(e)* is obtained by convolving *(a)* and *(b)*; in the frequency domain, its Fourier transform *(f)* is obtained by multiplying *(c)* and *(d)*.

12.2.2 Sampling

In a similar manner, sampling a signal consists of multiplying it by a sequence of unit impulses, separated by the sampling period T. From the convolution theorem, the Fourier transform of the sampled signal is the result of the convolution of the Fourier transform of the original signal by the dual sequence of impulses, separated by the sampling frequency $1/T$. According to the foregoing section, the Fourier transform of the sampled signal is the periodic continuation, in the frequency domain, of the Fourier transform $H(f)$ of the original signal. This is illustrated in Fig.12.3. Once again, we notice the duality between the time and frequency domains: *Sampling in one domain is equivalent to a periodic continuation in the dual domain.*

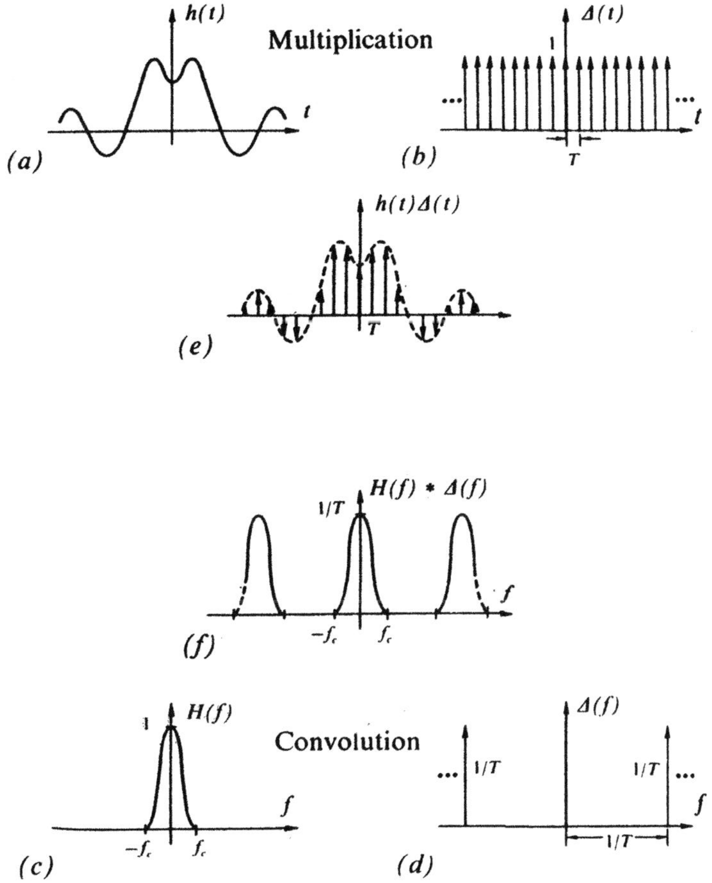

Figure 12.3: Sampling of a continuous signal. The sampled signal *(e)* arises from the product of the original signal *(a)* by the sequence of equi-distant impulses *(b)*. Its Fourier transform *(f)* is the convolution of *(c)* and *(d)*.

12.3 Shannon's theorem, Aliasing

Let $h(t)$ be a band-limited signal ($|f| < f_c$) sampled with a period T. According to the previous section, the Fourier transform of the sampled signal is the periodic continuation of the Fourier transform of the continuous signal, with a periodicity $1/T$. As a result, if

$$2f_c \leq \frac{1}{T} = f_s \tag{12.4}$$

as in Fig.12.3, the waveforms appearing in the frequency domain do not overlap and *the central lobe is not distorted.* In principle, it is possible to recover the original signal from the sampled one by low-pass filtering.

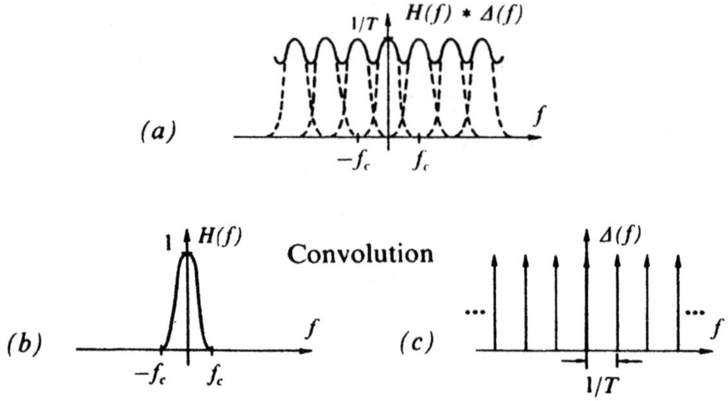

Figure 12.4: Interpretation of aliasing in the frequency domain. When the sampling frequency decreases, the various lobes of the Fourier transform of the sampled signal overlap.

Figure 12.5: Aliasing in the time domain. The sampled values of the two sine waves are identical.

If the sampling period increases and condition (12.4) is violated, the situation tends to that shown in Fig.12.4: The waveforms generated by the periodic continuation in the frequency domain tend to overlap and the central lobe is distorted. This phenomenon is known as ***aliasing***. Low-pass filtering cannot recover the original signal any longer.

The physical interpretation in the time domain is shown in Fig.12.5: Any sine wave at a frequency above $f_s/2 = 1/2T$ is aliased into another sine wave at a frequency below $f_s/2$. These two functions cannot be distinguished from their sampled values. $f_s/2 = 1/2T$ is the highest frequency for which sampling does not introduce distortion; it is called the ***Nyquist frequency***.

Formally, a continuous, band-limited signal $h(t)$ can be reconstructed from the sampled values $h(nT)$ according to

$$h(t) = \sum_{n=-\infty}^{\infty} h(nT) \frac{\sin \frac{\pi(t-nT)}{T}}{\frac{\pi(t-nT)}{T}} \tag{12.5}$$

Indeed, each term in the sum is of the form

$$\frac{\sin \frac{\pi t}{T}}{\frac{\pi t}{T}}$$

One can readily check from Table 12.1 that such a signal has its entire frequency content below the Nyquist frequency ($|f| < 1/2T$). The translation in time does not affect the frequency content. Besides, since

$$\frac{\sin \pi(k-n)}{\pi(k-n)} = \delta(k,n)$$

where $\delta(k,n)$ is the Kronecker delta [$\delta(k,n) = 1$ if $k = n$ and $= 0$ if $k \neq n$], each term in the sum vanishes at the sample times $t = kT$, except for $k = n$. As a result, Equ.(12.5) goes through all the sample values $h(kT)$.

The lower bound (12.4) on the sampling frequency for a given frequency content of the continuous signal is known as *Shannon's theorem*. Some oversampling is often advisable and, in order to avoid the contamination of the signal by high frequency measurement noise, it is highly recommended that the continuous data be put through a low-pass filter before sampling.

12.4 Fourier series

12.4.1 Orthogonal functions

The set of functions $u_i(t)$ is orthogonal over the interval $[-T_0/2, T_0/2]$ if they satisfy the orthogonality condition

$$\int_{-T_0/2}^{T_0/2} u_m(t)u_n(t)\,dt = c\,\delta(m,n) \tag{12.6}$$

Let

$$x(t) = \sum_{n=0}^{\infty} a_n u_n(t) \tag{12.7}$$

be the series expansion of $x(t)$ over the interval; the unknown coefficients a_i can be readily obtained by multiplying both sides of Equ.(12.7) by $u_i(t)$ and integrating over the interval; one gets

$$\int_{-T_0/2}^{T_0/2} x(t)u_i(t)\,dt = \sum_{n=0}^{\infty} a_n \int_{-T_0/2}^{T_0/2} u_n(t)u_i(t)\,dt$$

and, from the orthogonality condition,

$$a_i = \frac{1}{c}\int_{-T_0/2}^{T_0/2} x(t)u_i(t)\,dt \tag{12.8}$$

Thus, the coefficients of the expansion can be computed *independently*; this is one of the attractive features of orthogonal functions.

An orthogonal expansion is in fact a *mean square fit* of $x(t)$ in the space of available orthogonal functions $u_n(t)$. Indeed, the coefficients $\hat{a}_n$ of the truncated expansion

$$\hat{x}(t) = \sum_{n=0}^{N} \hat{a}_n u_n(t) \tag{12.9}$$

which minimize the mean square error

$$\varepsilon = \frac{1}{T_0} \int_{-T_0/2}^{T_0/2} [x(t) - \hat{x}(t)]^2 \, dt \tag{12.10}$$

are exactly $\hat{a}_n = a_n$ (this is a direct consequence of the orthogonality condition). If $\varepsilon \to 0$ as N increases, the set of orthogonal functions is *complete*.

Using the orthogonality condition, we can write the mean square value of the signal as

$$\frac{1}{T_0} \int_{-T_0/2}^{T_0/2} x^2(t) \, dt = \frac{c}{T_0} \sum_{n=0}^{\infty} a_n^2 \tag{12.11}$$

This is a special form of Parseval's theorem. The square of the expansion coefficients, a_n^2, can be regarded as the part of the power distribution of $x(t)$ in the orthogonal component $u_n(t)$.

12.4.2 Fourier series

The complex exponentials $u_n = \exp(j2\pi nt/T_0)$ constitute a special set of orthogonal functions which satisfy the following orthogonality condition

$$\int_{-T_0/2}^{T_0/2} e^{j(m-n)2\pi t/T_0} \, dt = T_0 \delta(m,n) \tag{12.12}$$

The expansion of a periodic function in its harmonic contributions is called its Fourier series. It can be written in terms of sine and cosine contributions or, in a more compact manner, in terms of complex exponentials:

$$y(t) = \sum_{n=-\infty}^{\infty} \alpha_n e^{j2\pi nt/T_0} \tag{12.13}$$

From the orthogonality condition, the coefficients of the expansion are

$$\alpha_n = \frac{1}{T_0} \int_{-T_0/2}^{T_0/2} y(t) e^{-j2\pi nt/T_0} \, dt \qquad n = 0, \pm 1, \pm 2 \ldots \tag{12.14}$$

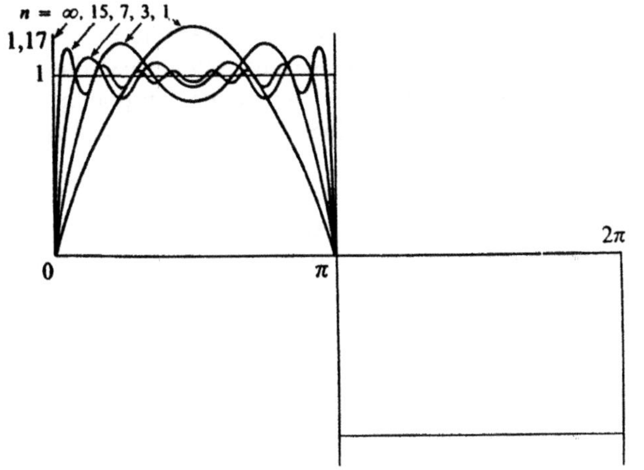

Figure 12.6: Illustration of the Gibbs phenomenon for a rectangular wave.

and Parseval's theorem reads (Problem P.12.1)

$$\frac{1}{T_0}\int_{-T_0/2}^{T_0/2} |y(t)|^2 dt = \sum_{n=-\infty}^{\infty} |\alpha_n|^2 \tag{12.15}$$

Since the harmonic functions are continuous functions of time, the function given by expansion (12.13) is also continuous. It converges towards the value of the function $y(t)$ wherever it is continuous, but it cannot match both values of the function at a point of discontinuity. Instead, the Fourier series converges towards the average value of the discontinuous function. Expanding discontinuous functions in terms of continuous orthogonal functions is the origin of a difficulty known as *Gibbs phenomenon.*

12.4.3 Gibbs phenomenon

A sequence of functions $S_n(t)$ tends to a limit $S(t)$ if, for any instant of time t and any given ε, one can find a value $N(t,\varepsilon)$ such that

$$|S_n(t) - S(t)| \leq \varepsilon \qquad \text{for} \qquad n \geq N(t,\varepsilon) \tag{12.16}$$

If it is possible to find a value of $N(\varepsilon)$ which is independent of t so that Equ.(12.16) applies for the complete interval, the convergence is *uniform.* Since an expansion in continuous functions of time is also continuous, the convergence cannot be uniform near a point of discontinuity. This is responsible for oscillations called *Gibbs phenomenon* as illustrated in Fig.12.6 for a rectangular wave. The figure shows the truncated expansions with an increasing number of terms ($n = 1, 3, 7, \ldots$). As n increases, the mismatch between the truncated expansion and the original function tends to concentrate near the points of discontinuity where the truncated expansions exhibit strong oscillations. The frequency of the oscillations is that of the first truncated harmonic component. As n increases, the overshoot near the discontinuity does not disappear and reaches a limiting value of 1.179 (Problem P.12.3).

A deeper insight into the phenomenon can, once again, be obtained from the convolution theorem: Truncating the Fourier series after n terms amounts to passing the signal through an ideal low-pass filter $H(f)$ whose cut-off frequency is anywhere between n and $n+1$ times the fundamental frequency $1/T_0$. In the time domain, this corresponds to convolving the original signal by the impulse response $h(t)$ of the ideal low-pass filter (see Table 12.1). The oscillations at the cut-off frequency of the filter which are present in $h(t)$ are transmitted into the filtered signal at every discontinuity as a result of the convolution.

12.4.4 Relation to the Fourier transform

We have seen that a periodic signal can be constructed by convolving one period of the signal, $h(t)$, by a sequence of unit impulses $x(t)$ separated by the period T_0:

$$y(t) = h(t) \star x(t) = h(t) \star \sum_{n=-\infty}^{\infty} \delta(t - nT_0) \tag{12.17}$$

From the convolution theorem,

$$Y(f) = H(f).X(f) = H(f).\frac{1}{T_0} \sum_{n=-\infty}^{\infty} \delta(f - \frac{n}{T_0})$$

or

$$Y(f) = \frac{1}{T_0} \sum_{n=-\infty}^{\infty} H(\frac{n}{T_0})\delta(f - \frac{n}{T_0}) \tag{12.18}$$

As illustrated in Fig.12.2, this is equivalent to sampling $H(f)$ in the frequency domain.

On the other hand, the Fourier series coefficients α_n are given by

$$\alpha_n = \frac{1}{T_0} \int_{-T_0/2}^{T_0/2} h(t) e^{-j2\pi nt/T_0}\, dt$$

and, since $h(t)$ vanishes outside the interval $[-T_0/2, T_0/2]$, the bounds of the integral can be changed to infinity:

$$\alpha_n = \frac{1}{T_0} \int_{-\infty}^{\infty} h(t) e^{-j2\pi nt/T_0}\, dt = \frac{1}{T_0} H(\frac{n}{T_0}) \tag{12.19}$$

Comparing Equ.(12.18) and (12.19), we see that *the Fourier transform of a periodic signal consists of a sequence of impulses whose intensities are the Fourier series coefficients.*

12.5 Graphical development of the *DFT*

In this section, we analyse the various steps leading from the continuous Fourier transform to its digital approximation; the presentation closely follows Brigham (1974). Since the numerical calculations can only be performed on finite sequences of numbers and for a finite set of control frequencies, the following operations must be carried out before computing the Digital Fourier Transform:

- ***sampling*** in the time domain (this transforms the continuous signal into a sequence of numbers).
- ***truncation*** to limit the size of the sequence.
- ***sampling in the frequency domain***: the *DFT* is calculated for a discrete set of control frequencies.

In this section, using the results of the previous sections, we examine graphically the relation between the *DFT* arising from the foregoing operations, and the continuous Fourier transform of the original signal. We consider the Fourier transform pair of Fig.12.7.a; for simplicity, we assume that $h(t)$ is even, so that $H(f)$ is real and even.

- **Sampling**

The first operation consists of transforming the continuous signal into a sequence of numbers. This amounts to multiplying $h(t)$ by a train of impulses $\Delta_0(t)$ separated by the sampling period T (Fig.12.7.b). As we have seen in section 12.2, the Fourier transform of the sampled signal, $H(f) \star \Delta_0(f)$ is the periodic continuation of $H(f)$. If the sampling frequency violates the condition of Shannon's theorem, some overlapping takes place during this operation and introduces aliasing. Without aliasing, the central lobe of $H(f) \star \Delta_0(f)$ is, to a constant factor, identical to $H(f)$.

- **Truncation**

Since the numerical calculations must involve sequences of finite length, the original signal is truncated after N samples, that is after a duration $T_0 = NT$. Truncation can be achieved by multiplying the sampled signal by a ***rectangular window*** $x(t)$ of duration T_0 (Fig.12.7.d). The corresponding Fourier transform is given by the convolution

$$[H(f) * \Delta_0(f)] * \frac{\sin \pi T_0 f}{\pi f} \tag{12.20}$$

This introduces some distortion in the Fourier transform, which is called *leakage*. Note that since

$$\lim_{T_0 \to \infty} \frac{\sin \pi T_0 f}{\pi f} = \delta(f) \tag{12.21}$$

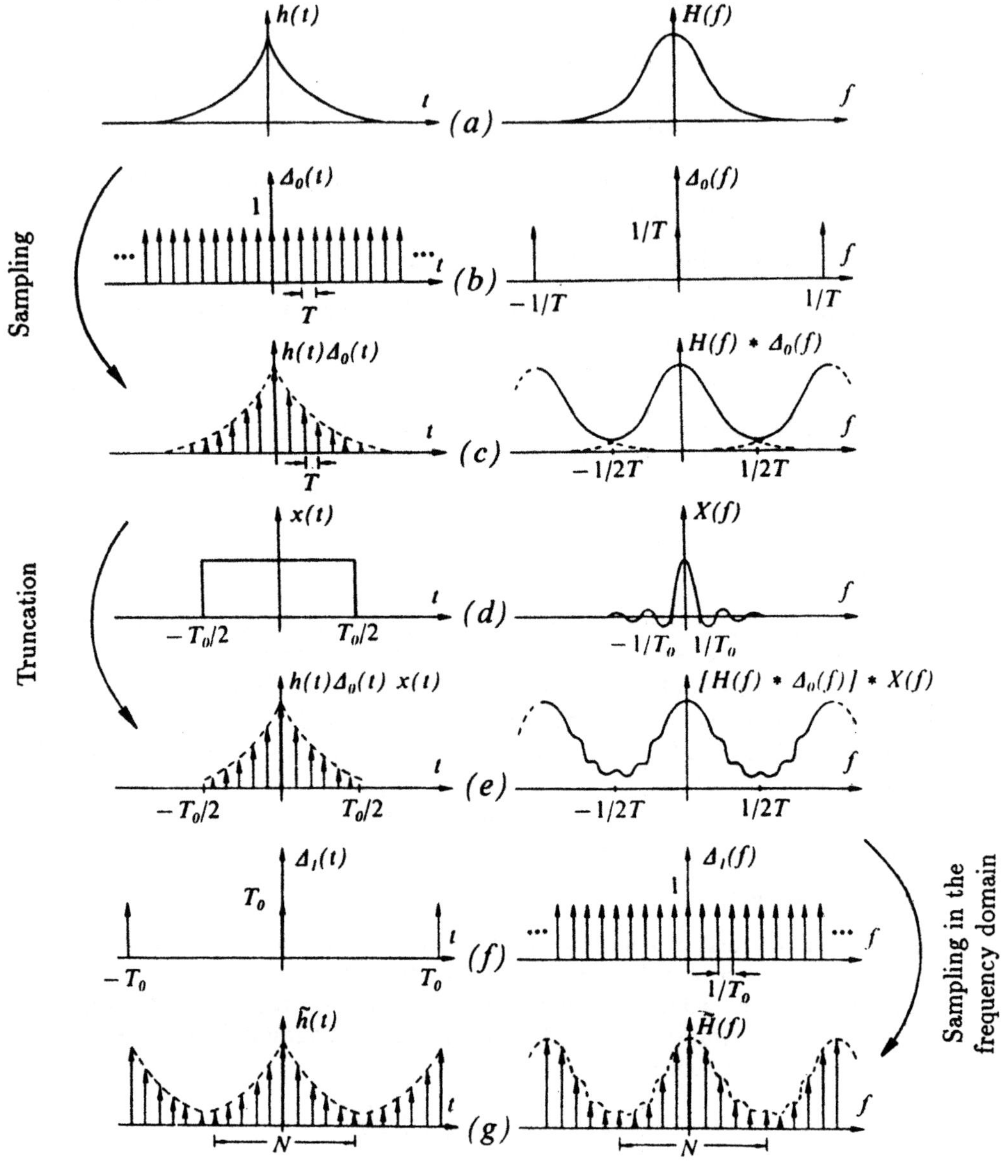

Figure 12.7: Graphical development of the *DFT*.

the distortion disappears as T_0 increases. Increasing the record length is not always practical, but a substantial reduction of the leakage can be achieved by using smoother windows which are better conditioned in the frequency domain. Leakage reduction is an important issue which will be addressed in more detail later.

- **Sampling in the frequency domain**

Since the numerical calculations can only be performed at a set of discrete frequencies, the Fourier transform is sampled in the frequency domain. This is equivalent to a periodic continuation in the time domain. If we adopt a frequency resolution $1/T_0$, the periodicity in the time domain is exactly the duration of the truncated records. The result is represented in Fig.12.7.g; it reads

$$\tilde{h}(t) = [h(t).\Delta_0(t).x(t)] \star \Delta_1(t) \tag{12.22}$$

$$\tilde{H}(t) = [H(f) \star \Delta_0(f) \star X(f)].\Delta_1(f) \tag{12.23}$$

where $h(t)$ is the original signal, $\Delta_0(t)$ is a sequence of impulses of unit intensity separated by the sampling period, $x(t)$ is the observation window and $\Delta_1(f)$ is a sequence of unit impulses in the frequency domain, separated by $1/T_0$.

Note that the final result is periodic in the time domain (due to sampling in the frequency domain) and in the frequency domain (due to sampling in the time domain).

12.6 Analytical development of the *DFT*

In the foregoing section, we have described graphically the operations which must be performed on a function to allow a numerical approximation of its Fourier transform. The various transformations have led from the original pair $[h(t), H(f)]$ to the modified one $[\tilde{h}(t), \tilde{H}(f)]$ defined by Equ.(12.22)-(12.23). In this section, we develop the mathematics behind each step of the transformation.

- **Sampling**

Sampling $h(t)$ amounts to multiplying it by a sequence of unit impulses $\Delta_0(t)$ separated by the sampling period T.

$$h(t)\Delta_0(t) = h(t) \sum_{k=-\infty}^{\infty} \delta(t-kT) = \sum_{k=-\infty}^{\infty} h(kT)\,\delta(t-kT) \tag{12.24}$$

- **Truncation**

Truncation after a finite duration $T_0 = NT$ amounts to limiting the above sum to N samples. If we use the rectangular window $x(t) = 1 \quad (-T/2 < t < T_0 - T/2)$, we get

$$h(t)\Delta_0(t)x(t) = \sum_{k=0}^{N-1} h(kT)\,\delta(t-kT) \tag{12.25}$$

- **Sampling in the frequency domain**

Sampling in the frequency domain is equivalent to a periodic continuation in the time domain, with a period T_0. This is achieved by convolving the signal obtained at the preceding step by:

$$\Delta_1(t) = T_0 \sum_{r=-\infty}^{\infty} \delta(t - rT_0) \tag{12.26}$$

The result,

$$\tilde{h}(t) = [h(t)\Delta_0(t)x(t)] \star \Delta_1(t) \tag{12.27}$$

is T_0 times the periodic extension of (12.25):

$$\tilde{h}(t) = T_0 \sum_{r=-\infty}^{\infty} \sum_{k=0}^{N-1} h(kT)\,\delta(t - kT - rT_0) \tag{12.28}$$

$\tilde{h}(t)$ is a periodic function and, according to section 12.4.4, its Fourier transform consists of a sequence of impulses:

$$\tilde{H}(f) = \sum_{n=-\infty}^{\infty} \alpha_n\,\delta(f - \frac{n}{T_0}) \tag{12.29}$$

with amplitudes α_n identical to the Fourier series coefficients. They can be calculated from one period of the function according to

$$\alpha_n = \frac{1}{T_0} \int_{-T/2}^{T_0-T/2} \tilde{h}(t) e^{-j2\pi nt/T_0} dt \qquad n = 0, \pm 1, \pm 2... \tag{12.30}$$

or

$$\begin{aligned}
\alpha_n &= \int_{-T/2}^{T_0-T/2} \sum_{k=0}^{N-1} h(kT)\delta(t - kT)e^{-j2\pi nt/T_0}\,dt \\
&= \sum_{k=0}^{N-1} h(kT) \int_{-T/2}^{T_0-T/2} e^{-j2\pi nt/T_0}\,\delta(t - kT)\,dt \\
&= \sum_{k=0}^{N-1} h(kT)e^{-j2\pi nkT/T_0}
\end{aligned}$$

Taking into account that $T_0 = NT$, we finally get

$$\tilde{H}(\frac{n}{T_0}) = \alpha_n = \sum_{k=0}^{N-1} h(kT)e^{-j2\pi nk/N} \qquad n = 0, \pm 1, \pm 2... \tag{12.31}$$

This equation is the starting point for defining the Digital Fourier transform of $h(kT)$.

12.7 Definition and properties of the *DFT*

12.7.1 Definition of the *DFT* and *IDFT*

Let $x(m), \quad m = 0, ..., N-1$ be a sequence of N numbers (complex in general). We define the *DFT* as

$$C_x(n) = \frac{1}{N} \sum_{k=0}^{N-1} x(k) W^{kn} \qquad n = 0, ..., N-1 \tag{12.32}$$

where W is an abbreviation for the complex exponential:

$$W = e^{-j2\pi/N} \tag{12.33}$$

With respect to $\tilde{H}(n/T_0)$ of Equ.(12.31), the reason for introducing the constant factor $1/N$ in the definition is that $C_x(0)$ is the average value of the sequence $x(m)$. The complex exponentials satisfy the following orthogonality relation:

$$\sum_{m=0}^{N-1} W^{km} W^{-lm} = N\delta(k, l) \tag{12.34}$$

It follows that the inverse transform (IDFT) reads

$$x(l) = \sum_{n=0}^{N-1} C_x(n) W^{-nl} \qquad l = 0, ..., N-1 \tag{12.35}$$

Since the complex exponential is *periodic* with respect to N:

$$W^{km} = W^{(k+N)m}$$

so are the sequences $x(k)$ and $C_x(n)$:

$$x(k) = x(k+N); \qquad C_x(n) = C_x(n+N)$$

This is in agreement with the discussion of the previous sections.

12.7.2 Properties of the *DFT*

- ***DFT* of a real function**

If $x(k)$ is real and N is even, then

$$C_x(N/2 + i) = C_x^*(N/2 - i) \qquad i = 0, ..., N/2 \tag{12.36}$$

It follows that $C_x(N/2)$ must be real, as also the average value $C_x(0)$. If $x(k)$ is even, $C_x(n)$ is real and even; if $x(k)$ is odd, $C_x(n)$ is imaginary and odd. Since an arbitrary real function can be decomposed into the sum of an even and an odd function, the real part of its DFT is even and its imaginary part is odd (Fig.12.8). Recall that $C_x(n+N) = C_x(n)$.

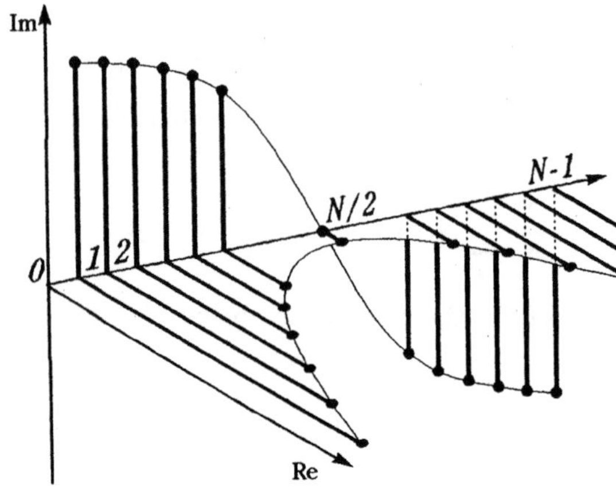

Figure 12.8: The DFT of a real function is such that its real part is even and its imaginary part is odd: $C_x(N/2+i) = C_x^*(N/2-i)$.

- **Translation theorems**

If $x(m)$ and $C_x(k)$ constitute a DFT pair, the translation theorems in the time and frequency domains read respectively

$$x(m-i) \Longleftrightarrow C_x(k)e^{-j2\pi ki/N} \tag{12.37}$$

$$x(m)e^{j2\pi im/N} \Longleftrightarrow C_x(k-i) \tag{12.38}$$

- **Parseval's theorem**

For a real sequence,

$$\frac{1}{N}\sum_{m=0}^{N-1} x^2(m) = \sum_{k=0}^{N-1} |C_x(k)|^2 \tag{12.39}$$

The theorem states that the square of the modulus of the DFT can be regarded as the frequency decomposition of the mean square value of the signal $x(m)$.

- **Relation to the continuous Fourier transform**

Comparing Equ.(12.31) and (12.32), we see that

$$C_h(n) = \frac{1}{N}\tilde{H}(\frac{n}{T_0}) \tag{12.40}$$

where $\tilde{H}$ has been derived according to Fig.12.7. Under the following conditions,

- The duration of the signal is smaller than T_0, so that *it is not altered by the truncation.*

- *The signal is band-limited and is sampled according to Shannon's theorem.*

$\tilde{H}$ is a faithful representation of H, because

$$\tilde{H}(\frac{n}{T_0}) = \frac{1}{T}H(\frac{n}{T_0}) \tag{12.41}$$

Combining the two previous equations with Equ.(12.19), we conclude that the DFT is related to the continuous Fourier transform of the original signal according to

$$C_h(n) = \frac{1}{N}\tilde{H}(\frac{n}{T_0}) = \frac{1}{T_0}H(\frac{n}{T_0}) = \alpha_n \tag{12.42}$$

Any departure from the two above conditions induces an error in Equ.(12.42) and, because the two conditions are in fact *contradictory* (a signal cannot be bounded simultaneously in the time and frequency domains), errors are indeed introduced in the approximation. The *unique* case where Equ.(12.42) applies rigorously is that where

- the signal is periodic;

- it is band-limited and sampled according to Shannon's theorem;

- the duration of the observation window, T_0, is a multiple of the period of the signal.

Under these circumstances, the periodic continuation of the truncated signal (Fig.12.7.g) restores the original signal, and Equ.(12.41) applies. Any departure from this situation leads to an error in Equ.(12.42).

Apart from the aliasing error which is easily dealt with by low-pass filtering before sampling and using a sampling rate fast enough, the most serious error is the leakage introduced by the truncation of the signal. Techniques for leakage reduction will be discussed in the next section.

- **Gibbs effect**

We know from Equ.(12.42) that, in the ideal conditions, $C_x(n) = \alpha_n$. Let us now perform a numerical experiment. We construct a digital approximation of a rectangular wave involving $N = 32$ samples by selecting the *DFT* coefficients identical to the Fourier series coefficients α_n for all the harmonic components below the Nyquist frequency ($n = 0, ..., 16$). The remaining coefficients are calculated according to Equ.(12.36). The *DFT* sequence is then inverse transformed into a time sequence $x(k)$ of 32 samples. The result is compared to the original rectangular wave in Fig.12.9. Strong oscillations occur at half the sampling frequency; their amplitude is maximum near the discontinuity. This is Gibbs phenomenon;

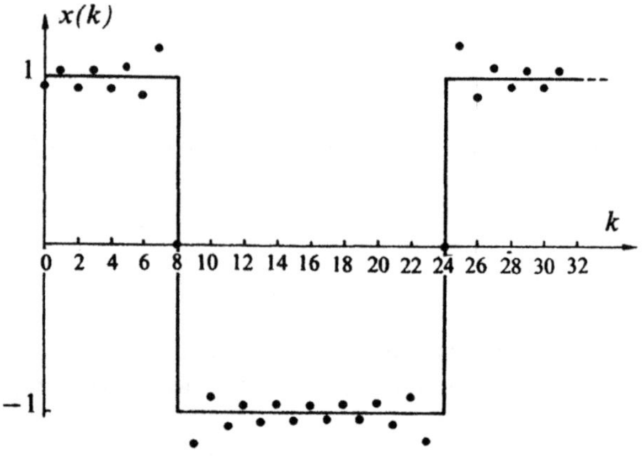

Figure 12.9: Gibbs phenomenon for a rectangular wave reconstructed from $C_x(n) = \alpha_n$ $(n = 0, ..., 16)$.

it can be explained as follows: The sequence that we have constructed consists of the sampled values of the truncated Fourier series expansion of the rectangular wave. The frequency of the oscillations is the Nyquist frequency (half the sampling frequency), at which the Fourier expansion has been truncated.

12.8 Leakage reduction

The leakage is the direct consequence of the finite length of the sequence and of the periodicity of the *DFT*. According to the previous section, a cosine function with exactly n cycles within the period of observation will give a *DFT* consisting of a single pair of non-zero components:

$$x(k) = \cos\frac{2\pi nk}{N} \Longleftrightarrow C_x(l) = \frac{1}{2}\delta(l,n) + \frac{1}{2}\delta(l, N-n)$$

On the contrary, if the harmonic function is not periodic within the observation window T_0, the *DFT* possesses a large number of non-zero components, even at frequencies very different from that of the original signal (Problem P.12.5). They arise from the convolution of the Fourier transform of the original cosine function by that of the observation window $f_R(t) = \Pi_{T_0/2}(t)$ (Fig.12.10)

$$f_R(t) = \Pi_{T_0/2}(t) \Longleftrightarrow \mathcal{F}_R(f) = \frac{\sin \pi f T_0}{\pi f} \tag{12.43}$$

$\mathcal{F}_R(f)$ consists of a main lobe of width $2/T_0$ and side lobes of width $1/T_0$, the amplitude falling off as $|f|^{-1}$. This slow decay rate is often unacceptable. A faster decay of the side lobes can be obtained at the expense of a wider main lobe by using a ***Hanning window*** (Fig.12.11).

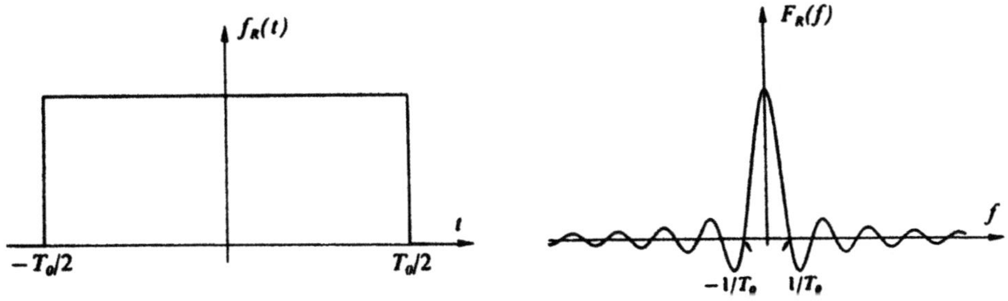

Figure 12.10: Centered rectangular window (also called *Box Car*).

$$f_H(t) = \frac{1}{2} + \frac{1}{2}\cos\frac{2\pi t}{T_0} \qquad t \in [-T_0/2, T_0/2] \tag{12.44}$$

Its Fourier transform can be obtained from the convolution theorem by noting that

$$f_H(t) = (\frac{1}{2} + \frac{1}{2}\cos\frac{2\pi t}{T_0}).f_R(t)$$

Therefore,

$$\mathcal{F}_H(f) = [\frac{1}{2}\delta(f) + \frac{1}{4}\delta(f - \frac{1}{T_0}) + \frac{1}{4}\delta(f + \frac{1}{T_0})] \star \mathcal{F}_R(f) \tag{12.45}$$

or

$$\mathcal{F}_H(f) = \frac{1}{2}\mathcal{F}_R(f) + \frac{1}{4}\mathcal{F}_R(f - \frac{1}{T_0}) + \frac{1}{4}\mathcal{F}_R(f + \frac{1}{T_0}) \tag{12.46}$$

The side lobes of the Hanning window fall off like $|f|^{-3}$ instead of $|f|^{-1}$, but the width of the main lobe is twice that of the rectangular window, which reduces the resolution. The Fourier transform of the signal after windowing is

$$x(t).f_H(t) \Leftrightarrow X(f) \star [\frac{1}{2}\delta(f) + \frac{1}{4}\delta(f - \frac{1}{T_0}) + \frac{1}{4}\delta(f + \frac{1}{T_0})] \star \frac{\sin \pi f T_0}{\pi f}$$

and, because the convolution is commutative and associative,

$$x(t).f_H(t) \Leftrightarrow \{X(f) \star \frac{\sin \pi f T_0}{\pi f}\} \star [\frac{1}{2}\delta(f) + \frac{1}{4}\delta(f - \frac{1}{T_0}) + \frac{1}{4}\delta(f + \frac{1}{T_0})] \tag{12.47}$$

The first convolution is the Fourier transform of the original signal with a rectangular window. If we consider the sampled values of the above expression for $f = k/T_0$, we find that the *DFT* coefficients $C_x^H(k)$ with a centered Hanning window can be obtained from those with a rectangular window $C_x^R(k)$ by the convolution

$$C_x^H(k) = \frac{1}{4}C_x^R(k-1) + \frac{1}{2}C_x^R(k) + \frac{1}{4}C_x^R(k+1) \tag{12.48}$$

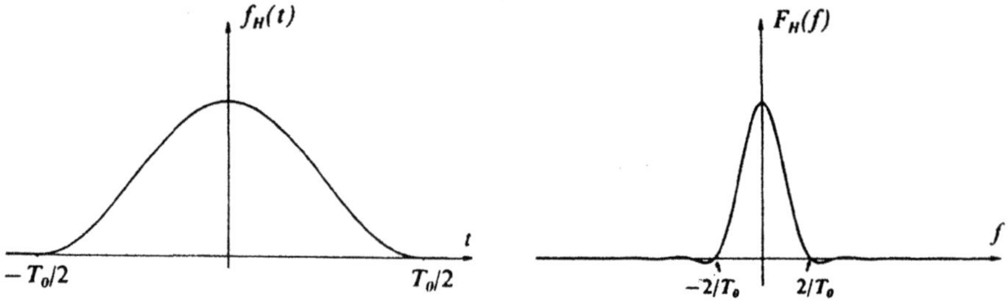

Figure 12.11: Centered Hanning window (also called *Cosine bell*).

If we use a non centered Hanning window from 0 to T_0, we can show (Problem P.12.6) that

$$C_x^H(k) = -\frac{1}{4}C_x^R(k-1) + \frac{1}{2}C_x^R(k) - \frac{1}{4}C_x^R(k+1) \tag{12.49}$$

In Fig.12.10 and 12.11, we see that the width of the central lobe of the Hanning window is twice that of the rectangular window and that the side lobes have opposite signs. This suggests that further reduction of the side lobes can be achieved by linearly combining the two windows. It can be shown that the window minimizing the maximum amplitude of the side lobes is

$$0.54 + 0.46\cos\frac{2\pi t}{T_0}$$

It is called the *Hamming window*; its highest side lobe is about one third of that of the Hanning window. It can be shown (Problem P.12.7) that the corresponding coefficients in Equ.(12.48) and (12.49) are $\{0.23, 0.54, 0.23\}$ instead of $\{0.25, 0.5, 0.25\}$.

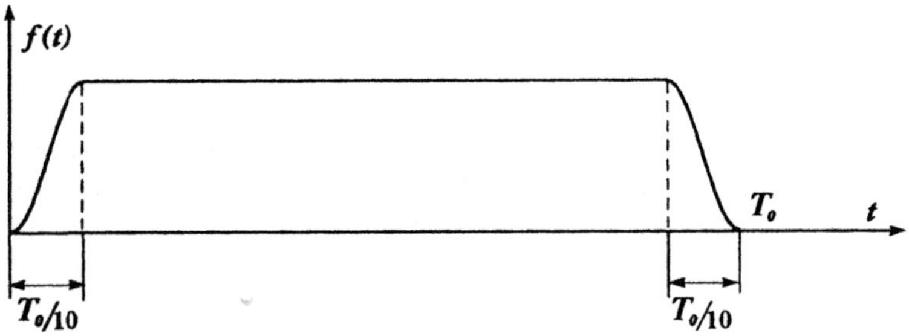

Figure 12.12: Cosine taper window.

In general, a window should satisfy the conflicting requirements of being flat, to concentrate the main lobe near $f = 0$, and smooth and slowly changing, to reduce the side lobes. A good compromise is represented in Fig.12.12. It consists of a flat central part with cosine tapered ends. The duration of the smoothing sections at both ends is 10% of the total duration of the signal.

12.9 Power spectrum estimation

Since the *FFT* algorithm has been introduced, the practical estimation of the spectral densities is based on Equ.(3.48) and (3.68). Combining these with Equ.(3.53), we can write the one-sided spectra as

$$G_{xx}(f) = \lim_{T_0 \to \infty} \frac{2}{T_0} \mathrm{E}[|\mathcal{X}(f, T_0)|^2] \tag{12.50}$$

$$G_{xy}(f) = \lim_{T_0 \to \infty} \frac{2}{T_0} \mathrm{E}[\mathcal{X}(f, T_0)\mathcal{Y}(f, T_0)^*] \tag{12.51}$$

In practice, the expected value operation is replaced by an ensemble average on a collection of sample records of finite duration acquired sequentially. If m sample records $x_k(t)$ and $y_k(t)$ of duration T_0 have been recorded, the auto and cross spectral density estimates are respectively

$$\hat{G}_{xx}(f) = \frac{2}{mT_0} \sum_{k=1}^{m} |\mathcal{X}_k(f, T_0)|^2 \tag{12.52}$$

$$\hat{G}_{xy}(f) = \frac{2}{mT_0} \sum_{k=1}^{m} \mathcal{X}_k(f, T_0)\mathcal{Y}_k(f, T_0)^* \tag{12.53}$$

where $\mathcal{X}_k(f, T_0)$ and $\mathcal{Y}_k(f, T_0)$ are estimates of the Fourier transform of the finite duration sample records $x_k(t)$ and $y_k(t)$. They can be obtained from the *DFT* according to Equ.(12.42):

$$\mathcal{X}_k(\frac{n}{T_0}, T_0) = T_0 C_{x_k}(n) \tag{12.54}$$

Substituting in the foregoing equations, we get

$$\hat{G}_{xx}(\frac{n}{T_0}) = \frac{2T_0}{m} \sum_{k=1}^{m} |C_{x_k}(n)|^2 \tag{12.55}$$

$$\hat{G}_{xy}(\frac{n}{T_0}) = \frac{2T_0}{m} \sum_{k=1}^{m} C_{x_k}(n) C^*_{y_k}(n) \tag{12.56}$$

These equations apply up to the Nyquist frequency provided that the sampling rate is such that there is no aliasing. The estimates are given at discrete frequencies separated by $1/T_0$. The maximum achievable resolution of the spectral estimate is thus the reciprocal of the record length. Increasing the record length T_0 improves the resolution, and increasing the number of records m reduces the scatter of the spectral estimate.

In computing the *DFT*, we must use the techniques for leakage reduction discussed in the previous section. Of course, tapering the record at the end reduces the *RMS* value of the signal. According to Parseval's theorem (12.15), the loss factor of the Hanning window is

$$\frac{1}{T_0}\int_{-T_0/2}^{T_0/2} f_H^2(t)\,dt = (\frac{1}{4})^2 + (\frac{1}{2})^2 + (\frac{1}{4})^2 = \frac{3}{8}$$

If a Hanning window is used, the spectral estimates should be multiplied by the scale factor 8/3. Because it affects equally the auto and the cross *PSD*, the loss factor does not affect the transfer functions, but it does affect the signal to noise ratio.

12.10 Convolution and correlation via *FFT*

An important application of the *FFT* is the approximation of the convolution and correlation integrals. A difficulty arises from the fact that, unlike the continuous Fourier transform, the *DFT* is periodic. Before addressing this important issue, we define the *periodic* convolution and correlation.

12.10.1 Periodic convolution and correlation

The discrete convolution of the periodic sequences $x(k)$ and $h(k)$ is defined by

$$y(k) = x(k) \star h(k) = \frac{1}{N}\sum_{i=0}^{N-1} x(i)h(k-i) = \frac{1}{N}\sum_{i=0}^{N-1} x(k-i)h(i) \qquad (12.57)$$

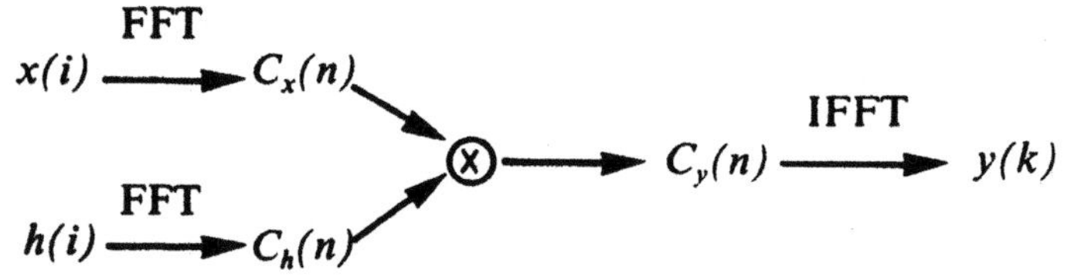

Figure 12.13: Periodic convolution via *FFT*.

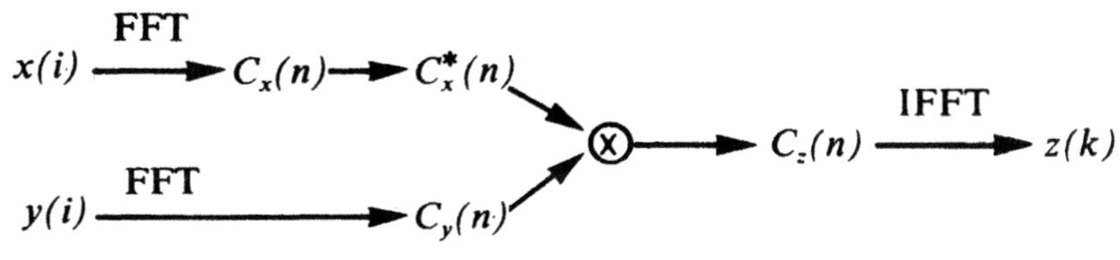

Figure 12.14: Periodic correlation via *FFT*.

It is called *periodic* convolution because the sequence $y(k)$ is also periodic. Its *DFT* satisfies the convolution theorem

$$C_y(n) = C_x(n)C_h(n) \tag{12.58}$$

This can be demonstrated as follows, with the usual notation $W = e^{-j2\pi/N}$,

$$C_y(n) = \frac{1}{N}\sum_{m=0}^{N-1} y(m)W^{nm}$$

$$= \frac{1}{N^2}\sum_{m=0}^{N-1}\sum_{i=0}^{N-1} x(i)h(m-i)W^{nm}$$

$$= \frac{1}{N}\sum_{i=0}^{N-1} x(i)\frac{1}{N}\sum_{m=0}^{N-1} h(m-i)W^{nm}$$

According to the translation theorem (12.37), the second sum is equal to $C_h(n)W^{ni}$ and we get

$$C_y(n) = \frac{1}{N}\sum_{i=0}^{N-1} x(i)W^{ni}C_h(n) = C_x(n)C_h(n)$$

An efficient method for computing periodic convolutions with a *FFT* algorithm is described in Fig.12.13. If the sequences are real,$C_x(n)$ and $C_h(n)$ can be calculated in a single *FFT*.

In a similar way, the discrete correlation of the periodic sequences $x(m)$ and $y(m)$ reads

$$z(k) = \frac{1}{N}\sum_{i=0}^{N-1} x(i)y(k+i) \tag{12.59}$$

$z(k)$ is also periodic; it is called the *periodic correlation*. Its *DFT* satisfies the correlation theorem

$$C_z(n) = C_x^*(n)C_y(n) \tag{12.60}$$

The demonstration is identical to that of the convolution theorem. Fig.12.14 describes the use of the *FFT* algorithm for the efficient computation of the periodic correlation.

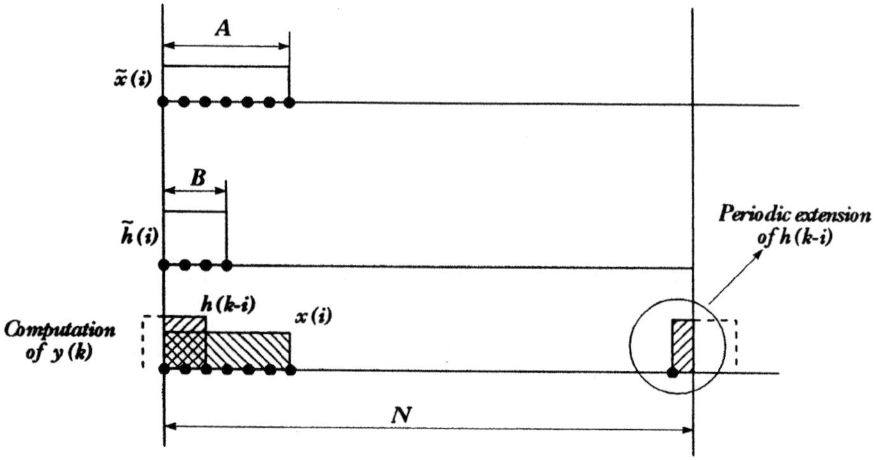

Figure 12.15: Under the condition $N \geq A+B-1$, the periodic and non-periodic convolutions are identical.

12.10.2 Approximation of the continuous convolution

The approximation of the continuous convolution by the discrete convolution leads to an efficient numerical method based on the *FFT*. However, the discrete convolution deals with *periodic* signals and, without special care, this may lead to substantial errors. This section describes methods which eliminate the periodicity error. All the discussion applies also to the correlation integral.

- **The two signals have a finite length**

This is the simplest situation. Let us denote by $\hat{x}(t)$ and $\hat{h}(t)$ the continuous signals of duration a and b respectively, by $\tilde{x}(kT)$ and $\tilde{h}(kT)$ the sampled values of the signal, involving respectively A and B samples, and by $x(kT)$ and $h(kT)$ the periodic sequences whose sample values in one period are identical to $\tilde{x}(kT)$ and $\tilde{h}(kT)$ (so far, the period NT is not specified). Furthermore, let $\hat{y}(t)$ be the continuous convolution $\hat{y}(t) = \hat{x}(t) \star \hat{h}(t)$, $\tilde{y}(kT)$ be the sampled values of $\hat{y}(t)$, $y(kT)$ be the periodic convolution $y(kT) = x(kT) \star h(kT)$ and finally, let $y^\star(kT)$ be defined as

$$y^\star(kT) = \frac{1}{N} \sum_{i=0}^{\infty} \tilde{x}(iT)\tilde{h}[(k-i)T] \tag{12.61}$$

$NTy^\star(kT)$ constitutes a numerical estimate of $\tilde{y}(kT)$, obtained by assuming a rectangular approximation of the integral over each sampling period. $y^\star(kT)$ has $A+B-1$ samples which differ from zero.

The condition on N under which the periodic convolution $y(kT)$ is identical to the numerical approximation $y^\star(kT)$ is

$$N \geq A+B-1 \tag{12.62}$$

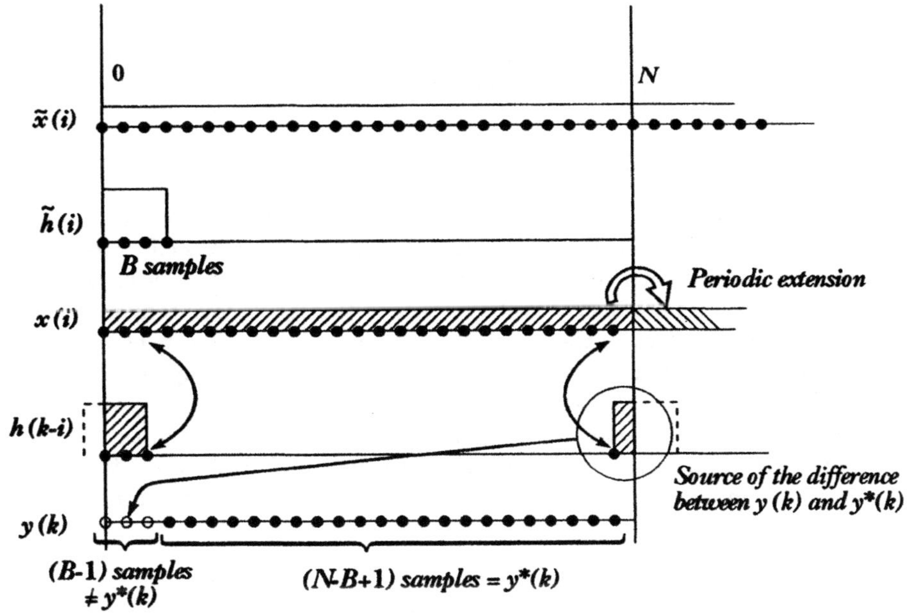

Figure 12.16: Convolution of a signal of infinite length $x(t)$ by one of finite length $h(t)$ (B samples). The first $B-1$ samples of the periodic convolution differ from the continuous convolution because of the extremity effect.

Indeed, under this condition, the infinite sum in Equ.(12.61) can be restricted to $N-1$ and, in the computation of the cyclic convolution, the periodic extension does not affect the value of the convolution, as illustrated in Fig.12.15. This makes $y(kT) = y^\star(kT)$.

Condition (12.62) states that the periodicity should be chosen in such a way that it is at least as long as the duration of the non-zero part of the continuous convolution ($NT \geq a + b$). If this is the case, the cyclic convolution $y(kT)$ can be regarded as a good approximation of $y^\star(kT)$, and the fast computation of the convolution with a *FFT* algorithm is a direct application of Fig.12.13.

- **One of the signals is of finite length**

Next, consider the case where the signal $\hat{x}(t)$ is infinite while $\hat{h}(t)$ is finite [$\tilde{h}(kT)$ has B samples different from zero]. Since the periodic convolution deals with periodic sequences, we truncate $\tilde{x}(kT)$ after N samples and define $x(kT)$ as its periodic extension. The terms involved in the construction of the periodic convolution $y(kT)$ are shown in Fig.12.16. The part of $h(k-i)$ close to $N-1$ arises from the periodic extension of $\tilde{h}(kT)$, and does not appear in the corresponding continuous convolution. It is responsible for a difference between $y(kT)$ and $y^\star(kT)$ for the first $B-1$ sample values; the remaining ones are identical. Two methods of eliminating the $B-1$ erroneous samples of the cyclic convolution are described in the following sections.

12.10.3 Sectioning Overlap-save

When computing the convolution with a *FFT* algorithm, we substitute a cyclic convolution for the original one. The signal of infinite length is cut into sections of N samples and each section is processed separately. For each of them, the $B-1$ initial values are in error with respect to the continuous convolution, due to the extremity effect in the cyclic convolution. The errors can be eliminated by sectioning in such a way that the various sections ***overlap*** by $B-1$ samples, as indicated in Fig.12.17. Each section of N samples is analysed separately, and the $B-1$ erroneous values of the cyclic convolution are deleted; the $N-B+1$ remaining values are saved and stored sequentially to produce the desired approximation. In this way, all the errors are eliminated, except for the $B-1$ initial values of the first section.

12.10.4 Sectioning Overlap-add

An alternative way of sectioning the record is described in Fig.12.18. Each section $x^k(i)$ is constructed from $N-B+1$ samples of the original record, supplemented by $B-1$ zeros. For each section, condition (12.62) is fulfilled and the cyclic convolution approximates the continuous convolution. Since the sum of the sections $x^k(i)$ restores the original signal, and the convolution is a linear operation, the convolution of the original signal is obtained by ***adding*** the partial convolutions of all the contributing sections, including the part relative to the $B-1$ overlapping samples.

12.11 *FFT* simulation of Gaussian processes with prescribed *PSD*

For many applications, it is necessary to generate sample records of stationary, ergodic Gaussian processes with a prescribed *PSD*. Examples are shown in Fig.10.10; they have been genèrated with a *FFT* subroutine according to the procedure described below. The two key parameters are the ***duration*** T_0 which controls the frequency resolution of the *DFT* ($f_0 = 1/T_0$) and the ***number of samples***, N, which controls the sampling period ($T = T_0/N$). The idea is to generate a sequence of *DFT* coefficients $C_x(k)$ with amplitude and phase such that the *IDFT* will produce the desired time sequence $x(i)$. Recall that $x(i)$ consists of the sampled values of a band-limited signal of period T_0 and of cut-off frequency equal to half the sampling frequency, $f_c = N/2T_0$. If the sequence $x(i)$ is real, the *DFT* coefficients must satisfy Equ.(12.36).

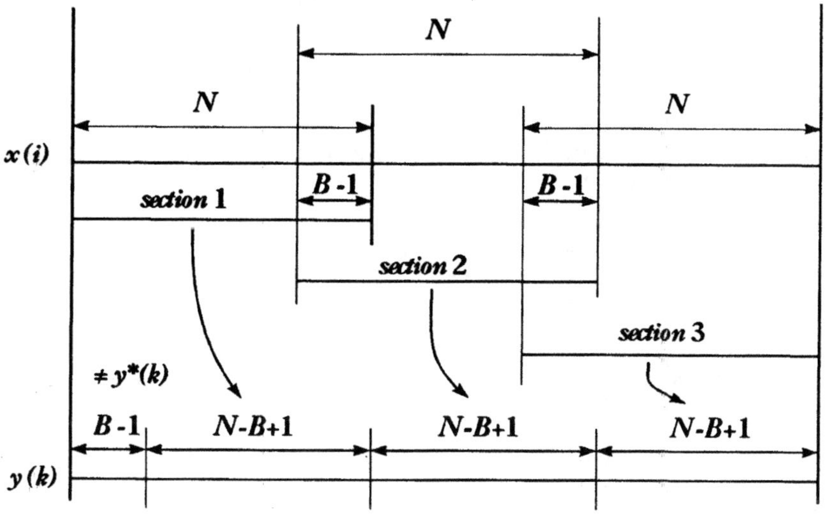

Figure 12.17: Sectioning Overlap-save.

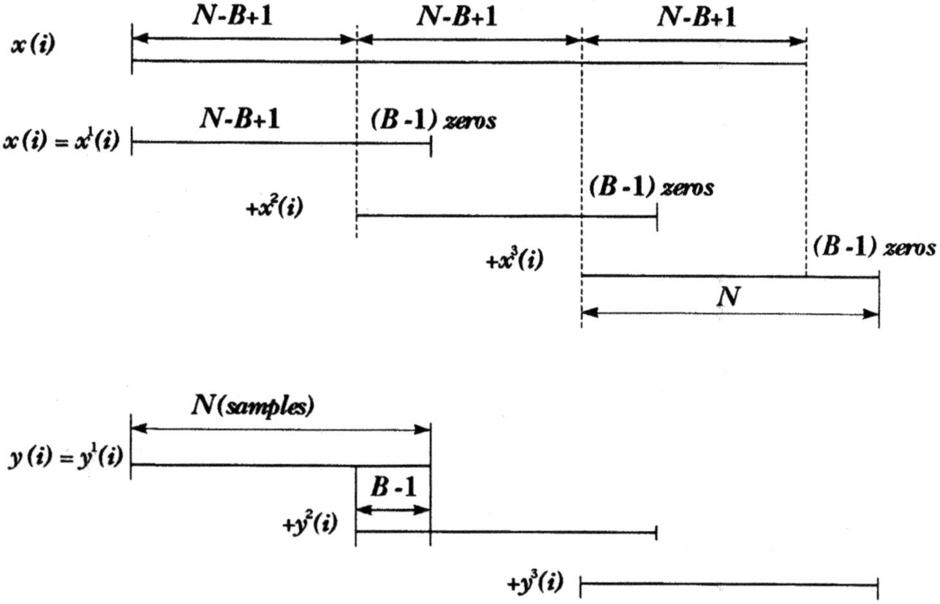

Figure 12.18: Sectioning Overlap-add. $y(k) = \sum_i y^i(k)$.

The amplitude of the *DFT* coefficients, $|C_x(k)|$, $k = 1, ..., N/2$ must be chosen in order to match the desired power distribution. For an ergodic process,

$$\frac{1}{T_0}\int_0^{T_0} x^2(t)\,dt \simeq \int_{-\infty}^{\infty} \Phi_{xx}(\omega)\,d\omega$$

Discretizing with time and frequency increments respectively $\Delta t = T_0/N$ and $\Delta\omega = \omega_0 = 2\pi/T_0$, we get

$$\frac{1}{N}\sum_{m=0}^{N-1} x^2(m) \simeq 2\sum_{k=1}^{N/2} \Phi_{xx}(k\omega_0)\,\omega_0$$

after taking into account that $\Phi_{xx}(\omega)$ is even and band-limited. From Parseval's theorem (12.39) and Equ.(12.36), this sum can be transformed into

$$2\sum_{k=1}^{N/2} |C_x(k)|^2 \simeq 2\sum_{k=1}^{N/2} \Phi_{xx}(k\omega_0)\,\omega_0$$

(the term relating to $k = 0$ has been omitted in both sides of this equation). According to the foregoing relationship, the proper power distribution will be met if we choose $|C_x(k)|$ according to

$$|C_x(k)| = [\Phi_{xx}(k\omega_0)\,\omega_0]^{1/2} \qquad k = 1, ..., N/2 \tag{12.63}$$

The accuracy of this equation is directly related to the frequency resolution. This point may become important if we want to simulate very narrow-band processes such as the response of an oscillator with very light damping. If the *PSD* $\Phi_{xx}(\omega)$ experiences sharp variations within the frequency discretization interval ω_0, this relation must be replaced by

$$|C_x(k)|^2 = \int_{(k-1/2)\omega_0}^{(k+1/2)\omega_0} \Phi_{xx}(\omega)\,d\omega \tag{12.64}$$

Let us now consider the phase distribution of the *DFT*. If we choose the phases $\theta_k = \arg[C_x(k)]$ as ***independent*** random variables with uniform distribution in $[0, 2\pi)$, for any time t, all the harmonic components $x^k(t) = A_k \sin(k\omega_0 t + \theta_k)$ will be independent random variables with probability distribution given by Equ.(2.39). According to the central limit theorem, the sum of a large number of independent random variables with arbitrary distribution is Gaussian at the limit. Note that the uniform phase distribution is not necessary to achieve a Gaussian process, but this choice is convenient.

In summary, after the duration T_0 and the record size N have been selected to achieve the appropriate frequency resolution and sampling rate, sample records of a zero mean, stationary, ergodic Gaussian process of prescribed spectral content can be obtained by inverse transformation of a *DFT* sequence generated as follows:

$$C_x(0) = 0$$

$$|C_x(k)| = [\Phi_{xx}(k\omega_0)\,\omega_0]^{1/2} \qquad k = 1, ..., N/2$$

$$\theta_k = \arg[C_x(k)] = \text{Uniformly distributed in } [0, 2\pi)$$

$$C_x(N/2 + i) = C_x^*(N/2 - i) \tag{12.65}$$

Any new set of statistically independent random phases θ_k produces a new sample record with the same spectral content, but statistically independent of the previous record. Computer independent random number generators are widely available. Finally, in contrast to section 3.8.4, each sample record generated according to Equ.(12.65) is representative of the frequency content of the process. This constitutes the ergodicity property.

12.12 References

N.AHMED & K.R.RAO, *Orthogonal Transforms for Digital Signal Processing*, Springer Verlag, 1975.

J.BENDAT & A.PIERSOL, *Random Data: Analysis and Measurement Procedures*, Wiley-Interscience, 1971.

J.BENDAT & A.PIERSOL, *Engineering Applications of Correlation and Spectral Analysis*, Wiley-Interscience, 1980.

G.D.BERGLAND, A guided tour of the Fast Fourier Transform, *IEEE Spectrum, July, 1969.*

R.B.BLACKMAN & J.W.TUKEY, *The Measurement of Power Spectra from the Point of View of Communications Engineering*, Dover, 1958.

R.N.BRACEWELL, *The Fourier Transform and its Applications*, McGraw-Hill, 1978.

E.O.BRIGHAM, *The Fast Fourier Transform*, Prentice Hall, 1974.

J.W.COOLEY & J.W.TUKEY, An algorithm for the machine calculation of complex Fourier series, *Math. Computation, Vol.19, pp.297-301, April 1975.*

B.GOLD & C.RADER, *Digital Processing of Signals*, McGraw-Hill, 1969.

R.HAMMING, *Digital Filters*, Prentice Hall, 1977.

H.F.HARMUTH, *Transmission of Information by Orthogonal Functions*, Springer Verlag, 1969.

IEEE Transactions on Audio and Electroacoustics, Special Issue on the *FFT*, *Vol.AU-15, No 2, June 1967.*

A.OPPENHEIM & R.SCHAFER, *Digital Signal Processing*, Prentice Hall, 1975.

A.PAPOULIS, *The Fourier Integral and its Applications*, McGraw-Hill, 1962.

12.13 Problems

P.12.1 *Show that for the Fourier series expansion, Parseval's theorem reads*

$$\frac{1}{T_0}\int_{-T_0/2}^{T_0/2} |y(t)|^2 dt = \sum_{n=-\infty}^{\infty} |\alpha_n|^2$$

P.12.2 *Show that the Fourier series coefficients (12.14) of the rectangular wave of Fig.12.6 are*

$$\alpha_{2k+1} = \frac{2}{j\pi(2k+1)}$$

$$\alpha_{2k} = 0 \qquad k = 0, \pm 1, \pm 2, ...$$

Show that it can be written equivalently

$$y(t) = \frac{4}{\pi}\sum_{k=0}^{\infty}\frac{\sin(2k+1)t}{2k+1}$$

P.12.3 *Show that the truncated Fourier expansion with n terms can be obtained from the original signal by the convolution*

$$y_n(t) = 2f_c\{\frac{\sin 2\pi f_c t}{2\pi f_c t}\} \star y(t)$$

where $n/T_0 < f_c < (n+1)/T_0$. Using this interpretation for a rectangular wave, show that the limit of the first overshoot due to the Gibbs effect is

$$y_\infty(0) = \frac{2}{\pi}\int_0^\pi \frac{\sin x}{x} dx \simeq 1.179$$

P.12.4 *Using a FFT subroutine (e.g. in MATLAB), show that the following DFT sequence*

$$C_x(2k) = 0 \qquad k = 0, ..., 8$$

$$C_x(2k+1) = \frac{2}{j\pi(2k+1)} \qquad k = 0, ..., 7$$

$$C_x(16+k) = C_x^*(16-k) \qquad k = 1, ..., 15$$

produces the sample values of the truncated expansion of a rectangular wave with $n = 15$ terms (compare with problem P.12.2). What would be the result if we enforce $C_x(i) = 0$ for $8 < i < 16$?

P.12.5 *Consider the sequence*

$$x(i) = \cos\frac{2\pi i a}{32} \qquad i = 0, ..., 31$$

(a) Using a FFT subroutine, compute the DFT coefficients for $a = 4$. Show that there are only two non-zero components $[C_x(4) = C_x(28) = 1/2]$.

(b) Do the same computations for $a = 3.9$. Comment on the leakage phenomenom.
(c) Do the same calculations with a Hanning window.
(d) Check that the DFT of the signal with the Hanning window can be obtained from that of the original signal (with a rectangular window) by

$$C_x^H = -0.25C_x(k-1) + 0.5C_x(k) - 0.25C_x(k+1)$$

(e) What would produce a Hanning window for $a = 4$.
P.12.6 *Show analytically that the DFT coefficients with a Hanning window from 0 to T_0 are related to those without window according to*

$$C_x^{Hanning} = -0.25C_x(k-1) + 0.5C_x(k) - 0.25C_x(k+1)$$

[Hint: Start from Equ.(12.47) and use the translation theorem.]
P.12.7 *Show that the DFT coefficients with a Hamming window from 0 to T_0 can be obtained from those without window according to*

$$C_x^{Hamming} = -0.23C_x(k-1) + 0.54C_x(k) - 0.23C_x(k+1)$$

Bibliography

M.ABRAMOWITZ & I.STEGUN, *Handbook of Mathematical Functions*, Dover, 1972.

R.J.ADLER, On the envelope of a gaussian random field, *J. of Appl. Prob. 15, pp.502-513, 1978.*

N.AHMED & K.R.RAO, *Orthogonal Transforms for Digital Signal Processing*, Springer Verlag, 1975.

A.ANGOT, *Complément de Mathématiques*, 5 ème édition, Editions de la Revue d'Optique, Paris, 1965.

S.T.ARIARATNAM & H.N.PI, On the first-passage time for envelope crossing for a linear oscillator, *Int. Journal of Control, Vol.18, No 1, pp.89-96,* 1973.

J.D.ATKINSON, Eigenfunction expansions for randomly excited non-linear systems, *Journal of Sound and Vibration 30(2), pp.153-172,* 1973.

G.AUGUSTI, A.BARATTA & F.CASCIATI, *Probabilistic Methods in Structural Engineering,* Chapman & Hall, 1983.

J.BENDAT & A.PIERSOL, *Random Data: Analysis and Measurement Procedures*, Wiley-Interscience, 1971.

J.BENDAT & A.PIERSOL, *Engineering Applications of Correlation and Spectral Analysis*, Wiley-Interscience, 1980.

G.D.BERGLAND, A guided tour of the Fast Fourier Transform, *IEEE Spectrum, July, 1969.*

L.A.BERGMAN & J.C.HEINRICH, On the moments of time to first passage of the linear oscillator, *Earth. Eng. Struct. Dyn., Vol.9, pp.197-204, 1981.*

A.T.BHARUCHA-REID, *Elements of Theory of Markov Processes. and their Applications*, McGraw-Hill, 1960.

J.BIETRY, C.SACRE & E.SIMIU, Mean wind profiles and change of terrain roughness, *Proc. ASCE, Vol.104, ST 10, October 1978.*

J.BIGGS, *Introduction to Structural Dynamics*, McGraw-Hill, 1964.

R.B.BLACKMAN & J.W.TUKEY, *The Measurement of Power Spectra from the Point of View of Communications Engineering*, Dover, 1958.

A.BLANC-LAPIERRE & R.FORTET, *Théorie des Fonctions Aléatoires*, Masson, Paris, 1953.

R.D.BLEVINS, *Flow-Induced Vibration*, Van Nostrand Reinhold Co, 1977.

V.V.BOLOTIN, *Statistical Methods in Structural Mechanics*, Holden-Day, 1969.

R.N.BRACEWELL, *The Fourier Transform and its Applications*, McGraw-Hill, 1978.

E.O.BRIGHAM, *The Fast Fourier Transform*, Prentice Hall, 1974.

A.E.BRYSON & Y.C.HO, *Applied Optimal Control (Optimization, Estimation and Control)*, J.Wiley, 1975.

D.E.CARTWRIGHT & M.S.LONGUET-HIGGINS, The statistical distribution of the maxima of a random function, *Proc. Roy. Soc. Ser. A, 237, pp.212-232,* 1956.

T.K.CAUGHEY, Classical Normal Modes in Damped Linear Dynamic Systems, *Trans. ASME, J. of Applied Mechanics, Vol.27, No 2, pp.269-271, 1960.*

T.K.CAUGHEY & H.J.STUMPF, Transient response of a dynamic system under random excitation, *J. of Applied Mechanics, Vol.28, pp.563-566, 1961.*

T.K.CAUGHEY, Derivation and application of the Fokker-Planck equation to discrete nonlinear dynamic systems subjected to white noise excitation, *J. of Acoustical Society of America, Vol.35, No 11, November 1963.*

T.K.CAUGHEY, Nonlinear theory of random vibrations, *Advances in Applied Mechanics 11, pp.209-253,* 1971.

G.K.CHAUDHURY & W.D.DOVER, Fatigue analysis of offshore platforms subject to sea wave loadings. *J. Fatigue, 7 ,No 1, Mar., 13-19,* 1982.

R.W.CLOUGH & J.PENZIEN, *Dynamics of Structures*, McGraw-Hill, 1975.

J.W.COOLEY & J.W.TUKEY, An algorithm for the machine calculation of complex Fourier series, *Math. Computation, Vol.19, pp.297-301, April 1975.*

R.B.COROTIS, E.H.VANMARCKE, & C.A.CORNELL, First passage of non-stationary random processes, *ASCE J. of Eng. Mech. Div. ,EM2, April 1972.*

R.B.COROTIS & E.H.VANMARCKE, Time dependent spectral content of system response, *ASCE J. of Eng. Mech. Div., Vol.101, October 1975.*

H.CRAMER & M.R.LEADBETTER, *Stationary and Related Stochastic Processes*, Wiley, 1967.

S.H.CRANDALL & W.D.MARK, *Random Vibration in Mechanical Systems*, Academic Press, 1963.

S.H.CRANDALL, Zero crossings, peaks, and other statistical measures of random responses, *The Journal of the Acoustical Society of America, Vol.35, No 11, pp.1693-1699, November* 1963.

S.H.CRANDALL, K.L.CHANDIRAMANI & R.G.COOK, Some first passage problems in random vibration, *ASME Journal of Applied Mechanics, Vol.33, pp.532-538, September* 1966.

S.H.CRANDALL, The role of damping in vibration theory, *Journal of Sound and Vibration 11(1), pp.3-18, 1970.*

S.H.CRANDALL, First crossing probabilities of the linear oscillator, *Journal of Sound and Vibration 12(3), pp.285-299,* 1970.

R.R.CRAIG Jr., *Structural Dynamics*, Wiley, 1981.

T.DAHLBERG, Optimization criteria for vehicles travelling on a randomly profiled road - a survey, *Vehicle System Dynamics, 8, pp.239-252, 1979.*

T.DAHLBERG, Comparison of ride comfort criteria for computer optimization of vehicles travelling on randomly profiled roads, *Vehicle System Dynamics 9, pp.291-307, 1980.*

A.G.DAVENPORT, The application of statistical concepts to the wind loading of structures, *Proc. Inst. Civ. Eng., Vol.19, pp.449-471, August 1961.*

A.G.DAVENPORT, Note on the distribution of the largest value of a random function with application to gust loading, *Proc. Inst. Civ. Eng., Vol 28, pp.187-196,* 1964.

A.G.DAVENPORT, The treatment of wind loading on tall buildings, ***Proceedings of the Symposium on Tall Buildings***, University of Southampton, Pergamon Press, London, 1966.

W.B.DAVENPORT, ***Probability and Random Processes***, McGraw-Hill, 1970.

W.B.DAVENPORT & W.L.ROOT, ***An Introduction to the Theory of Random Signals and Noise***, McGraw-Hill, 1958.

M.DEL PEDRO & P.PAHUD, ***Mécanique Vibratoire***, Presses Polytechniques et Universitaires Romandes, Lausanne, 1989.

A.DER KIUREGHIAN, Structural response to stationary excitation, ***ASCE J. of Eng. Mech. Div., Vol.106, EM6, pp.1195-1213, 1980.***

A.DER KIUREGHIAN, A Response spectrum method for random vibration analysis of MDF systems, ***Earth. Eng. Struct. Dyn., Vol.9, pp.419-435, 1981.***

J.L.DOOB, ***Stochastic Processes***, Wiley, 1953.

I.ELISHAKOFF, A.Th. VAN ZANTEN, & S.H.CRANDALL, Wide-band random axisymmetric vibration of cylindrical shells, ***ASME J. of Applied Mechanics, Vol.46,No 2, pp.417-422, June 1979.***

I.ELISHAKOFF, ***Probabilistic Methods in the Theory of Structures***, Wiley, 1982.

F.ELLYIN & K.GOLOS, Multiaxial fatigue damage criterion. ***Trans. ASME, J. of Engineering Materials and Technology, Vol.110, January, 63-68***, 1988.

B.ETKIN, ***Dynamics of Atmospheric Flight***, Wiley, 1972.

D.J.EWINS, ***Modal Testing: Theory and practice***, Wiley, 1984.

B.FRAEJIS de VEUBEKE, Influence of internal damping on aircraft resonance, AGARD report, November 1959.

Y.C.FUNG, ***An Introduction to the Theory of Aeroelasticity***, Dover, 1969.

M.GERADIN & D.RIXEN, ***Mechanical Vibrations, Theory and Application to Structural Dynamics***, Wiley, 1993.

R.J.GIBERT, ***Vibrations des Structures***, Eyrolles, 1988.

B.GOLD & C.RADER, ***Digital Processing of Signals***, McGraw-Hill, 1969.

D.J.GORMAN, An analytical and experimental investigation of the vibration of cylindrical reactor fuel elements in two-phase parallel flow, ***Nuclear Science and Engineering, 44, pp.277-290, 1971.***

A.H.GRAY, First passage time in a random vibrational system, ***ASME J. of Applied Mechanics, pp.187-191, March 1966.***

E.J.GUMBEL & P.G.CARLSON, Extreme values in aeronautics, ***J. of Aeronautical Sciences, 21, pp.389-398, 1954.***

E.J.GUMBEL, ***Statistics of Extremes***, Columbia University Press, 1958.

R.HAMMING, ***Digital Filters***, Prentice Hall, 1977.

J.K.HAMMOND, On the response of single and multidegree of freedom systems to non-stationary random excitations, ***Journal of Sound and Vibration 7(3), pp.393-416, 1968.***

J.K.HAMMOND, Evolutionary spectra in random vibrations, ***The Journal of the Royal Stat. Soc., series B, Vol.35, No 2, pp.167-188, 1973.***

J.K.HAMMOND & R.F.HARRISON, Nonstationary response of vehicles on rough ground, a state space approach. *ASME J. of Dyn. Syst. Meas. & Cont., Vol.103, Sept. 1981.*

H.F.HARMUTH, *Transmission of Information by Orthogonal Functions*, Springer Verlag, 1969.

A.M.HASOFER & PETOCZ, Extreme response of the linear oscillator with modulated random excitation, *pp.503-512 of Statistical Extremes and Applications*, J.TIAGO de OLIVEIRA (ed.), 1984.

J.C.HOUBOLT, Atmospheric turbulence, *AIAA Journal, Vol.11, No 4, April 1973.*

IEEE Transactions on Audio and Electroacoustics, Special Issue on the FFT, *Vol.AU-15, No 2, June 1967.*

W.D.IWAN, & P-T.D.SPANOS, Response envelope statistics for nonlinear oscillators with random excitation, *ASME J. of Applied Mechanics, Vol.45, March 1978.*

M.KAC, Random walk and the theory of Brownian motion; *American Mathematical Monthly 54, No 7, pp.369-391*, 1947. Reprinted in *Selected Papers on Noise and Stochastic Processes*, N.WAX ed., Dover, 1954.

J.C.P.KAM & W.D.DOVER, Fast fatigue assessment procedure for offshore structures under random stress history. *Proc. Instn Civil Engrs, Part 2, 85, Dec., 689-700*, 1988.

P.KREE & C.SOIZE, *Mécanique Aléatoire*, Dunod, 1983.

S.KRENK, Nonstationary narrow-band response and first-passage probability, *ASME Journal of Applied Mechanics, Vol.46, pp.919-924, December* 1979.

R.S.LANGLEY, On various definitions of the envelope of a random process, *J. of Sound and Vibration, 105(3), pp.503-512, 1986.*

R.S.LANGLEY, Structural response to non-stationary non-white stochastic ground motions, *Earth. Eng. Struct. Dyn., Vol.14, pp.909-924, 1986.*

J.H.LANING & R.H.BATTIN, *Random Processes in Automatic Control*, McGraw Hill, 1956.

C.E.LARSEN & L.D.LUTES, Predicting the fatigue life of offshore structures by the single-moment spectral method. *Probabilistic Engineering Mechanics, Vol.6, No 2*, 1991.

M.C.LEE & J.PENZIEN, Stochastic analysis of structures and piping systems subjected to stationary multiple support excitations, *Earth. Eng. Struct. Dyn., Vol.11, pp.91-110, 1983.*

W.C.LENNOX & D.A.FRASER, On the first-passage distribution for the envelope of a nonstationary narrow-band stochastic process. *ASME Journal of Applied Mechanics, Vol.41, pp.793-797, September* 1979.

Y.K.LIN, *Probabilistic Theory of Structural Dynamics*, McGraw-Hill, 1967.

Y.K.LIN, First-excursion failure of randomly excited structures, *AIAA Journal, Vol.8, No 4, pp.720-725*, 1970.

Y.K.LIN, First-excursion failure of randomly excited structures II, *AIAA Journal, Vol.8, No 10, pp.1888-1890*, 1970.

Y.K.LIN & W.F.WU, Along wind response of tall buildings on compliant soil, *ASCE J.of Eng. Mech. Div., Vol.110, EM1, January 1984.*

M.LIVOLANT, F.GANTENBEIM & R.J.GIBERT, Méthodes statistiques pour l'estimattion de la réponse des structures aux séismes, *Mécanique, Matériaux, Electricité, No394-395, Oct.-Nov. 1982.*

R.M.LOYNES, On the concept of the spectrum for non-stationary processes, *J. Roy. Stat. Soc., Series B,30(1), pp.1-30, 1968.*

L.D.LUTES & C.E.LARSEN, Improved spectral method for variable amplitude fatigue prediction. *J. Struct. Div., ASCE, 116 (4), 1149-1164*, 1990.

R.H.LYON, On the vibration statistics of a randomly excited hard-spring oscillator, *J. Acoust. Soc. Am., Vol.32, pp.716-719, 1960.*

R.H.LYON, On the vibration statistics of a randomly excited hard-spring oscillator II, *J. Acoust. Soc. Am., Vol.33, No 10, pp.1395-1403, October 1961.*

P.H.MADSEN & S.KRENK, Stationary and transient response statistics, *ASCE J. of Eng. Mech. Div., Vol.108, EM4, pp.622-635, 1982.*

W.D.MARK, On false-alarm probabilities of filtered noise, *Proceedings IEEE, Vol.54, pp.316-317, February 1966.*

W.D.MARK, Spectral analysis of the convolution and filtering of non-stationary stochastic processes, *Journal of Sound and Vibration 11(1), pp.19-63, 1970.*

L.MEIROVITCH, *Computational Methods in Structural Dynamics*, Sijthoff & Noordhoff, 1980.

P.G.MERTENS & A.PREUMONT, Improved generation of *PSD* functions, artificial accelerograms and spectra, fully compatible with a design response spectrum, *SMIRT-12*, paper K 13/3, Stuttgart 1993.

D.MIDDLETON, *An Introduction to Statistical Communication Theory*, McGraw-Hill, 1960.

J.W.MILES, On structural fatigue under random loading, *J. of Aeronautical Sciences, 21, pp.753-762, 1954.*

L.D.MITCHELL, Improved methods for the Fast Fourier Transform (FFT) calculation of the frequency response function, *ASME J. Mech. Design, Vol.104, pp.277-279, April 1982.*

D.E.NEWLAND, *Random Vibrations and Spectral Analysis*, Longmans, 1975.

N.M.NEWMARK & E.ROSENBLUETH, *Fundamentals of Earthquake Engineering*, Prentice Hall, 1971.

N.C.NIGAM, Phase properties of a class of random processes, *Earth. Eng. Struct. Dyn., Vol.10, pp.711-717, 1982.*

N.C.NIGAM, *Introduction to Random Vibration*, MIT Press, 1983.

M.NOVAK, Random vibrations of structures, Proceedings of *ICASP-4*, Florence, pp.539-550, June 1983.

Y.OHSAKI, On the significance of phase content in earthquake ground motions, *Earth. Eng. Struct. Dyn., Vol.7, pp.427-439, 1979.*

M.D.OLSON & G.M.LINDBERG, Jet noise excitation of an integrally stiffened panel, *Journal of Aircraft, Vol.8, No 11, pp.847-855, November 1971.*

A.OPPENHEIM & R.SCHAFER, *Digital Signal Processing*, Prentice Hall, 1975.

A.PAPOULIS, *The Fourier Integral and its Applications*, McGraw-Hill, 1962.

A.PAPOULIS, *Probability, Random Variables and Stochastic Processes*, McGraw Hill, 1965.

J.PAQUET, Etude experimentale in situ de l'effet du vent sur la tour Maine-Montparnasse, *Annales de l'Institut Technique du Batiment et des Travaux Publics, No 376, October 1979.*

E.PARZEN, *Stochastic Processes*, Holden Day, 1962.

G.PETIT BOIS, *Tables of Indefinite Integrals*, Dover, N-Y, 1961.

A.G.PIERSOL, Optimum resolution bandwidth for spectral analysis of stationary random vibration data, *Shock and Vibration, Vol.1, No 1, pp.33-43, 1993.*

A.POWELL, On the fatigue failure due to vibrations excited by random pressure fields. *J. Acoust. Soc. Am., Vol.30, No 12, pp.1130-1135, December 1958.*

A.POWELL, On the approximation to the *infinite* solution by the method of normal modes for random vibrations. *J. Acoust. Soc. Am., Vol.30, No 12, pp.1136-1139, December 1958.*

G.H.POWELL, *Missing mass* correction in modal analysis of systems, *SMIRT-5*, paper K10/3, Berlin, 1979.

H.PRESS, Athmospheric turbulence environment with special reference to continuous turbulence, *AGARD* Report 115, 1957.

A.PREUMONT, The generation of non-separable artificial earthquake accelerograms for the design of nuclear power plants, *Nuclear Engineering and Design, 88, pp.59-67, 1985.*

A.PREUMONT, On the peak factor of stationary Gaussian processes, *Journal of Sound and Vibration 100(1), pp.15-34*, 1985.

A.PREUMONT, Bayesian analysis of the extreme value of a time-history, *Reliability Engineering and System Safety, 20, pp.165-172.*

M.B.PRIESTLEY, Evolutionary spectra and non-stationary processes, *J. Roy. Stat. Soc., Series B, 27(2), pp.204-237, 1965.*

M.B.PRIESTLEY, Power spectral analysis of random processes, *Journal of Sound and Vibration 6(1), pp.86-97, 1967.*

M.B.PRIESTLEY & H.TONG, On the analysis of bivariate non-stationary processes, *J. Roy. Stat. Soc., Series B, 35(2), 1973.*

R.L.RACICOT & F.MOSES, A first-passage approximation in random vibration, *ASME Journal of Applied Mechanics, pp.143-147, March 1971.*

S.O.RICE, Mathematical analysis of random noise, *Bell System Technical Journal, 23, pp.282-332*, 1944; *24, pp.46-156*, 1945. Reprinted in *Selected Papers on Noise and Stochastic Processes*, N.WAX ed., Dover, 1954.

J.B.ROBERTS, Probability of first passage failure for stationary random vibration, *AIAA Journal, Vol.12, No 12, pp.1636-1643, December 1974.*

J.B.ROBERTS, First passage time for the envelope of a randomly excited linear oscillator, *Journal of Sound and Vibration, 46 (1), pp.1-14*, 1976.

J.B.ROBERTS, Response of nonlinear mechanical systems to random excitation, Part 1: Markov methods, *Shock Vib. Dig., 13 (4), pp.17-28, April 1981.*

J.B.ROBERTS, Response of nonlinear mechanical systems to random excitation, Part 2: Equivalent linearization and other methods, *Shock Vib. Dig., 13 (5), pp.15-29, May 1981.*

E.ROSENBLUETH J.I.BUSTAMENTE, Distribution of structural responses to earthquakes, *Proc.ASCE, J. Eng. Mech. Div., Vol.88, EM3, pp.75-106, 1962.*

B.SAHAY & W.LENNOX, Moments of the first-passage time for narrow-band process, *J.of Sound and Vibration, 32(4), pp.449-458, 1974.*

D.J.SAKRISON, *CommunicationTheory: Transmission of Waveforms and Digital Information*, Wiley, 1968.

J.E.SHIGLEY & L.D.MITCHELL, *Mechanical Engineering Design*, McGraw-Hill, 1983.

S.SHIHAB & A.PREUMONT, Non-stationary random vibrations of linear multidegree-of freedom systems, *J. of Sound and Vibration 132(3), pp.457-471, 1989.*

M.SHINOZUKA, Random processes with evolutionary power, *Proc. ASCE, J. Eng. Mech. Div., Vol.96, EM4, pp.543-545, 1970.*

E.SIMIU & R.H.SCANLAN, *Wind effects on structures*, Wiley, 1978.

G.SINES & G.OHGI, Fatigue criteria under combined stresses or strains. *Trans. ASME, J. of Engineering Materials and Technology, Vol.103, April, 82-90,* 1981.

C.SOIZE, Gust loading factors with nonlinear pressure terms, *Proc ASCE, Vol.104, ST6, pp.991-1007, June 1978.*

G.P.SOLOMOS & P-T.D.SPANOS, Solution of the Backward-Kolmogorov equation for nonstationary oscillation problem. *ASME Journal of Applied Mechanics, Vol.49, pp.923-925, December* 1982.

G.P.SOLOMOS & P-T.D.SPANOS, Oscillator response to nonstationary excitation, *J. of Applied Mechanics, Vol.51, pp.907-912, December 1984.*

R.L.STRATONOVICH, *Topics in the Theory of Random Noise*, 1, Gordon & Breach, N-Y, 1963.

R.L.STRATONOVICH, *Topics in the Theory of Random Noise*, Vol.2, Gordon & Breach, N-Y, 1967.

A.A.SVESHNIKOV, *Applied Methods of the Theory of Random Functions*, Pergamon Press, 1966.

D.H.TACK, M.W.SMITH & R.F.LAMBERT, Wall pressure correlations in turbulent airflow. *J. Acoust. Soc. Am., Vol.33, No 4, pp.410-418, April 1961.*

J.TAYLOR, Manual of Aircraft loads, *AGARDograph 83*, 1965.

R.H.TOLAND & C.Y.YANG, Random walk model for first passage probability, *Proc. ASCE, J. Eng. Mech. Div., EM3, pp.791-807, June 1971.*

E.H.VANMARCKE, Properties of spectral moments with applications to random vibration, *ASCE J. Eng. Mech. Div., EM2, pp.425-446, April* 1972.

E.H.VANMARCKE, On the distribution of the first-passage time for normal stationary random processes, *ASME Journal of Applied Mechanics, pp.215-220, March* 1975.

E.H.VANMARCKE, Structural response to earthquakes, Ch.8 of *Seismic Risk and Engineering Decisions*, C.LOMNITZ & E.ROSENBLUETH, Eds., Elsevier, 1976.

E.H.VANMARCKE, *Random Fields: Analysis and Synthesis*, The MIT Press, 1983.

M.C.WANG & G.E.UHLENBECK, On the theory of Brownian motion II, *Review of Modern Physics, Vol.17, No 2 and 3, April-July, pp.323-342*, 1945. Reprinted in *Selected Papers on Noise and Stochastic Processes*, N.WAX ed., Dover, 1954.

G.B.WARBURTON, *The Dynamical Behaviour of Structures*, Pergamon Press, 2nd Edition, 1976.

W.WEDIG, A critical review of methods in stochastic structural dynamics, *Nuclear Engineering and Design, 79, pp.281-287, 1984.*

R.G.WHITE, & J.G.WALKER (Eds.) *Noise and Vibration*, Ellis Horwood Publishers, 1982.

P.H.WIRSCHING & E.B.HAUGEN, Probabilistic design for random fatigue loads. *Journal of Engineering Mechanics Division, ASCE, Vol.99, EM6, December, 1165-1179*, 1973.

P.H.WIRSHING & M.C.LIGHT, Fatigue under wide band random stress. *J. Struct. Div., ASCE, 106 (7), 1593-1607*, 1980.

J.M.WOZENCRAFT & I.M.JACOBS, *Principles of Communication Engineering*, Wiley, 1965.

C.Y.YANG, *Random Vibration of Structures*, Wiley, 1986.

J.N.YANG & M.SHINOZUKA, On the first excursion probability in stationary narrow-band random vibration, *ASME Journal of Applied Mechanics, pp.1017-1022, December* 1971.

J.N.YANG & M.SHINOZUKA, On the first excursion probability in stationary narrow-band random vibration, II, *ASME Journal of Applied Mechanics, pp.733-738, September* 1972.

J.N.YANG, Simulation of random envelope processes, *J. of Sound and Vibration, 21 (1), pp.73-85, 1972.*

J.N.YANG, First excursion probability in non-stationary random vibration, *J. of Sound and Vibration, 27 (2), pp.165-182, 1973.*

J.N.YANG & Y.K.LIN, Along-wind motion of multistory building, *Proc. ASCE, J. Eng. Mech. Div., Vol 107, EM2, pp.295-307, April 1981.*

Index

9 780792 330363

GPSR Compliance

The European Union's (EU) General Product Safety Regulation (GPSR) is a set of rules that requires consumer products to be safe and our obligations to ensure this.

If you have any concerns about our products, you can contact us on ProductSafety@springernature.com

In case Publisher is established outside the EU, the EU authorized representative is:

Springer Nature Customer Service Center GmbH
Europaplatz 3
69115 Heidelberg, Germany

Batch number: 10755292

Printed by Printforce, the Netherlands